Perspectives on Animal Behavior

A First Course

Ten Original Essays by

John Alcock
Assistant Professor of Zoology
Arizona State University, Tempe

Jarvis Bastian
Associate Professor of Psychology
University of California, Davis

Gordon Bermant
Center Fellow
Battelle Seattle Research Center
and Affiliate Professor of Psychology
University of Washington, Seattle

Robert C. Bolles
Professor of Psychology
University of Washington, Seattle

Thomas Harrington
Associate Professor of Psychology
University of Nevada, Reno

Benjamin L. Hart
Associate Professor of Anatomy
School of Veterinary Medicine
University of California, Davis

Peter H. Klopfer
Professor and Director
Field Station for Animal Behavior Studies
Duke University, Durham

Dale F. Lott
Associate Professor of Psychology
University of California, Davis

G. Mitchell
Associate Professor, Department of Psychology
and Associate Research Psychologist
Department of Behavioral Biology, Primate Research Center
University of California, Davis

Robert M. Murphey
Assistant Professor of Psychology
University of California, Davis

Benjamin D. Sachs
Associate Professor of Psychology
University of Connecticut, Storrs

Robert P. Scobey
Assistant Professor of Behavioral Biology
University of California, Davis

Perspectives on Animal Behavior

A First Course

Edited by Gordon Bermant

Scott, Foresman and Company
Glenview, Illinois   Brighton, England

Library of Congress Catalog Card Number 72-95049
ISBN: 0-673-07577-X

Regional offices of Scott, Foresman and Company are located
in Dallas, Texas; Glenview, Illinois; Oakland, New Jersey; Palo
Alto, California; Tucker, Georgia; and Brighton, England.

PREFACE

One of the great pleasures in a teacher's life is the successful transmission of his own intellectual enthusiasm to a student or a class. The rewards of academic research in psychology and biology are often subtle and refined, and students sometimes have difficulty understanding why anyone would spend a lifetime engaged in such pursuits. Indeed the teacher may lose sight of "the point" of his research, that is, the intellectual scheme in which his efforts have meaning. When this occurs students are likely to cry for relevance because the teacher has failed to supply the contexts or perspectives which explain and justify his public commitment to the esoteric facts and theories of academic life. Such failure of communication can easily occur in the field of animal behavior.

This book was designed to allow each of its authors to transmit his own enthusiasm and perspective about the field of animal behavior to students with relatively little background in psychology or biology. Each author was asked to write about those ideas which excited him the most and which he believed would be most interesting to bright students in the first course in animal behavior, comparative psychology, ethology, or related courses. The editor placed no other limitations upon the author's choice of content. As far as possible, each author was also allowed to speak in the voice of his own particular style. Diversity of form and content, accurately reflecting the diversity of views that exist in this exciting field of study, and allowing the student to come to grips with areas of disagreement as well as consensus, has been the goal from the outset of the project.

The book is divided into four sections. The first section contains just one chapter, which carries the same title as the book. Here the fundamentals of the zoological, physiological, and psychological perspectives on animal behavior are introduced and elaborated sufficiently to motivate the material in the later chapters. Section

Two contains three chapters. One deals with the evolution of behavior, one with the genetic analysis of behavior, and one with the
development of social and emotional behaviors within individuals.
Placing genetics between evolutionary development (phylogeny)
and individual development (ontogeny) is appropriate because an
individual's genetic endowment is the linkage between its own
history and the history of its species. Only in man does history have
any other sense.

Section Three contains two chapters given over explicitly to
physiological considerations. The first of these describes the fundamentals of sensory systems, particularly in nonhuman and human
primates. The second introduces and expands upon the concept
of reflex. The intention in both chapters was to provide all the
material necessary for at least an introductory understanding. The
authors of both chapters have been particularly concerned to convey the fascination inherent in the physiology of behavior, while
avoiding the introduction of any more technical jargon than was
absolutely necessary.

The final section, with four chapters, considers four functional
behavior categories: courtship and mating, parental behavior,
learning, and communication. In these chapters, as in all the others,
every attempt has been made to present the material as simply as
possible without doing damage to the inherent complexity of important ideas. It is, as every writer knows, much more difficult to make
one's material simple than to make it complex. Everyone associated
with this book has worked hard to present his ideas as simply as
the ideas and his grasp on them would permit. What we have come
up with finally, I think, is not a survey of the field, nor a handbook,
nor an orthodox textbook, but rather an invitation to the field issued
by twelve of its practitioners.

This book was begun as a result of conversations I had with

George S. Reynolds of the University of California, San Diego. The organization and list of authors were developed primarily in Davis, California, during early 1969; Gayle McFadgen assisted at that point. Officers of the Battelle Memorial Institute, in particular R. S. Paul, E. R. Irish, and T. W. Ambrose, have been most generous and encouraging. Mark Starr and Mary McGuire have been extremely helpful in keeping track of references, permissions, deadlines, and so on. Barbara Prittie stuck to her typewriter at times when, I suspect, others would not. John Alcock did much more than it appears at first glance. And Roberta Bermant helped me maintain perspective on *Perspectives.*

Gordon Bermant
Seattle, Washington

CONTENTS

TO THE ANIMALS—ALL OF US

1 Perspectives on Animal Behavior

Gordon Bermant and John Alcock

From the mosquito that makes us miserable at a picnic to the lamb we slaughter for chops to the dog that brings in the paper to the elephant that does handstands in the circus, animals are very much a part of our everyday lives. In one way or another each of us shows an interest in animals and their behavior. We surround ourselves with pets, go to zoos and parks to see strange animals, and spend hours at the movies and more hours seated numbly before a television set, staring at all sorts of programs about animals—from Mickey Mouse and the Pink Panther to National Geographic specials on ants or chimpanzees. Either directly or indirectly, even the most urban of us is often exposed to the behavior of nonhuman animals.

To show interest in a Walt Disney animal-life special or in the antics of the family mutt is normal and readily understandable; but to turn this interest into a profession—investigating the behavior of beetles or the nerve cells of a sea slug or the way rats copulate—seems, perhaps, a bit curious and worthy of some explanation if not justification. Are these just the eccentric activities of ivory-tower academics with no particular significance or relation to anything else? The authors of this book would not have written it if they believed the answer was yes. Each of us holds the view that, in one way or another, understanding the behavior of animals is an important scientific task. But we also appreciate that the reasons for our views may not be obvious to everyone, and we will begin by trying to remedy this.

Animals have fascinated men throughout all of human history. Some of the earliest art known, paintings on walls of caves in France and Spain, are of animals. In prehistoric times the fascination of man with animal was obviously related to the central role hunting played in human existence. Both food and clothing, or the lack of them, were the direct result of a man's success in the hunt. From the beginning there has been a deep interest in animals based on their *utility* for the promotion of human welfare.

As important as animals were for the survival and basic economy of early men, one suspects that the artists among them had something other than utility in mind when they painted their representations of the bull on the wall of the cave; that, like us, they found the forms and motions of animals fascinating in themselves: sometimes beautiful, sometimes repulsive, but provoking a response that was immediate, apparently untrained, and often quite strong. It seems natural for man to turn to the representation of animals for *aesthetic value* and *symbolism*.

Our language is filled with phrases relating the characteristics of animals and men: sly as a fox, busy as a bee, blind as a bat, quick as a cat, and so on. Just what some of these phrases really mean is not too clear: after all, just how sly *is* a fox? When you stop to think about it, these kinds of phrases are more the result of placing characteristic descriptions of human behavior onto animals rather than the other way around. But the point remains: we see in animals enough which reminds us of ourselves that we may be inclined to believe that understanding animal behavior will give us some insight into human behavior. It wasn't until the latter part of the nineteenth century that there was any systematic or scientific foundation for this belief. It was well before that time, however, that the characteristics of animals were widely used as comparisons and contrasts with human behavior; we will present some examples later. Since the time of Darwin, most biological scientists have at least admitted the possibility that principles emerging from the study of animal behavior may have direct bearing upon the explanation or deeper understanding of human behavior. Among these scientists there have been some who have taken that possibility as a fact and made claims for a strong similarity between the behavior of men and, for example, albino rats, or monkeys and apes, or greylag geese. In general, then, a third deep interest in animal behavior comes from the increase in human *self-understanding* that may arise from its study.

Utility, aesthetics and symbolism, and self-understanding: these three topics encompass much of our concern with animals and their behavior. Now we will consider each of them in some more detail.

Utility

In the modern world our utilitarian interest in animals and their behavior may be grouped under three headings: agricultural, medical, and ecological.

Agricultural Concerns: Feeding, Reproduction, and Pest Control

Agriculturalists have two basic reasons for becoming involved with the

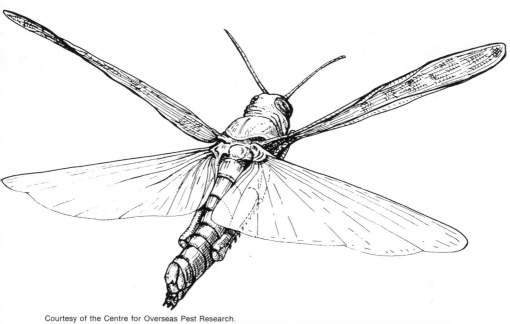

Courtesy of the Centre for Overseas Pest Research.
Figure 1-1 A desert locust in flight.

study of animal behavior. First, their success depends upon knowledge of the feeding and reproductive behavior of their livestock and poultry. Beef cattle, swine, and lambs that eat well and remain quiet will yield more and better meat than nervous, finicky animals. Modern husbandrymen will go to substantial lengths to arrange the environments of their animals so as to maximize food intake and minimize outside disturbances (Cole and Ronning, 1973). Animals that are particularly desirable in regard to their commercial characteristics (meat quality and quantity, milk production, wool production, etc.) are used as breeding stock for the next generation. Analyses of the reproductive behavior of domestic stock can lead to innovative techniques in sperm collection and insemination (Hafez, 1962).

The second area of agricultural concern has to do with the control or destruction of agricultural pests. The scope of this problem cannot be underestimated: insects make off with a very large percentage of all the food grown in the world each year, and rats and insects together may consume fifty percent annually of all stored foodstuffs in some underdeveloped countries (Ehrlich and Ehrlich, 1970). Although at one time broadly-acting pesticide applications may have appeared to be an answer to insect depredations, it is becoming increasingly obvious that insecticides have a number of serious drawbacks associated with their use. Sprayings often do not succeed in permanently lowering pest populations, and there is even evidence that they may sometimes produce *larger* pest populations than would have occurred without spraying. The target insect usually evolves into an insecticide-resistant species over a period of years. Moreover, pesticides affect the pest's natural enemies adversely, destroy pollinating insects, and have totally unexpected and disastrous effects on some bird populations (Peakall, 1970). Humans themselves may not be immune to prolonged exposure to low dosages of a variety of these toxic substances. In any event, there is growing consensus that a biologically sounder approach to pest control is necessary.

Perhaps the single most important agricultural pests are the several species of large grasshoppers, especially one—the desert locust. They are extremely dramatic plant consumers. These animals are found in scattered low density populations during most years. However, at irregular intervals, their numbers increase sharply and the locusts band together and fly great distances, descending at times to strip the earth of its vegetation. A single swarm may contain ten billion individuals (Evans, 1968); it requires little imagination to appreciate the damage a group of this size could inflict on a farmer's crops. The Centre for Overseas Pest Research located in London has workers in various parts of the world trying to learn what initiates swarming behavior, what factors control the movements of locust bands in outbreak areas, and what strategies might be effective in stopping the animals before they set off on their aerial journey; it may be impossible to control them afterwards. The results of this behavioral research have not as yet led to a solution of the problems posed by locusts, but it is certain that we are closer to the solution than we would have been without this important work.

In the case of another insect pest, the screw-worm fly, biobehavioral research has led to a complete solution to an agricultural problem. This insect, which roughly resembles a housefly or bluebottle fly, has the unsavory habit of laying three or four hundred eggs in the wounds and sores on livestock. A single wound may carry one thousand eggs. The eggs hatch into wormlike larvae that set about consuming the flesh of their hosts. This debilitates or kills cattle and makes the screw-worm fly a very unpopular animal species with cattle owners.

Fortunately for both the livestock and their owners in the southeastern United States, the United States Department of Agriculture was able to eradicate the fly entirely from the area. They did so with a program based on the knowledge that the male fly's mating behavior was not affected by sterilizing irradiation. Over a series of fly generations planes released a total of more than *two billion* male flies that had been reared on tons of whale and horse meat and then sterilized. The sterile males competed with fertile males for females. Once a female fly has mated she will not mate again. This means that if she copulates with a sterile male her reproductive career is over. Eventually there were so few fertile males relative to the number of sterile males that the female flies found only sterile males and produced only sterile eggs; in this way the fly was driven to extinction in Florida (Knipling, 1960).

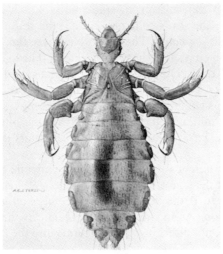

Courtesy of the Trustees of the British Museum (Natural History).

Figure 1-2 A human body louse.

Although biological methods of pest control based on knowledge of reproductive behavior and natural enemies are becoming better known and more popular, it is unlikely that many will be as completely successful as the screw-worm control program. However, biological control programs in general should eventually do a better job than most pesticide programs; additionally, of course, they are likely to have far fewer deleterious environmental side-effects.

Medical Concerns: The Control of Human Disease Vectors

Although most readers of this book are likely to feel remote from diseases in which animals play a direct or facilitating role, most of the people who have lived in the world have not been so lucky. Hans Zinsser's fascinating book *Rats, Lice, and History* (1935) makes this point quite clearly as it describes the hideous effects of epidemic plagues on human populations in the past. The "hero" of Zinsser's book is typhus fever, a disease produced by several species of microorganisms belonging to the genus *Rickettsia*. The disease is characterized by very high fever, pains in the limbs, and severe headache. On the fourth or fifth day an eruption of small pink spots spreads over the body. The spots soon become purplish, then brown. Death rates in epidemics of typhus may run as high as 40 percent.

Epidemic typhus, the most severe form of the disease, was a standard feature of European history during the major military campaigns of the past. More recently, during the five-year period 1917–1921, twenty-five million persons contracted the disease in Russia and Poland, and two to three million died.

The mode of transmission of epidemic typhus was not discovered until 1909, when Charles Nicolle proved that the responsible vector was a louse, *Pediculus humanis.* The louse contracts the disease from an infected person when it sucks the person's blood. The louse will die within approximately 9 days. During that time its feces contain the microorganism *Rickettsia prowazekii* (named for von Prowazek, a Polish scientist who died of typhus while studying it). If the louse moves to an uninfected person and bites him, and the person scratches the bite and breaks the skin, then the *Rickettsia* contained in the feces enters the human bloodstream.

A somewhat milder form of the disease, endemic typhus, is carried by rats and their fleas, not by lice. Fleas ingest *Rickettsia mooseri* that have infected a rat. If the flea then deposits feces on a man, there is a chance of infection. As in the case of epidemic typhus, it is infection through the feces, not direct injection of germs into the bloodstream, that produces infection.

The major controls of typhus are immunization, rat control, and DDT campaigns. The behavior of rats, lice, and fleas has not been put directly to use against them in the fight against typhus. However, as Zinnser has pointed out, before people understood the role of these animals in the production of epidemics, the disease was totally mysterious and thus doubly terrifying. The terrible death rates produced by the epidemic were due in part to the panic that stemmed from the unknown nature of the disease's cause. People by the thousands would attempt to flee an affected area and thereby spread typhus more widely, disrupting agriculture and other activities over huge regions. In more ways than one, knowledge of the louse's role in transmitting typhus was fundamental for its control.

Similarly, understanding of the life cycles of mosquitoes and the protozoan responsible for malaria (the most serious debilitating disease in the world prior to the advent of DDT) was an essential foundation for successful anti-malaria programs. In fact, it was not until 1897 that Ronald Ross showed that some mosquito species could contract the malaria parasite from a malarial patient and that the parasites matured in their mosquito hosts (Oldroyd, 1965).

The stories of typhus and malaria and the associated control programs are by now classic examples of the usefulness of behavioral biology for public health. We will complete this section by describing two more cases, less well known in general, that illustrate the important issues involved. The two diseases in question are *schistosomiasis* and *Chagas' disease.*

Schistosomiasis is a very unpleasant disease caused by an infection of liver flukes, a kind of flatworm (*Phylum Platyhelminthes*). The life cycle of the fluke has several stages. One form, called *cercaria,* lives in fresh water such as anti-malaria drainage ditches and irrigation canals. Humans who use this water for drinking, washing, or swimming are likely to become infected by the *cercaria,* which can penetrate through the skin.

Once within the human bloodstream the fluke lives for a while in the lungs, then migrates to the liver, and finally ends up in the veins supplying the bowel, the bladder, or the small intestine (depending upon the species of fluke involved). Nurtured by human blood, the flukes begin to lay eggs—as many as 3000 per day. A few of these eggs escape the human body through the urine or feces, but most of them stay trapped within the body and produce the symptoms of the disease. Some of the eggs that escape will reach fresh water eventually, at which time they mature into a form of the fluke called *miracidia.* Each *miracidium* is motile, and in its journeys it may reach and infect a particular species of fresh-water snail (different species of flukes are associated with different species of snails). Within the snail the fluke undergoes a series of sexually productive phases which result, finally, in the production of the *cercaria* which bore out of the snail and into the water, ready to infect humans—and the cycle is complete (Figure 1-3).

The interaction of the fluke's complicated life cycle with changing styles of human existence may be seen by the consequences of the massive Aswan Dam project in southern Egypt. In earlier times *schistosomiasis* was relatively rare in that part of Egypt because of the relative lack of bodies of still water: still, fresh water is required by the snails who serve as intermediate hosts. However, with the advent of the dam and the massive complex of irrigation ditches that accompanied it, the snail has spread and with it the opportunity for *schistosomiasis.* The rate of infection has increased substantially: the enormous productive potential of the fluke and the abundance of snails in the proper habitat make *schis-*

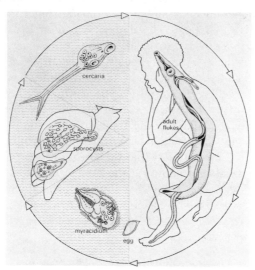

From *Lower Animals* by Martin Wells. Copyright 1968 by McGraw-Hill Book Company. Used with permission of McGraw-Hill Book Company.

Figure 1-3 The various stages of the life-cycle of the liver fluke, and their relationships with human and snail hosts.

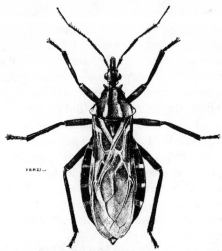

Figure 1-4   An assassin bug capable of transmitting Chagas' disease to human beings.

tosomiasis a very difficult disease to control. This is one example of the unfortunate higher-order consequences of massive environmental interventions.

Finally, we turn to Chagas' disease, which is characterized by fever, swelling, heart malfunctions, and, in children particularly, nervous system damage that is often fatal. The disease is caused by an infection of protozoans (*Trypanosoma cruzi*). The trypanosome is carried by a large insect known as the assassin bug (genus *Triatoma*), which lives comfortably in or near human habitation, particularly in the cracks and crevices in adobe brick dwellings (Figure 1-4). The assassin bug is active at night; should it encounter a human it may insert its beak under the skin and draw blood, thereby allowing the trypanosome to move in the other direction and enter the person's bloodstream. The protozoan multiplies rapidly within the human body and the disease process begins; there is no satisfactory treatment known at the present time.

Chagas' disease has been a serious problem in South America. Incidentally, Charles Darwin encountered the assassin bug over 100 years ago during his travels in western Argentina. In his book, *The Voyage of the Beagle* Darwin wrote:

At night I experienced an attack (for it deserves no less a name) of the Benchuca, . . . , the great black bug of the Pampas. It is most disgusting to feel soft wingless insects about one inch long, crawling over one's body. (Darwin, 1860, p. 331)

It may be that Darwin contracted Chagas' disease at that time, because he was chronically ill with a mysterious malady upon his return to England and until his death.

It has been clear for many years that one way to prevent or reduce the incidence of the disease was to make human homes less accessible to the assassin bug. Modern, relatively expensive methods of construction or fumigation are out of the question for most of those who are affected by the problem. However, observations of the building habits of a South American bird, the *hornero* or ovenbird, suggested a solution. The bird gets its name from the large melon-shaped mud nest it builds, which looks something like the outdoor ovens South Americans often used to bake bread. The nest does not crack and become fissured like adobe; it is a remarkably durable piece of work. The ovenbird makes its nest out of a mixture of cow dung and sand, which upon drying is odor free. Public health officials in Brazil reasoned that they could copy the ovenbird's formula and use the resulting cement to cover the inside walls of adobe houses. This would seal the walls and prevent the development of new cracks and crevices, thereby depriving assassin bugs of available hiding places. In 1958 two hundred thousand homes were plastered with "ovenbird brand" cement, and officials hoped to be able to apply the cement to ten times that number within the decade. Apparently this procedure has been effective in reducing the incidence of Chagas' disease; man has benefited by borrowing the technology of a bird (Welty, 1963).

Animal Conservation

There are a variety of reasons for wishing to preserve animal species in the natural environment by creating national parks and preserves. One practical justification for such parks is that they can provide information about the ecology of truly natural areas—information that may be critical in understanding factors that contribute to ecological breakdown in agricultural and industrial areas. However, a policy of preservation of animals and space leads to problems in the management of these areas; some of these problems can be solved only by the study of animal behavior.

A well-publicized example of this sort of research involves the grizzly bears of Yellowstone and Glacier National Parks. Almost every summer we read that some visitors to the parks have been maimed, and in some cases killed, by the bears. One way to prevent further tragedies of this sort would be to kill the grizzlies; this would not be a particularly difficult task. But if one believes that the parks exist, in part, for the protection of the grizzlies, very few of which remain in the United States, and that these animals are an important component of the region's ecology and worth preserving for this reason, then killing the bears is an unsatisfactory solution.

If campers *and* bears are to be saved, it is critical to know as much as possible about why the bears are behaving the way they are. This means finding out how far individual grizzlies roam, the size of their total population, under what circumstances they come close to human camps, and so on. Over the past several years, thorough studies by Park Service officials and the Craighead brothers, John and Frank, have shown that the "grizzly problem" is due in good measure to the location and utilization of garbage dumps in the park. A very large proportion of the parks' grizzly population has been attracted to and learned to exploit the ready sources of food provided by human garbage. The garbage has lured the bears into areas packed with campers and, worse yet, continuing contact with the garbage has habituated the bears to human odors, which they might otherwise avoid. Instead they have learned to associate these odors with food.

The obvious solution to the problem is immediately to remove the garbage dumps and make other arrangements for campers' waste disposal. This has been the approach taken by the Park Service. But the Craigheads believe that a gradual elimination of the dumps would be the wiser course: by weaning the bears away from the dumps gradually, the bears may come to spend progressively longer periods of time in more remote park regions. The Craigheads are concerned that the immediate clearing of the dumps will lead to aggressive food searching by the bears in areas where the density of campers is high. (See also Johnson, 1972.) One way or another, we should have the answer to this debate within the next several years.

African national parks are faced with other, equally complex problems that require studies of animal behavior. For example, antelopes and other grazing species often migrate enormous distances during the year. If they are to be protected effectively, it is crucial to know the extent of their movements so that park boundaries can be properly drawn. Similarly, the question of how to maintain a proper balance of the many game and predatory species living in an area can be answered only through the study of the behavior and ecology of these species. For example, recent observations in Serengeti Park in Tanzania have revealed that each of the three most abundant grazing antelopes (wildebeest, zebra, and Thompson's gazelle) consumes a different part of the grasses that grow on the plains. The feeding patterns of each of these species seems to facilitate the others; thus the way the zebra eats makes it possible for the gazelle to forage easily. Moreover, all three species prefer to forage in short as opposed to long grass.

These simple findings have important management implications. First, any policy or action that affects one species is sure to affect the others; their close interdependence in feeding activity assures this. Second, it may be possible to develop a more cooperative policy with African pastoralist tribesmen who are accustomed to grazing cattle in the areas bordering the park. It had been previously thought that cattle must be kept out of the park for fear they would overgraze the grassland and

a

b

Picture a), photograph by L. H. Barnard. Picture b) from Godey's, November 1883.

Figure 1-5 Some examples of the use of feathers as ornaments: a) A *luluai,* of New Guinea wears the plumed "golden wig." b) Feather-trimmed bonnets were the fashion for American women in 1883.

thus menace the food supply of the wild species. Now it appears that the cropping of tall grass by cattle might act to the advantage of the antelopes, who prefer shorter grass anyway (Bell, 1971). In general, one can be certain that throughout the world the conservation of animal species will depend to a large extent on how well their behavior and ecology are understood.

Aesthetics and Symbolism

Paradoxically, perhaps, human appreciation for the natural beauty of other animals often leads to the animals' destruction. New Guinea tribesmen have killed birds of paradise for their plumes, probably for thousands of years, because these feathers make such spectacular ornaments; and at the end of the nineteenth century the demand by European and American women for the feathers of egret, gull, tern, hummingbird, and bird of paradise, was so great that some of these species were nearly exterminated (Figure 1-5).

Fortunately, our aesthetic sensibilities need not always be satisfied only by the death of the object of our admiration. Humans have probably always found flying birds and butterflies attractive. The fascination of birds in flight has expressed itself in many forms, one of which is the very ancient sport of falconry. A successful falconer must learn a great deal about the behavior of his birds (White, 1951). Other species have been specially bred because of their aesthetic value; canaries, parakeets, and several species of exotic fish are in this category. Species once valued for their utility, for example, horses, are now kept, in this country at least, only for their aesthetic value (or the monetary gain that may be associated with it). Highly refined breeds of dogs have arisen as a result of dog fanciers' concern for the fine points of conformation or obedience; unfortunately, this concern does not always act in the best interests of the health or reproductive success of the breed. The varieties of purebred domestic cats, although not so numerous as dogs, nevertheless also reflect human interest in the refinements of animal beauty. In other countries, for example, in Southeast Asia, fish and gamecocks are bred both for beauty and for fighting.

In addition to the admiration we show for the form or grace of some other species, humans are also quick to fasten upon special traits of some animals and elevate these to symbolic status. In the case of medieval English falconry there was a direct relationship between the rank of a man and the kind of bird he might legitimately use in his sport. The eagle was considered the most majestic of the species, and hence could only be flown by the king. The peregrine falcon, a somewhat smaller bird, but still magnificent, could be flown by an earl, while flying the still smaller kestrel, which could hardly be trained, was left to the knaves. Some other examples of symbolic birds are shown in Figure 1-6.

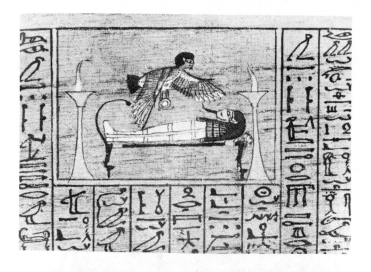

From *The Folklore of Birds* by Edward A. Armstrong. Dover Publications, Inc., New York, 1958, 1969. Reprinted through permission of the publisher.

Figure 1-6 Two examples of birds playing a symbolic role in human burials: a) The Egyptians believed that a soul bird carried the soul to the next world. b) Canadian Eskimos believed that a diving bird, the Arctic loon, carried the dead person's spirit with it on a final dive to the land of the dead.

Worship or symbolization of animals has by no means been restricted to birds. The ancient Egyptians revered a host of animals, including the scarab beetle. Some scarab beetles collect balls of dung and push them about with their hind legs until they reach a hole in the ground where they bury themselves and their collection. There they either feast on the ball of dung or lay an egg upon it. The egg hatches into a larva that

From *Fabre's Book of Insects* by J. Henri Fabre. Dodd, Mead & Company, New York, 1921.

Figure 1-7 Two dung beetles rolling a ball of dung to a nest site.

consumes the dung before pupating. The Egyptians believed the scarab represented the agency that moved the sun across the sky each day. Figure 1-7 shows two of these beetles.

Snakes, by virtue perhaps of their resemblance to the human penis, appear as symbols in the myths of many cultures including our own: for instance, in the story of the Garden of Eden. Similarly, the sexual and muscular power of bulls have made them favorite symbols. The Minoans of ancient Crete (2500–1500 B.C.) held elaborate rituals in which acrobats performed spectacular feats of agility in a ring with bulls; and today, at Sir Arthur Evans' reconstruction of the Minoan capitol of Knossos, there stands a huge set of horns, representing the importance of the bull in Cretan culture.

There are literally hundreds of examples of the human proclivity to convert animals into symbols on the basis of their appearance and behavior. Readers interested in this aspect of human nature would enjoy reading Laurens van der Posts' *The Heart of the Hunter* (1961), which provides a beautiful treatment of the mythical world of the Bushmen of the Kalahari desert. This is a world whose creator is a praying mantis, whose source of wisdom is the large and graceful eland antelope, and whose inhabitants are the mongoose, the porcupine, the hyena, the lynx, and many other animals, each with its own personality and destiny.

Self-Understanding

The use of animals and animal forms in decoration and myth is closely related to our third introductory theme: the relation between animals and human self-understanding. In very recent years there has been a spate of books and articles written for general audiences which attempt to show that modern studies of animal behavior give us direct and useful insights into the current human situation: Robert Ardrey's *African Genesis* (1961) and *The Territorial Imperative* (1966), Konrad Lorenz' *On Aggression* (1966), Desmond Morris' *The Naked Ape* (1967), Lionel Tiger's *Men in Groups* (1970), and Lionel Tiger and Robin Fox's *The Imperial Animal* (1971) are perhaps the best-known examples of this genre. Each of these books develops a theme concerning modern humans and links it with facts and/or speculation about human prehistory and the behavior of currently living nonhuman animals. These themes are usually provocative and the books well written enough that large numbers of people buy, read, and talk about them. But it would be an error to suppose that these books represent the first attempt to bring the behavior of animals to bear upon human self-understanding, for there is a long history in this tradition. Originally, and to a surprisingly large extent still today, the behavior of animals has been used to make decisions about human character and worth: in other words, to make value judgments about man by comparing and contrasting him with animals. Comparative psychology was originally comparative *moral* psychology, and some of the moralizing flavor of the prescientific work remains in the modern popularizations of animal behavior study.

A nice example of early comparative moral psychology is found in the work of the famous French essayist Michel de Montaigne (1533–1592). In the essay *Apology for Raymond Sebond* Montaigne makes use of personal observations and Roman texts to chastise men for their pride and sense of elevation above the animals:

. . . there is no apparent reason to judge that the beasts do by natural and obligatory instinct the same things that we do by our choice and cleverness. . . . Why do we imagine in them this compulsion of nature, we who feel no similar effect? Besides, it is more honorable, and closer to divinity, to be guided and obliged to act lawfully by a natural and inevitable condition, than to act lawfully by accidental and fortuitous liberty; and safer to leave the reins of our conduct to nature than to ourselves. . . . it is apparent that it is not by a true judgment, but by foolish pride and stubbornness, that we set ourselves before the other animals and sequester ourselves from their condition and society. (Montaigne, 1948, p. 336–337; 358)

During the seventeenth and eighteenth centuries there was a flowering of satire, much of which used comparisons between man and animals for telling effect. The theme expressed above by Montaigne was often reiterated. The tendency to denigrate man by claiming him to be no better than animals reached a climax during the middle of the eighteenth century, at which time a reaction began. Subsequent essayists and philosophers, while not quite so misanthropic, were nevertheless concerned to describe the irrational in man. And, to the extent that irrationality was considered an incorrigible part of human nature, the comparison of human behavior with that of nonhumans was considered apposite (Lovejoy, 1961).

This theme submerged in nineteenth-century philosophy but soon reappeared in the work of some physicians and anthropologists of the time as they worked their way through to a conception of the unconscious mind. These ideas culminated at the end of the century with the work of Freud and the development of psychoanalysis (Ellenberger, 1970). Although the Freudian and neo-Freudian positions represent but one general viewpoint out of several in contemporary American psychology, their significance in reawakening a concern with human irrationality and instinct was and still is enormous. Freud himself had relatively little to say about the relations between men and nonhuman animals. It remained for the students of Charles Darwin's ideas to connect natural selection and the descent of man to a post-Freudian view of human nature.

It is essentially this perspective that characterizes a good deal of the current popular extrapolations of animal-behavior study. And, lest the reader let the apparent scientific foundation for these extrapolations confuse him concerning their status as moral statements, he should compare the sixteenth-century statement of Montaigne, given above, with the following twentieth-century statements of Robert Ardrey:

We must know that while the human brain exceeds by far the potentialities of that possessed by any other animal species, its psychological processes probably differ not at all from those of other higher animals, and from those of lower animals as well. . . . while our evolutionary inheritance seems to place a limitation on human freedom, an eternity of natural selection has presented us as its legacy with the foundation of human strength. . . . . man no different from any other animal is a complex of expressions, frequently conflicting, in which no single determinant . . . holds exclusive permanent domain. . . . Whether morality without territory is possible in man remains as our final, unanswerable question. (Ardrey, 1966, p. 350–351)

In the preceding discussion we ignored the development of the ideas which led to Darwin's espousal of organic evolution through the principle of natural selection. We will turn to these ideas now, beginning with the ideas of René Descartes (1596–1650), with whom much of modern Western science and philosophy begins. It is sometimes stated that Descartes claimed animals to have no minds (souls), and that this lack is what distinguished them definitively from humans (Boring, 1950). However, other scholars have pointed out that Descartes was himself confused or uncertain about how to treat the mental or spiritual aspect of nonhuman life (Klein, 1970; Peters, 1965). Although his early view was that animals could be viewed simply as machines (seventeenth-century machines), his later writing was more ambiguous on this point. Nevertheless, he never totally gave up the view that animals could be considered totally as *bodies,* even though these bodies might have concomitant, temporary minds or souls. And it was this distinction between animals as bodies and men as bodies plus immortal souls (dualism), codified and strengthened by Descartes, that later evolved into the position on the separation between men and other animals that the Darwinian revolution overturned.

The issues at stake in relating men to other animals arose from the very old theme of the *Great Chain of Being,* which was

. . . a conception of the universe as . . . composed of an immense (or infinite) number of links ranging in hierarchical order from the meagerest kind of existents, which barely escape nonexistence, through 'every possible' grade up to the . . . highest possible kind of creature. . . . (Lovejoy, 1960, p. 59)

Until the eighteenth century the Chain of Being was a static conception: the universe was perfect and full, the expression of its Creator. Thus when Descartes distinguished between man and other animals, he was making a judgment which he took to be true without regard to time: neither man nor animals would ever change. The implications of this static scheme, when scrutinized by eighteenth-century philosophers, were sufficiently troublesome to call the entire thesis into question. Voltaire and Samuel Johnson, among others, wrote highly critical essays against the concept. Voltaire, for example, could not believe that the organisms then living constituted all possible organisms. In 1764 he asked,

Is there not visibly a gap between the ape and man? Is it not easy to imagine a featherless biped possessing intelligence but having neither speech nor the human shape, who would answer to our gestures and serve us? And between this new species and that of man can we not imagine others? (Lovejoy, 1960, p. 252)

Such critical attacks were part of the movement toward a more dynamic conception of nature which culminated in the middle of the nineteenth century with Charles Darwin and the theory of natural selection. The century that elapsed between Voltaire's questions and the first edition of Darwin's *The Origin of Species* (1859) saw, among other achievements in biology, the completion of the great classificatory system of Linnaeus (1758) and the publication of Lamarck's evolutionary theory (1809). Although Lamarck's theory of the inheritance of acquired characters was overthrown by the theory of natural selection, it nevertheless contributed a great deal to the intellectual process and climate which prompted the work of Darwin.

Charles Darwin, Natural Selection, and Evolutionary Biology

The concept of natural selection which Darwin presented and supported with examples and argument in the *The Origin of Species by Means of Natural Selection, Or the Preservation of Favored Races in the Struggle for Life* (1859) is so fundamentally a part of the biological approach to human self-understanding that we must be sure to have a proper appreciation of it before proceeding. It is not an exaggeration to say that natural selection is *the* central idea for all of modern biology and is basic to any biological question.

Natural selection is the outcome of differences in the reproductive success of individuals. Reproductive success is a relative matter: a successful individual is one that has more offspring reaching reproductive age and having offspring themselves, than some other individual. One must understand that the essence of the concept is not survival *per se,* nor strength, nor intelligence, but simply reproductive ability. "Survival of the fittest," a phrase which seems to be aligned with Tennyson's phrase "Nature, red in tooth and claw," in fact refers to the survival in populations of those hereditary factors, genes, that contribute to relative reproductive success. Survival itself is important only to the extent that it brings about the production of reproductively viable offspring. A person who lives to be 100 years old but remains childless is an evolutionary failure. A woman who dies in childbirth but whose child reproduces is, relative to the first person, an evolutionary success. There are cases, in fact, in which rapid post-reproduction death may be advantageous; if the individual is not capable of reproducing again (which is true for the members of many animal species) and takes no role in the care of its offspring (similarly true for many if not most animal species), then life after reproduction serves no direct evolutionary function and may exhaust food supplies, attract predators to younger animals, and so on. Among the insects which comprise approximately eighty percent of all the animal species now living, there are a large number in which death follows rapidly upon the laying or fertilizing of eggs.

Another subtle but critical point to understand about natural selection is that it is not *concerned* with the survival of animal species in that sense of concern which connotes "caring about" or "intentionally producing." Selection is not a supernatural entity or guiding force capable of steering a species along the path of preservation. It is simply the fact that individuals differ in how many copies of their genes they can get into the next generation. These reproductive differences between individuals are the result of differences in how well adapted they are to the *current* environment. Should the environment change rapidly relative to the species' reproductive rate and genetic variability, extinction may well result. Natural selection does not and cannot plan for future environments; ninety-eight percent of all the animal species that ever graced the earth are now extinct. We emphasize this point because it is very easy to find philosophers and even some biologists who, intentionally or not, present evolution through selection as a progressive or directed phenomenon in the world, as if evolution implied continuous betterment for the preservation of species. In our opinion, there are no nontheological reasons to accept this view.

The Natural Selection of Behavior

All bodily structures and functions, without exception, are products of heredity realized in some sequence of environments. So also are all forms of behavior, also without exception. (Dobzhansky, 1972, p. 530)

This assertion, made by one of America's most eminent geneticists, serves well as the basic axiom for all modern studies of animal behavior. As we shall see below, zoological and psychological perspectives on animal behavior have traditionally differed in regard to the relative emphasis placed on heredity and environment, but we shall also see that a new synthesis of these historically divergent views may be emerging.

When we speak about the evolution or natural selection of behavior, it is important that we keep in mind the distinction between structural and functional descriptions of behavior. A structural description describes the motion of an animal, or part of an animal, in space over time; it is a characteristic of an animal that may be reasonably expected to have a precisely describable physiological correlate. A functional description, on the other hand, describes what the consequences of the animal's motions are such as defending a territory, reproducing, or escaping from a predator. There is an almost indefinitely large number of ways, structurally speaking, that an animal might behave to fulfill any one function. In fact, however, research into animal behavior reveals time and again that a single animal species shows only a limited variety of "structural solutions" to its "functional problems" of food acquisition and intake, reproductive behavior, defensive behavior, and so on. We need not provide examples here; later chapters are devoted entirely to these topics. For now it is important to note only that the *species specific* (or *species-typical* or *species-characteristic*) nature of the behavior gives us confidence that it, like the muscles, nerves, and other structures which embody it, is a product of natural selection. The behavioral capacities of animals are evolutionary products.

This fact is disturbing to many people who apparently believe that if behavioral adaptations are the result of natural selection they must be "genetically determined" or fixed in some invariable manner. Extrapolating to humans, this would mean that our behavior is predetermined as well—since we are animals with an evolutionary history—and this is repugnant to many of us, largely for philosophical reasons. However, such thinking is based on a misunderstanding of the role of genes in development (see Chapter 3). The fact that selection has operated on behavior in the past simply means that living animals have the kind of genetic information that tends to interact favorably with certain environments. These gene-environment interactions lead to the development of animals which can behave in adaptive ways in an environment typical of the species. However as Dobzhansky (1972) points out with special reference to human beings,

Environments are infinitely variable, . . . and new ones are constantly invented and added. . . . It would require not a scientific but something like divine knowledge to predict how much the stature, or IQ, or mathematical ability of any individual or population could be raised by environmental or educational modifications or improvements. . . . (p. 530)

In other words, although there are most definitely limitations and restrictions placed on the behavioral flexibility of animal species, humans included, it does not mean that behavioral development occurs without regard to the environment. Animals often have a substantial degree of developmental flexibility with a variety of behavioral options depending on the kind of environment they find themselves in. Perhaps more important still, the concern of the anti-determinists overlooks the extremely positive aspect of selection which acts to favor individuals whose behavior happens to be adaptive, able to meet the functional problems of a particular environment. The end products of the evolutionary process are populations of animals wonderfully adapted to a way of living and reproducing in their specific habitat.

Three Perspectives on Animal Behavior

Professional students of animal behavior typically receive advanced training in one or more of three disciplines: zoology, physiology, and psychology. It is one of the themes of this book that rather general differences between these three disciplines have prompted and promoted three separable perspectives on the behavior of animals and the best ways to study it. A second theme is that in recent years increased communication and training across these disciplines have led to the beginnings of a synthesis of viewpoints, so that an integrated discipline which studies animal behavior, a discipline well grounded in evolutionary and functional biology and acute to the subtleties of psychological analysis, is emerging and gaining strength. A number of different names might be suggested for this emergent discipline: comparative animal behavior, behavioral biology, psychobiology, and biopsychology are some of the labels currently in use in America and England. We will not be concerned with choosing among these names, but rather with 1) describing briefly the differences that characterized the three perspectives on animal behavior and 2) suggesting how the new synthesis appears to be developing. The other chapters in the book will provide ample evidence of both the similarities and dissimilarities introduced here.

Zoological Perspective: The Development and Foundations of Ethology

Ethology is a natural science, a branch of biology, from which it took the comparative method for the study of behavioral morphology and the analytic method for the causal analysis of behavioral physiology. Its philosophical base is a critical realism. Its orientation is neo-Darwinistic and it enjoys a fruitful exchange of ideas with other schools of behavior, especially with behaviorism. . . . (Eibl-Eibesfeldt, 1970, p. 9)

This characterization of ethology, which we will take to be equivalent to the zoological perspective on animal behavior, emphasizes both its similarities and dissimilarities with physiological and psychological (behaviorist) approaches to the study of animal behavior. A very brief history of the development of ethology will help us to understand how the modern field acquired its current character. Readers interested in other, more complete treatments of this material may consult Eibl-Eibesfeldt (1970), Hess (1962), Hinde (1959), Lockard (1971) or Lorenz (1970).

As we have already pointed out, modern animal behavior study begins with Darwin and the theory of natural selection. Several authors have noted that in addition to *The Origin of Species* (1859), it was Darwin's later work, including *The Descent of Man* (1871) and *The Expression of the Emotions in Man and Animals* (1872), that provided immediate impetus for the development of animal behavior study (Lockard, 1971; Lorenz, 1965b). Some of the men Darwin influenced went immediately in search of "higher mental faculties" in animals, confident in their search because of the continuity in mental life that Darwin claimed must exist between nonhuman and human forms. This research effort consisted, for the most part, of collecting information about farm animals and pets considered by their owners to be particularly intelligent. As long as one believed the anecdotes owners and trainers told, then one could believe that there were few mental feats performable by people that at least one dog, horse or pig had not also accomplished. When, however, these stories were checked out and the animal in question tested, the result was likely to be somewhat less astonishing. The best-known case of this sort involved a German horse, "Clever Hans," who, it had been demonstrated time and again, could do arithmetic by tapping out answers to his trainer's questions with his forefoot. Close observation of horse and trainer revealed that when Hans arrived at the proper number of taps the trainer would, unconsciously, make a move of one sort or another that Hans responded to by stopping his tapping. The trainer then would reward Hans with food or affection. Thus, although Hans had learned a subtle discrimination that his trainer reinforced regularly, he was not doing arithmetic (Watson, 1914). By approximately the end of the century, interest in the anecdotal approach to the study of animal intelligence had ceased. The new view of the matter was precisely stated by C. Lloyd Morgan in his textbook of 1894: "In no case may we interpret an action as the outcome of the exercise of a higher psychical faculty, if it can be interpreted as the outcome of the exercise of one which stands lower in the psychological scale." This rule is now commonly known as *Lloyd Morgan's Canon;* it is the application of the more general principle of parsimony in science to the interpretation of the behavior of animals.

The development of the study of animal intelligence during the twentieth century is more the story of psychology than of ethology, and we will return to that story below. There was another development at the turn of the century that led directly to the modern field of ethology: two ornithologists, one American and one German, recognized that the Darwinian view of evolution held for behavior as well as for organs and other structural characters. This insight, which we have already described above, came almost simultaneously to Charles Otis Whitman of the University of Chicago and Oskar Heinroth of the Berlin Zoological Gardens. Both men realized that closely related species could be differentiated on a behavioral basis: in other words, that some forms of behavior could serve as a useful taxonomic indicator. Heinroth in particular was concerned to demonstrate that phylogenetic relationships could be established by behavioral analysis, and that the concept of *homology* ("resemblance due to inheritance from a common ancestry"—Simpson, 1961), so useful to anatomists and embryologists, would be applicable in behavior study as well.[1]

Both Whitman and Heinroth trained students who furthered this new field of ethology. Whitman's student was Wallace Craig, who introduced the important distinction between *appetitive behavior* and a *consummatory act*. A consummatory act is the last act of a behavioral chain; when the act is complete the animal rests or engages in some other obviously different kind of activity. The behavior which precedes it operates to place the animal in a situation where the consummatory act can take place. Craig called this sort of behavior appetitive on the grounds that it expressed the animal's "appetite" for the stimulus situation in which the consummatory response could occur. As Hess (1962) has pointed out, Craig's analysis of the distinction between appetitive behavior and consummatory acts avoided the philosophical difficulty of maintaining that animals strive purposefully to behave in ways that are adaptive or beneficial for survival. By suggesting that the goal of appetitive behavior is the discharge of the consummatory act, Craig gave us an idea which is central to much of our modern theory of the motivation of animal behavior. This idea will return, in various guises, throughout this book.

Among the students of Heinroth was the person who, more than any other, stimulated the development of modern ethological theory: Konrad Lorenz. Lorenz has spanned more than forty years with almost continuous

---

[1] Not all modern zoologically-oriented students of animal behavior believe that the concept of homology applies legitimately to behavior. The arguments have been summarized by Atz (1970). For other kinds of criticism, see Callan (1970) and Eisenberg (1972).

scientific research and writing. One of his early papers, "Contributions to the study of the ethology of social Corvidae" was published in 1931; yet he is probably best known, at least to American readers, for his recent popular book *On Aggression* (1966). Much of the flavor and character of early ethological theory stemmed from Lorenz's attempts to apply the ideas of Heinroth, Craig, the zoologist H. S. Jennings, and the physiologists Charles Sherrington and Erich von Holst to his own enormous fund of observational knowledge about animals (Lorenz, 1970).

Early in his career Lorenz was joined by the Dutch zoologist Niko Tinbergen. Together they performed some of the experiments that illustrated basic ethological concepts, including the famous analysis of egg-rolling in the greylag goose (Lorenz and Tinbergen, 1938, 1957) to which we will return shortly. Tinbergen's subsequent career, no less than Lorenz's is virtually definitive of the development of modern ethology; it was his 1951 textbook *The Study of Instinct* which first made the ethological perspective available to a large number of American readers, and his other books, including *Social Behaviour in Animals* (1953b), *Curious Naturalists* (1958), and *The Herring Gull's World* (1960b, 1953a), have provided elegant descriptions of animal behavior which set very high standards for other animal behaviorists to follow.

The fundamental ideas of ethological theory may be approached by referring to the study of egg-retrieving in the greylag goose that Lorenz and Tinbergen described first in 1938. The basic observations are these: a goose is sitting on a shallow nest containing a clutch of eggs. An egg is placed a short distance away from the nest. When the goose first sees the egg she does not reach toward it nor does she continue to look at it; rather she looks away and then returns her glance to it, now for a longer period. Then, after a series of brief and relatively rapid movements of her neck, she stretches her neck toward the egg as far as she can. The egg has been placed too far away for her to reach the egg without leaving the nest. But she does not get up immediately; instead she remains with her neck outstretched for several seconds. Then, still in the outstretched position, she arises from the nest and steps toward the egg with a way of walking that is observed only near the nest. She stops walking when the underside of her lower beak has contacted the egg. At that time the muscles in her outstretched neck begin to quiver with a characteristic tension, and, slowly, her neck contracts, with the result that the egg is rolled slowly toward her body. Of course the egg, being egg-shaped, has a tendency to roll off to one side or another depending on the terrain over which it is moving; thus the goose also moves her beak from side to side, thereby keeping the egg on a more or less straight course. In this way the goose returns the egg to the nest.[2]

---

[2] This description is a highly condensed paraphrase of the English translation of Lorenz and Tinbergen's original work (Lorenz and Tinbergen, 1957); it fails to capture the elegance and detail of their prose, to which the reader is directed for a fuller appreciation of ethological description.

On some occasions the egg being brought into the nest slips away and rolls back down the slope of the nest. Sometimes when this happens the goose is observed to continue the saggital (back-and-forth) movements of her neck but not the lateral movements. Moreover, the goose may now roll the eggs already in the nest around for a short period of time, until at last she settles down once again over them—at which time she rests until she sees the egg that has not been retrieved and begins the neck-stretching behavior again.

The egg-rolling behavior of the greylag goose has a number of features that illustrate quite beautifully the basic ethological concepts. We will discuss each of these briefly.

Fixed Action Pattern

These are movements which are highly stereotyped and once released are "played back" with little reference to environmental influences. The drawing-back movement of the greylag is a *fixed action pattern* (FAP). It is stereotyped; it is done with little variation from performance to performance by the same individual or among different individuals; and once begun the behavior is carried out whether or not the egg or any other object is actually being retrieved. Should the egg slip away or be taken away by an experimenter in the course of the movement, the animal does not stop its action but instead completes it. This component of the consummatory response does not, then, depend upon external stimulation for completion but appears to be "driven" by internal excitation. It is as if the goose had an internal program for this movement; once the program starts it continues until completion.

This action is easily distinguished from the side-to-side movements that balance the egg on the underside of the beak while rolling it back toward the nest. The lateral movements are flexible, responsive to patterns of changing tactile stimulation provided by the moving egg. This *taxis* or steering component of the retrieval behavior is guided by external stimulation, unlike the straight back movement of the tucked-down head. Thus it is possible to distinguish between an innate motor pattern and the steering movements that are superimposed on it.

Releasers (Sign Stimuli)

The stimuli that elicit fixed action patterns often constitute only a small fraction of the total stimulus information potentially available. The key or effective qualities of the total stimulus situation that release a behavior pattern are called *sign stimuli* or *releasers*. Lorenz and Tinbergen were able to show that very simple cues cause the greylag to exercise its retrieval responses. Almost anything with smooth rather than broken outline would elicit the stretching-out-of-the-neck response. And anything that provided certain tactile stimuli would be drawn back toward the nest. Because of the extraordinarily simple nature of the relation between stimulus and response, the two researchers could elicit neck stretching by placing a small toy dog or a large yellow balloon just outside the nest. The roundedness of the object was sufficient to trigger a response regardless of its size, color, and other properties. Lorenz and Tinbergen easily induced the greylags to retrieve wooden blocks and giant dummy eggs very unlike the bird's own eggs but which provided the proper tactile cues for the behavior.

Ethologists have discovered an enormous number of releasers; Eibl-Eibesfeldt (1970) provides many examples. A great deal of animal behavior is triggered by a few key aspects of an external stimulus; this fact makes it possible to mimic the essential parts of an object and elicit behavioral responses to what appears to a human to be a very incomplete model of the real thing. It should be noted, moreover, that it is possible to conduct similar studies with human beings. A two-month-old infant will often smile if a person, any person, bends closer to the child while directly facing it. The same response, muscles pulling the lips back in the expression we label a "smile," can be elicited by lowering a two-dimensional mask toward the infant provided there are two dark circles where eyes might be on a more complete model of a human face. A mask with just one or no "eyes" is largely ineffective in releasing a baby's smile, even if it is in other respects a better imitation face (Spitz and Cobliner, 1965). In ethological terminology, a smile is a fixed action pattern triggered by a releaser (visual stimulus of an approaching face-sized object with two dark, contrasting spots).

Innate Releasing Mechanism

In order to account for the stereotyped response of animals to releaser stimuli Lorenz hypothesized that there must be an *innate releasing mechanism* (IRM). This physiological mechanism is responsible for receiving neural messages stimulated by the releaser and for issuing specific messages to the muscles involved in the performance of a FAP. For example, the greylag goose must according to the Lorenzian model have in its central nervous system a structure or system that orders the appropriate response (pulling back the neck) when it receives a specific key message from tactile receptors on the undersurface of the beak. This mechanism ignores or blocks out other patterns of stimulation from these receptors. The result is that when and only when the goose is in certain situations and confronted with certain stimuli does it pull its head back. Almost always, under natural conditions, an egg is drawn back to the nest. The biologically appropriate outcome occurs, but totally without foresight or intention.

The concept of an IRM has been revised somewhat since its original formulation but it seems almost certain that it is a fundamentally correct idea. Recently physiologists have been providing evidence that nervous systems are designed to play out certain motor responses when a key or triggering input is relayed to certain neurons (see especially Willows, 1971). It is a measure of the insight of Lorenz and Tinbergen that they were able to develop the concept of the IRM in the 1930's and 1940's at a time when there was almost no physiological evidence in support of the idea.

The reader will find more detailed explanations of ethological concepts in Tinbergen (1958) and Eibl-Eibesfeldt (1970). For our current purposes it is enough to have some rudimentary feel for the ethological view of animal behavior. A classical ethologist sees animals as collections of behavior patterns, a very large percentage of which are innate fixed action patterns (remember that when an ethologist says "animal" he isn't automatically thinking about human beings or other mammals, but may be thinking about one of the 300,000 species of beetles, or a flatworm, or a rotifer, or a starfish, or a sea cucumber, or a jumping spider). These behavior patterns are viewed as having ecological functions, a role to play in meeting environmental demands placed on the individual in its natural environment. And finally, these behavioral units are considered to be the product of evolution every bit as much as the structural features that characterize the individual members of an animal species.

Physiological Perspective

As we have already mentioned, the work of Descartes during the first half of the 17th century provided a foundation for much subsequent scientific development. Descartes was not primarily an experimentalist, but he provided a rich source of ideas with which experimentalists could profitably work. His contributions to biology included his arguments that mathematics could fruitfully be brought to bear in physiology and his insistence that the lives of animals could be understood completely on materialist principles. Some of his more famous conjectures, for example that the mind (soul) interacted with the body primarily at the locus of the pineal gland in the brain, were designed to solve the problem of the relation between human mind (soul) and body that plagued him and many subsequent philosophers. He also held that muscles were innervated by nerves on a hydraulic basis; "animal spirits" flowed through tubular nerves to fill the cavities of muscles and cause them to act. In fact the notion that nerves were hollow and allowed the passage of active fluid lasted for a long time after Descartes. A contemporary of Descartes, William Harvey, had in 1628 overturned earlier theories of the motion of blood in the body by showing the pumping nature of the heart and the closed nature of the circulatory system. The powerful implications of this discovery for the understanding of circulatory physiology carried over, by analogy, to considerations of the workings of the nervous system, so that the idea of nervous "fluids" which ran in nerves and affected muscles lasted into the eighteenth century. Toward the end of the seventeenth century, in 1674, Leeuwenhoek had used his newly invented microscope to examine the nerves of cows to see if they contained cavities, but his results were not clear enough to dismiss the theory (Brazier, 1959).

It was not until the end of the eighteenth century that the electrical nature of nervous activity became firmly established. Some of the figures directly involved in elucidating the physics of electricity were also important in showing the role of electrical processes in living tissue: late in the century, for example, Volta and Galvani became involved in a dispute concerning the proper interpretation of Galvani's experimental results on the effects of electrical stimulation upon frog's legs. The dispute was partially resolved by another giant of science in that period, Alexander von Humboldt (Brazier, 1959).

One of the most interesting developments of neural sciences in the nineteenth century began with the development of phrenology in the early 1800's. Phrenology was the product of Franz Josef Gall (1758–1828), who, along with his student J. C. Spurzheim, created a storm of controversy with their theory on the relation between the brain, the skull, and behavior. From his days as a student Gall had believed that there were definite relationships between certain behavioral tendencies or capacities, and rather obvious features of the face or head. One of his earliest conjectures, for example, was that persons with particularly protuberant eyes had special facility as memorizers. During the early part of his career, which was reasonably orthodox and successful, Gall worked on this basic notion. He finally evolved it to the point where it included the assertion that behavioral predispositions, tendencies, or capacities (classically called human faculties) were associated in a one-to-one fashion with particular areas of the brain. If a particular faculty were highly developed in an individual then the brain would be swollen in the corresponding locus; furthermore, the skull would be relatively elevated over that portion of the brain in order to accommodate its relatively greater size. From this it followed, Gall argued, that the character or personality of an individual could be measured by careful measurement of the skull's configuration.

The extreme nature of Gall's hypothesis, and his unwillingness to accept counterexamples as indicative of its weakness, cast him and his work into substantial disrepute in many scientific circles. Phrenology became very popular nevertheless, which only deepened the establishment's antipathy to Gall's work. Interestingly enough, many well-known scientists and other influential figures were drawn to some parts of the phrenological viewpoint; although they may have discounted the general validity of cranioscopy (the use of skull conformation to infer facts about the brain or character), they recognized that Gall had made definite contributions to contemporary biology.

Young (1970) has outlined three of Gall's major achievements: first, Gall established as finally acceptable the idea that the brain was the organ associated with mental events, and that scientific study of mentality was possible by analysis of brain function. Second, Gall broke from traditional ways of listing human faculties, which he claimed with some validity to be remnants of medieval philosophy, not based upon careful observation. Unfortunately Gall replaced the old list with a much longer one of his own, which was not superior in all regards. Nevertheless, the break with the past was an important step. And third, Gall stressed the similarity between men and animals in regard to their faculties. Although he was firmly in the tradition of the static Great Chain of Being, he called attention to human behavioral characteristics as parts of nature, not apart from it.

A good deal of the scientific debate surrounding Gall's work had to do with whether or not the surface of the brain was electrically excitable. This was a very important issue to the physiologists involved, for upon its outcome seemed to rest some very basic assumptions about the nature of man and of mind. As mentioned above, the electrical nature of nervous activity was finally convincingly demonstrated at the end of the eighteenth century, just as Gall was promoting his views. Some of his critics, including the famous French physiologist Pierre Flourens, rejected Gall's theories entirely, in part because they believed that the uppermost portion of the brain was electrically passive, not a part of the sensory-motor, reflex-like operation of the rest of the nervous system. According to Flourens the cerebral hemispheres acted as a unit, indirectly, to produce voluntary movement. He believed that all mental activity took place in the hemispheres; this portion of brain was the seat of the mind (soul), and attempts such as Gall's to break up its functions by analysis entailed fatalism and denial of the soul's existence. These views of Flourens were very influential, in part because of the well-deserved prestige he had received for his work in experimental brain physiology; however, as Young (1970) has pointed out, Flourens' judgment in the case of Gall and localization of function in the hemispheres was clouded by his philosophical prejudice and by his unwarranted generalization from the forebrain of birds, with which he had done much of his experimental work, to the cortex of man.

Much brain physiology during the remainder of the nineteenth century centered on the question of localization of function and the nature of the cortex. Several strong suggestions from clinical investigators, including Pierre Broca who advocated the localization of speech function in a particular area of the human brain and whose name is now used to refer to that area, were insufficient to convince the many followers of Flourens and the other scientists who argued against localization. It was not until 1870 that two young Germans, Gustav Frisch and Eduard Hitzig, published the results of experiments in which they had successfully produced predictable movements in the face, neck, and legs of dogs by electrical stimulation of different, specifiable portions of the cortex. With this one paper they had demonstrated three fundamental principles: the electrical excitability of the cortex, the importance of the cortex in the direct elicitation of movements, and the localization of cortical functioning in the production of these movements. The objections of Flourens and his followers to these three ideas were overruled; the idea of cortical localization was rapidly accepted and expanded upon, particularly by the

Englishman David Ferrier, whose experimental results using monkeys were the first to confirm the generality of Frisch and Hitzig's findings (Young, 1970). The cortex of the brain was the last stronghold against a completely materialistic analysis of mental and behavioral functioning. From approximately 1870 on, there has been no serious objection to the view that the brain operates completely as an electrochemical system; however, as we shall mention now and again briefly, there is still much uncertainty and some disagreement as to the best way to characterize this system.

Until the twentieth century all analysis of the relation between brain function and behavior had been relatively crude. The idea that behavior could be measured and related quantitatively to physiological variables was a product of the early years of this century. Two men were preeminent in analyzing relatively simple behavioral events and relating them to relevant physiology: the Englishman Charles Sherrington and the Russian Ivan Pavlov.

In 1906 Sherrington, then a Professor of Physiology at the University of Liverpool, presented the Silliman Lectures at Yale University under the title *The Integrative Action of the Nervous System.* In these lectures Sherrington summarized the results of a large number of experiments he had conducted on posture, movement, and brain function. Most of the work was conducted with dogs whose spinal cords had been severed, thereby removing the influence of the brain from spinal function. The basic unit of his analysis was the *reflex arc,* a "chain of structures" including a *receptor* (e.g. a sense organ in the skin), an *effector* (e.g. the cells of a gland or muscle), and a *conductor* (the neural pathway between receptor and effector). He defined the "simple reflex" as a relationship wherein the effector responds to a change in the receptor while the rest of the organism is unchanged. However, he was quick to point out that:

A simple reflex is probably a purely abstract conception, because all parts of the nervous system are connected together and no part of it is probably ever capable of reaction without affecting and being affected by various other parts, and it is a system certainly never absolutely at rest. But the simple reflex is a convenient, if not a probable, fiction. (Sherrington, 1961, p. 7)

The first three of Sherrington's lectures were given over to an analysis of the properties of this "convenient fiction": latency of response after stimulation, amplitude of response, response "after discharge," response threshold and refractory period, and response inhibition were all defined and explicated by means of careful observation and inference. All of these concepts have now become the foundations of our understanding of spinal cord functioning.

It is important to note that a feature of the nervous system that we currently take for granted, namely that it is made up of discrete neuronal units, had been demonstrated anatomically only a few years before

Sherrington's lectures. The Nobel Prize of 1906 was awarded to two histologists whose efforts led to the development of the neuron theory: Golgi and Ramon y Cajal. Golgi's contribution had been to develop the staining techniques which made possible the discoveries of Ramon y Cajal and others. However, Golgi remained in 1906 antagonistic to the neuron theory and continued to espouse the classical view that the central nervous system was a reticulum or net of fibers. Ramon y Cajal's Nobel address, on the other hand, was a recounting of the evidence which led him to the general conclusion that "nerve elements possess reciprocal relationships in *contiguity* but not in *continuity*" (Ramon y Cajal, 1967, p. 220). Thus, when Sherrington gave his 1906 lectures he showed a certain scientific reserve when he described the neuron theory and the way in which neurons communicated: "In view, therefore, of the probable importance physiologically of this mode of nexus between neurone and neurone it is convenient to have a term for it. The term introduced has been *synapse*" (Sherrington, 1961, p. 17).

In the remainder of his lectures Sherrington dealt with increasingly complex aspects of neural functioning. He ended his lectures by emphasizing the extreme importance of the cerebrum in vertebrate life and the relationship between heredity and experience in the adaptations of the individual organism:

These adjustments, though not transmitted to the offspring, yet in higher animals form the most potent internal condition for enabling the species to maintain and increase in sum its dominance over the environment in which it is immersed. A certain measure of such dominance is its ancestral heritage. . . . but the creature has to be partially readjusted if it is to hold its own in the struggle. Only by continual modification of its ancestral powers to suit the present can it fulfill . . . its life purpose, namely, the extension of its dominance over its environment. For this conquest its cerebrum is its best weapon. It is then around the cerebrum, its physiological and psychological attributes, that the main interest of biology must ultimately turn. (Sherrington, 1961, p. 390)

We have mentioned that Golgi and Ramon y Cajal won the Nobel Prize for Physiology or Medicine, 1906. Two years earlier the same prize had been awarded to Ivan Pavlov, then 55 years old and Professor of Physiology at the Russian Military Academy at St. Petersburg, for his research into the physiology of digestion. Pavlov was honored for his advances in technique as well as his substantive findings, for he had pioneered the use of surgical procedures which allowed gastric processes to be observed in the otherwise normal animal. In his acceptance speech Pavlov suggested that it was his view of the organism that allowed him, for example, to succeed with certain surgical procedures where others had failed: "A striking proof of the power of science that regards the organism as a machine!" Pavlov went on to conclude his address with a description of the work which was at the time in its relatively early stages but which subsequently served as a basis for much of American experimental and physiological psychology: the conditioned salivary reflex. Pavlov was insistent that the fact of "conditioned stimulation" was completely explainable by reference to physiological principles. His final statement suggested his optimism for the role of physiological sciences in the explanation of behavior:

Essentially only one thing in life interests us: our psychical constitution, the mechanism of which was and is wrapped in darkness. All human resources, art, religion, literature, philosophy, and historical sciences, all of them join in bringing light in this darkness. But man has still another powerful resource: natural science with its strictly objective methods. This science . . . is making huge progress every day. The facts and considerations [about conditioned reflexes] are one out of numerous attempts to employ a consistent, purely scientific method of thinking in the study of the mechanism of the highest manifestations of life in the dog, the representative of the animal kingdom that is man's best friend. (Pavlov, 1967, p. 154–155)

Modern Physiological Concepts
We have dealt with the history of behavioral physiology in order, hopefully, to provide the reader with a sense of how the physiologist has approached his task. Unlike the zoologist, his primary concern is with the mechanism: The physiologist's question is always "How does it work?" However, it turns out time and time again that there are intimate connections between the ways animals "work" and their evolutionary histories and current environmental circumstances. One of the most exciting aspects of modern animal behavior study is the discovery of physiological mechanisms particularly attuned to species-specific needs. For example, the structure and operating characteristics of the eye of the frog have been shown to be exquisitely well-suited for fly-catching (Lettvin, Maturana, McCulloch, and Pitts, 1959). Similarly, the ear of the moth, which contains only two cells, nevertheless does all the work required to detect the approximate whereabouts of an onrushing bat (Roeder, 1966). These and other related examples indicate that the ethological notion of innate releasing mechanism, which Lorenz originally assumed to be a central nervous system process, may in some cases be a feature of peripheral sensory specialization (Marler, 1961; Marler and Hamilton, 1966). Much of the work done by the nervous system in filtering the enormous amount of energy at the surface of the organism is accomplished in the periphery.

Increases in surgical skills and electronics technology have allowed physiologists to probe deep into the brain to record electrical activity, stimulate electrically and chemically, or remove discrete amounts of brain tissue. Most of this work has been done with a few species of laboratory mammals (rats, dogs, and cats), but there is increasing interest in applying these techniques to a broader spectrum of species and relating the results to ethological and comparative psychological theories. For example, Glickman and Schiff (1967) presented an argument based upon behavioral and physiological evidence that the subcortical brain loci directly involved in the organization of species-specific responses are also part of the neural systems involved with positive and negative reinforcement. These authors suggest that the neural basis of reinforcement is the "activation of neural systems located in the brainstem which mediate the expression of . . . species-specific responses." This physiological theory ties in nicely with the classical ethological view, discussed above, that it is consummatory responding *per se* that is rewarding for an animal.

Finally, developments in the fields of endocrinology and neuroendocrinology have served as the basis for a deepened understanding of the relations between hormones and behavior. It has been a matter of practical knowledge for hundreds of years that castrating male farm animals at a young age prevented adult sexual behavior. But the scientific investigation of the relations between hormones and behavior did not begin until approximately 1900. Interest in the hormone-nervous system interactions was spurred and maintained by interests in psychosomatic medicine. Psychiatrists were at one time prescribing the removal of endocrine glands, or the injection of their secretions, in patients with particular disorders, well before the physiology or chemistry of these manipulations were understood (Scharrer and Scharrer, 1963).

Advances in organic chemistry and related technology have allowed rapid advances in the study of hormones and their behavioral effects. For many years the quality and quantity of hormones circulating in the bloodstream could be measured only by *bioassay,* i.e., the treatment of a standardized biological preparation with the substance to be measured. For example, one indicant of the male sex hormone testosterone was the extent to which comb growth could be stimulated in a castrated rooster by injection of the unknown substance. Although some of these bioassays were quite sensitive, they were also tedious and approximate; more recently it has become possible to make direct chemical measurements of certain hormones in the bloodstream (e.g., Resko, Feder, and Goy, 1968), and to implant very small amounts of hormone directly into the brain (Davidson, 1969). Most studies of this sort are concerned with sexual and maternal behavior, but there is also a growing literature in the relation between endocrine secretions and aggressive behavior and learning (Bronson and Eleftheriou, 1965; Levine, 1968).

In summary, then, the physiological perspective views behavior as the *result* or *observable by-product* of biophysical processes. The physiologist's task is to describe these processes in terms that are, in so far as possible, in the languages of physics and chemistry.

The Psychological Perspective

In an earlier section we discussed the philosophical tradition of using animals and their behavior as analogues or comparative baselines for the evaluation of human conduct. We also pointed out that since the impact of Darwin, approximately the last one hundred years, it has been possible to argue, if not prove, that important aspects of human behavior are related to the behavior of nonhumans in a more directly biological way, perhaps homologically. The discipline of psychology has been the home for most of the controversy about this and related issues. This is as it should be, for virtually all psychologists, regardless of their differences, would agree that psychology's task *includes* the explanation of behavior; indeed for many psychologists this is the definitive task of psychology. But there is at the roots of psychological thinking uncertainty and disagreement about priorities, i.e., what behavior we ought to be studying, and also about the kinds of explanatory models or principles we ought to bring to bear upon behavior. These disagreements surface in different ways at different times; so for example, one recent commentator has said

Throughout the history of the study of man there has been a fundamental opposition between those who believe that progress is to be made by a rigorous observation of man's actual behavior and those who believe that such observations are interesting only in so far as they reveal to us hidden and possibly fairly mysterious underlying laws that only partially and in a distorted form reveal themselves to us in behavior. (Searle, 1972, p. 16)

Two other commentators, in despair over what they take to be the pretentiousness of much modern psychological theorizing, have argued that

The over abstract character of the concept of behavior (and that of stimulus) tends finally to produce the illusion that a conceptually homogeneous set of lawful relationships *has been achieved or is achievable in psychology. (Zener and Gaffron, 1962, p. 541, also quoted in Koch, 1964, p. 33)*

For many students of psychology, who have been taught that careful attention to behavior is exactly what will prevent psychology from becoming overabstract and lost in a sea of fuzzy ideas, it may seem curious to find behavior criticized for being too big for its scientific britches. But it is, generally speaking, a careful consideration of this possibility that is required in order to gain understanding of the problems involved in relating the behavior of animals to the behavior of men, i.e., doing a systematic comparative psychology that includes our own species.

Historically, it has been mental life and its attributes that have been the major issues to be dealt with whenever psychologists have tried to develop a systematic comparative psychology or otherwise to show how human conduct can be related to the behavior of nonhuman animals. In other words, given the enormous apparent differences between humans and other animals, particularly in regard to those aspects of our lives that have to do with our linguistic capacities (and that turns out to be virtually everything interesting about us), what reasons do we have for believing that behavioral studies of nonhuman animals will shed any light on our own situations?

The first experimental psychologists were not directly concerned with this question. We can date the beginning of experimental psychology with the formal establishment of the first psychological laboratory at the University of Leipzig in 1879. The founder of the laboratory, Wilhelm Wundt (1832–1920), was an enormously productive physiologist and physician whose growing interests in human conduct and consciousness led him to become Professor of Philosophy in Leipzig in 1875. Within four years he had combined his philosophical interests with his training in observation and experimentation to found his *Psychologisches Institut.* The work of the laboratory fell for the most part into the areas of modern psychology called sensation, perception, feeling, and attention. Wundt maintained that the subject matter of psychology was immediate experience and that its method was introspection. The goals of the field were to analyze consciousness into its elements, determine how the elements were connected, and to formulate general laws that would describe these connections. The concern in all cases was with the mind of the normal adult human being. Thus, although Wundt was well-trained in physiology, having worked with both Johannes Müller and Hermann von Helmholtz, and although he wrote voluminously on a number of physiological, philosophical, and social topics as well, his primary bequest to psychology was a viewpoint and method for which the question of the behavior and mentality of nonhuman animals was irrelevant (Boring, 1950).

Among Wundt's students was the Englishman Edward B. Titchener (1867–1927), who came to America in 1892 after receiving his Ph.D. at Leipzig. Titchener spent his entire academic career at Cornell University attempting to establish the Leipzig viewpoint in America. Early in his career he gave this viewpoint a name: *structural psychology* since it was concerned with the structure of consciousness. He also provided the name that has ever since been used to characterize the main opposition to his views: *functional psychology* (Boring, 1969; Heidbreder, 1969).

At least in part, functional psychology was a reaction against structuralism, or at least against the proposition that the only legitimate way to do psychology was as Wundt and Titchener prescribed. In addition to Harvard's William James, the most important functionalists included the philosopher John Dewey, James Angell, and Harvey Carr, all of whom were at the University of Chicago in the early years of this century. The functionalists did not oppose the study of consciousness or introspection as a scientific method; rather they believed that psychology needed to adopt a broader view of its mission, which included understanding the functions of consciousness as well as its structure. The methods for the development of this understanding included, in addition to introspection, the observation of other persons and of animals. This broadening of legitimate methodology was justified on the grounds that consciousness was an important feature of the biological adaptiveness of animal life, i.e. consciousness had evolutionary significance.

The "Chicago School" of functionalists and other functionally oriented psychologists, in particular Edward L. Thorndike of Columbia, provided the intellectual matrix for the development of American experimental psychology during the early years of this century. It is important to note that all of these men, as part of or in addition to their scientific interest, had deep interests in education and the educational process. The case is particularly clear for Thorndike, who did some of the earliest animal experimental psychology in America. He began by doing experiments with chicks in the basement of William James' home while he was a graduate student at Harvard. From Harvard he went to Columbia and completed his doctoral dissertation, entitled *Animal Intelligence: An Experimental Study of the Association Processes in Animals,* in 1898; thirteen years later, in 1911, he published his book *Animal Intelligence.* In this work he formulated some of the basic principles of learning in animals, including his well-known *Law of Effect:* "When a response or series of responses leads to success or to a satisfying state of affairs, the connection between the situation and this response is strengthened, while other responses not so satisfying (i.e. annoying) are weakened and hence rendered less probable of recurrence" (Boring, 1950). The relationship between this formulation and more modern concepts of learning by reward or reinforcement is obvious enough. As we shall see in Chapter Nine, another of Thorndike's formulations, the Selective Association Principle, has recently enjoyed a renaissance of importance in the study of animal behavior. But now we want to emphasize that *it is the predominating interest in the concept of learning, or more generally the modifiability of behavior, that has historically characterized American or psychological studies of animal behavior and provided a strong contrast with the European or ethological perspective.* It has been only in the past ten to twenty years that there has been much overlap between American experimental animal psychology and ethology, although both fields developed with an appreciation of Darwin's message (a different appreciation admittedly), and, ironically enough, both fields drew considerable strength from founding members who were at the University of Chicago at the same time: Whitman in zoology and Dewey, Angell, and Carr in philosophy and psychology. But between that time and this, American psychology went through a major revolution, during which its approach to animal behavior became technically sophisticated and theoretically elaborated while remaining very, very narrow in the numbers of species and problems investigated. It is to a very brief outline of that revolution, behaviorism, that we will now turn.

Behaviorism

We have already noted that the European followers of Darwin, George Romanes in particular, were concerned to find evidence for higher mental processes in nonhuman animals; we have also pointed out that American functional psychology was willing to admit of consciousness in animals. In both groups it was allowed that consciousness was a functionally significant entity, in other words that in some sense or another the consciousness of organisms was an evolutionary adaptation to environmental demands. The problem with investigating this idea is, of course, that it is impossible to observe conscious processes in animals directly. For that matter it is impossible to view it in other people directly, but at least people talk to each other in ways that give us confidence in arguing from our own experience to that of the persons that surround us. In the view of the founder of behaviorism, however, scientific advances in psychology were impossible so long as psychologists continued to pay attention to consciousness.

John B. Watson (1878–1958) received his Ph.D. from Angell at Chicago in 1903. He subsequently accepted a position at Johns Hopkins University and, ten years after receiving his degree, published the paper that began the behavioristic revolution: "Psychology as the Behaviorist Views It" (Watson, 1913). Watson was entirely unequivocal in his views:

Psychology as the behaviorist views it is a purely objective branch of natural science. Its theoretical goal is the prediction and control of behavior. Introspection forms no essential part of its methods, nor is the scientific value of its data dependent upon the readiness with which they lend themselves to interpretation in terms of consciousness. The behaviorist, in his efforts to get a unitary scheme of animal response, recognizes no dividing line between man and brute. The behavior of man, with all of its refinement and complexity, forms only a part of the behaviorist total scheme of investigation . . . . behaviorism is the only logical functionalism. . . . The consideration of the mind-body problem affects neither the type of problem selected nor the formulation of the solution of that problem. . . . I should like to bring my students up in the same ignorance of such hypotheses as one finds among the students of other branches of science. (Watson, 1913, reprinted in Sahakian, 1968, p. 450)

Watson's views and substantial personal presence combined to make behaviorism, the psychology of stimulus and response, enormously popular in a very brief period of time. As one psychological historian has said, "For a while in the 1920's it seemed as if all America had gone behaviorist" (Boring, 1950). Indeed Watson provided a stimulus for many psychologists of that time and subsequently. Some were reacting with positivist zeal against what they took to be the sloppiness of introspection. Others, in particular a young Chinese psychologist named Zing-Yang Kuo (1898–1970), were reacting against the concept of instinct then fashionable in American psychology, particularly as described and defended by the eminent Harvard psychologist William McDougall. In 1921, only three years after leaving China and while still an undergraduate student at the University of California, Kuo published his first paper, "Giving Up Instincts in Psychology" (Kuo, 1921). In that paper and subsequently he argued that the instinct concept tended to blind psychologists to the important task of describing the development of behavior. Kuo argued for an approach to the study of behavior that brought the full resources of physiology, anatomy, and embryology to bear upon understanding behavioral development. In some ways Kuo stands as the earliest member, if not the founder, of a tradition in biological psychology that also has included Theodore Schneirla, Daniel Lehrman, and other influential biobehavioral scientists. Readers interested in pursuing the history and/or current status of this very viable perspective on animal behavior may consult the work by Aronson, Tobach, Lehrman, and Rosenblatt (1970), Gottlieb (1972), and Kuo (1967).

The Age of Learning Theories

The evolution of behaviorism from its inception until approximately 1950 is in good part the history of the development of theory, experimental method, and evidence regarding learning. The outlines of this develop-ment have been traced by participants and critics (e.g. Hilgard, 1948; Kimble, 1961; Koch, 1964; Skinner, 1969). The most famous names in this history include Clark L. Hull, Edward C. Tolman, Edwin R. Guthrie, and B. F. Skinner. Of these names Skinner's is undoubtedly today the best known; indeed he is the best-known psychologist in the country. Part of Skinner's preeminence is due to the apparent applicability to practical human affairs of the basic principles of operant conditioning (Krasner and Ullmann, 1965; Skinner, 1948, 1953, 1968, 1969, 1971; Ulrich, Stachnik, and Mabry, 1966; Yates, 1970). The effectiveness of these procedures at least in some cases is unquestionable, but the relationship between the procedures and the more general behaviorist theory out of which they developed is being called into question (Locke, 1971; Wilkins, 1971).

During the years between approximately 1935 and 1955 the major learning theorists and their students, in particular the groups surrounding Hull and Tolman, developed their views in a spirit of competitiveness. The issues that were debated included, for example, the necessity and/or sufficiency of reinforcement for learning, the nature of secondary rein-forcement and its relation to primary reinforcement, the nature of dis-crimination learning (the "continuity vs. noncontinuity controversy"), the content of what is learned (the "place vs. response controversy"), and the number of "kinds of learning" that had been shown to exist (Osgood, 1953). The details of these several controversies were very intricate, and, as it seems from our current perspective, not solvable in the terms in which they had been stated. For our current purposes it is enough to point out that all of these controversies were generated in part in response to issues relevant to human behavior, yet the "crucial experiments" designed to test the competing hypotheses of the "Hullians" and "Tolman-ians" were conducted with albino rats. The justification for this approach to psychology stemmed from the assumption that, within rather broad limits, "behavior" was "behavior" and "organisms" were "organisms." The notion of species-specificity in behavior, of highly refined behavioral adaptations that might make generalization from rat to man questionable, simply was submerged throughout the heyday of American learning theory. In a paper he wrote late in his career, Tolman expressed his position in the following, rather disarming way:

I have always wanted my psychology to be as wide as the study of a life career and as narrow as the study of a rat's entrance into a specific blind. But, actually, I have for the most part studied only average numbers of entrances into specific blinds, or jumps to a particular type of door, and have hoped that the principles found there would have something to do with a really interesting piece of behavior—say, the choice and pursuit of a life career. (Tolman, 1959, p. 97)

Tolman's statement has the advantage of being straightforward and modest; with these exceptions it is characteristic of the psychological perspective on animal behavior during the period in question.

In 1950 Frank A. Beach published a paper entitled "The Snark was a Boojum," in which he documented how far from a genuine comparative base American psychology had moved. By cataloguing the articles published in the relevant psychological literature in regard to the problems investigated and the species utilized in the investigation, he was able to show the predominant fascination with studies of learning and conditioning employing the laboratory-bred rat. In 1948, for example, approximately fifty-five percent of the experiments published in the *Journal of Comparative and Physiological Psychology* dealt with conditioning or learning, and almost seventy percent utilized laboratory rats as subjects. In commenting upon this situation Beach suggested, as have we in this discussion, that a major reason for the narrow base of American psychology has been its anthropocentrism. The rat, and more recently the pigeon, provide no intrinsic interest to most psychologists. Rather their behavior is taken either as representative of the workings of very general and abstract behavioral principles which are species-unspecific or simply as heuristic material for the generation of ideas which will, hopefully, some day be tested against the realities of human conduct. As Beach put it,

There has been no concerted effort to establish a genuine comparative psychology in this country for the simple reason that with few exceptions American psychologists have no interest in animal behavior per se. (Beach, 1950, reprinted in McGill, 1965, p. 9)

A Unified Approach to the Study of Animal Behavior

The original reaction of American animal psychologists to ethological theory, especially to the classical formulations of Lorenz, was skeptical and in some cases hostile. It seemed to psychologists, even those with good biological training and understanding, that Lorenzian ideas about instinct implied too much in the way of "behavioral fixity" and were inadequate to account for the facts of learning as psychologists had developed them. For their part, the ethologists believed that the American approach to learning was premature; in his 1951 text, for example, Tinbergen commented upon American psychology as follows:

[American psychologists] do not concentrate on innate behavior; rather they specialize in the higher types of behavior. This specialization has developed as a consequence of 1) the general interest in man and human conduct, and 2) the general acceptance of man's evolutionary descent from ape-like ancestors. These factors have brought about a preoccupation with what is often called "prehuman" behavior in mammals. The result has been a certain neglect of innate behavior, which has led in some instances to entirely unwarranted generalizations. In my opinion, this disregard of innate behavior is due to the fact that it is not generally understood that learning and many other higher processes are secondary modifications of innate mechanisms, and that therefore a study of learning processes has to be preceded by a study of the innate foundations of behavior. (Tinbergen, 1951, p. 6)

One of the best-known disputes between ethological and psychological perspectives arose in 1953 when Daniel S. Lehrman published his paper "A Critique of Konrad Lorenz's Theory of Instinctive Behavior" (Lehrman, 1953). In that paper he raised a number of points against Lorenz's conceptualizations of instinctive behavior. We will not review his detailed arguments here. Instead we want to point out, first, that in a general sense Lehrman's criticisms of ethological theory were identical to the criticisms Kuo had leveled thirty years earlier against the earlier instinct theories of McDougall. Thus Lehrman began his concluding paragraph by saying "Any instinct theory which regards 'instinct' as immanent, preformed, inherited, or based on specific neural structures is bound to divert the investigation of behavior development from fundamental analysis and the study of developmental problems" (Lehrman, 1953, p. 358). Second, the criticisms of Lehrman and other psychologists did not go unheeded by Lorenz and other ethologists. In his 1965 book *Evolution and Modification of Behavior* Lorenz responded to what he called "behaviorist arguments" against his views and traced the modernization of ethological concepts. And third, in a more recent paper, Lehrman (1970) has reassessed his earlier critique and Lorenz's subsequent response to it; he concluded with the view that, from today's viewpoint, differences between ethological and psychological perspectives have decreased perceptibly. The easiest way to describe the remaining distinction is in terms of emphasis of interest: ethologists are primarily concerned to clarify the facts and principles of behavioral adaptation, while psychologists are more concerned to analyze the immediate causation and development of behavior. These tasks are not mutually exclusive or competitive, and each can benefit from the gains made by the other.

It would appear, then, that some synthesis, unification, or at least a *détente* is emerging. There are other signs of this healthy scientific development, including the publication of two integrative, high-level textbooks (Hinde, 1966, 1970; Marler and Hamilton, 1966). Similarly, there is very little remaining of the old-fashioned behaviorist's disdain for physiology; for example, at least some of the objections that Skinner raised against "neurological explanation" in 1938 no longer appear in his estimate of the physiological psychology of 1969 (Skinner, 1938, Chapter 12; Skinner, 1969, Chapter 9). Furthermore, operant conditioning techniques have become a very valuable tool in the hands of physiological psychologists, particularly in the evaluation of the effects of drugs and other chemical agents.

What is the appropriate framework in which to place the discipline of comparative animal behavior study that arises at the intersection of zoological, physiological, and psychological perspectives? The answer, it seems to us, is that some fundamental principles of evolutionary biology are sufficient to accommodate the unified approach to behavior. Here is a list of these principles:

1) Animal species are the product of a historical process controlled by the action of natural selection. In the last analysis selection acts at the genetic level since individuals influence future generations by the passage of genes from one generation to the next. Some individuals make a larger contribution than others. This *is* natural selection.

2) Genes are the foundation for all the traits, structural and behavioral, that characterize an animal. Each individual has a genotype, a set of genetic information acquired from its parents. In the absence of the genotype there can obviously be no development; the genotype controls the way in which the environment is utilized in constructing an individual.

3) Development for most but not all animals involves, among other things, the construction of a nervous system, an endocrine system and sets of muscles. These are the physiological foundations for behavior. Selection favors those animals whose genes interact with the environment to produce successful physiological foundations, those facilitating the behavioral processes that lead to reproductive success.

4) Each individual's behavior is tested at the ecological level. How an animal behaves when faced with a predator, or with the task of finding food, or with the problem of locating and securing a mate will determine the survival and reproductive success of that individual.

5) Some individuals will pass these ecological tests better than others. Their genes will be transmitted to future generations replacing genes that contributed to less successful physiological systems and behavioral tendencies and abilities.

It should be clear from these general principles that an evolutionary approach to behavior demands the synthesis of information about the genetics, physiology, and ecology of individual animals. It suggests that genetic analyses will reveal a direct link between the presence of certain genes and specific adaptive behavior patterns; the work of Dilger (1961) on nest-building in hybrid *Agapornis* parrots and Rothenbuhler (1964 a and b) on nest-cleaning in honey bees provides examples. The evolutionary approach also suggests that there will be discoverable relationships between the ecological demands placed upon a species and its neuro-anatomy and neurophysiology. There is in the work of Roeder (1963) and Dethier (1971) on insect behavior and physiology ample evidence to verify the fruitfulness of this hypothesis. In general, the evolutionary approach provides a framework for thinking about all aspects of animal behavior and their interrelationships.

The Nature-Nurture Problem

One of the enduring general problems in the study of behavior has been what is usually called the *nature-nurture controversy*. We have alluded to this issue many times already in this chapter without giving it a label. Thus the distinction between the classical ethological and psychological perspectives, or between the views of McDougall and Kuo, or Lorenz and Lehrman, can be seen as a difference in the relative importance in the determination of behavior attributed to "heredity" or "nature" on the one hand and "environment" or "nurture" on the other. In one form or another, the problem is mentioned in almost every chapter of this book. How does a unified perspective on animal behavior apply to the resolution of this problem?

To begin with, it is now agreed upon by all thoughtful students of the problem that both genes *and* environment contribute to all aspects of development and thus to all behavioral traits. Sometimes this truth is taken to imply that the nature-nurture issue is just a pseudoproblem. However, the questions remain, *how* are these contributions made and are there ways of measuring the *relative importance* of genotype and environment in regard to a particular trait?

One approach to the latter question is found in quantitative population genetics. This approach is outlined in some detail later in the book; for now we will point out only that it is an analysis of *populations*, not of individuals. Behavior-genetic analysis provides a way to sort out the behavioral variability in a population into components due to genotype, to environment, and to the interaction between them. The power of the analytic procedures are greatest when experiments have been carefully designed in advance to utilize them; *post hoc* assessment of the genetic component of some trait, for example human intelligence, is a relatively weak procedure. But, when the circumstances permit, this sort of analysis can provide reliable, quantitative estimates of the heritability of a behavioral trait.

There is, however, a limitation to this general approach, which is that it is restricted to measurements of differences between individuals of the same species. Questions about similarities within a species and differences and similarities between species are not answerable by these methods, as presently developed. Yet it is clear that there are many examples of behaviors (fixed action patterns) that are performed very uniformly within a species and which are very different from behavior patterns performed by members of other species. How are we to investigate the gene-environment interactions that lead to the development of fixed action patterns?

First, information about the ecology of the species and the function of the behavior pattern in question is indispensable. On the basis of this information one may be able to develop a working hypothesis. For example, if the behavior pattern were a defensive response (freezing), exhibited in the presence of a specific alarm call from another member of the species, and if the species lived in a fairly broad range of environments, we might predict that the development of this behavior would be relatively insensitive to many environmental situations or variable experiences the animal might have during its development. Why? 1) This is the kind of behavior pattern that must be performed properly the first time the animal hears the alarm call. Selection pressure, in the form of a predator, would act severely against an individual that had to learn how to act when this particular stimulus appeared in its environment. 2) Furthermore, because the young animals of this species are reared in a wide range of environments, selection might be expected to favor genes that "channeled" development along certain generally successful lines and that were therefore not sensitive to the peculiarities of each different habitat. It is advantageous only that the response appear reliably and completely at the appropriate moment; various modifications based on where and how the creature was reared are unlikely to improve the freezing response.

How might this hypothesis be tested? The classic "deprivation experiment" of ethologists is a reasonable test (Lorenz, 1965a). In this type of study the animal is deprived of its normal environmental conditions and reared, usually in social isolation, in some minimal, artificial situation. The animal would be tested after reaching the age at which the freezing response usually occurred or perhaps after an even more prolonged period of deprivation. This would be done by playing a recording of the alarm call to the experimental subject and observing its response. If it behaved more or less normally, the hypothesis would have been supported. Remember that the hypothesis did not state that the behavior would prove to be genetically determined and that the environment would prove to be totally irrelevant, but simply that most environmental variables would not have an effect on the development of the freezing response to the alarm call. What the results of the experiment would indicate is that natural selection has favored the evolution of a genetic system that more or less guarantees the development of a particular behavior in a very wide range of conditions given some absolutely minimal input of food, water, air to breathe, etc. Selection has favored such a developmental system because of the critical importance of a proper response in a wide range of habitats.

Thus innate behavior, behavior patterns that are species-specific, stereotyped, and appear in socially deprived animals, reflect one kind of developmental program, in which genetic instructions act to channel development rigidly along certain restricted lines.

The development of other kinds of behavior patterns may be more susceptible to subtle environmental influences. For example, an animal that is moderately omnivorous and lives in a variety of habitats might be expected to take advantage of food items abundant in one place or time but rare or absent in other places or at other times. Of course this learned behavior also has a genetic basis. In this case the genetic instructions lead to the development of a nervous system capable of storing information about food items, information that is used to alter preexisting dietary preferences or food-collecting movements. The environment provides the basic building blocks for the construction of a nervous system with special properties that are the foundation for a *specific* kind of learning (see Lorenz, 1970, and Bolles in this book). In addition the environment provides a key piece or pieces of information, in this case about food items, that in some way alters the action of the animal's nervous system and thus its future behavior toward food.

Thus the development of both instinctive behavior and learned behavior requires environmental input and genetic information, but how the genotype interacts with the environment differs in the two cases. Fixed action patterns depend upon the construction of a nervous system with specific IRMs that "recognize" certain stimuli on first encounter with them and order the animal to behave in a very specific way. Learned behavior depends upon the construction of a nervous system with a *learning mechanism* that can store certain information and use this information to change a behavior pattern of that individual animal. Certain behavioral responses are "built-in," innate, because the cost of making an initial mistake is so high, e.g., being eaten by a predator because the animal failed to stop moving when an alarm signal was given. Other responses are learned because the advantage of a flexible response to some environmental variability is high (as in learning to avoid a strange food or distinctively flavored liquid if you become sick after eating or drinking it—see Bolles, Chapter 9).

Table 1-1 A Comparison of the Behavior of Some Social Carnivores, Human Beings, and Primates[1]

	Lion	Wolf	Hunting Dog	Hyena	Man	Nonhuman Primates
Share Food[2]	yes	yes	yes	yes	yes	very rare
Group Defense of Territory	yes	yes	only during breeding	yes	yes	almost never
Killing of Individuals *not belonging to group*	yes	yes	?	yes	yes	very rare
Cooperative Behavior (Especially in Hunting)	yes	yes	yes	yes	yes	very rare

[1] Data from Schaller, 1972; Mech, 1967; Van Lawick-Goodall, 1970; Kruuk, 1972.
[2] Usually involves permitting others to feed at captured prey and the hunting dog and wolf acutally bring food back for cubs and adults left to guard them.

Advances in the Study of Human Behavior

We have tried to argue that evolutionary behavioral biology has something useful to say about the nature-nurture controversy. Here an effort will be made to explore briefly how this kind of thinking also permits new insights into the nature of human behavior. One of the key reasons given for studying other animals is to gain perspective about the action of natural selection on animals with a way of life *similar* to that of our ancestors. The "ecological method" (Lockard, 1971) is based on the premise that similar selection pressures lead to the evolution of similar behavioral adaptations. Convergent evolution (the evolution of similar traits in distantly related animal species) is a very well-documented phenomenon. A behavioral example is the independent evolution of sonar systems in bats and porpoises, two extremely different groups of mammals. Both, however, produce pulses of sounds and detect echoes reflected by obstacles and objects in their path. Both live in environments, nighttime and murky waters, where visual cues are not available.

A key evolutionary question about humans is this: what were the behavioral consequences of adopting a semicarnivorous mode of existence? Attempts to answer this question by studying our close relatives—apes and monkeys—are not likely to be productive because no other primate begins to approach the human dedication to meat eating and hunting. Instead, an evolutionary perspective on the issue suggests that social carnivores should be examined; and, in fact, field studies of the behavior of hyenas, hunting dogs, wolves, and lions have revealed some very interesting parallels between their behavior and our own.

This kind of comparison based on recent knowledge about the behavior of social carnivores suggests that the hunting way of life of our ancestors may have had profound behavioral effects. For example, since game animals, especially larger ones, are relatively rare and dispersed compared to vegetable foods, a very large area is required to support a group of humans, lions, hyenas, or wolves. Therefore it pays to defend this area against intruders that would undercut the base of support of the group. Territorial defense is more effective if carried out by a group. Because females may be burdened with young, or because the young animals cannot come hunting yet and must be protected by some adults while many are away, food-sharing may evolve. In general, various forms of cooperative behavior *within* such a group may arise because the welfare of individuals depends so heavily on coordinated group activities: area defense, hunting, protection of the young, etc. Some baboon species provide exceptions to the general rule that nonhuman primates do not engage in cooperative behavior. Interestingly enough, these species are savannah dwellers subject to selection pressures similar to those operating on our human ancestors. Cooperative behavior is highly developed in savannah-living baboons as a protection against large predators, leopards and lions, that are a special threat to primates walking on relatively exposed grasslands. Since there is no similar advantage to cooperating with or even tolerating strangers, these would be driven off or killed if possible by social carnivores. Our ancestors were subjected to selection pressures very different from those operating on most primates. An evolutionary biologist would contend that the result was a decidedly un-primate-like creature, *Homo sapiens,* with a set of behavioral adaptations similar to those of nonprimates living in similar ecological niches. Thus in many respects we are more like a hyena than a chimpanzee because the selection pressures operating on hyenas and humans have more in common than those operating on chimps and humans.

This interpretation is controversial (see Klopfer's chapter in this book) but we present it here to indicate how what may appear to be a rather odd source of behavioral information may be exploited in an effort to gain an evolutionary perspective on human behavior.

Hopeful Conclusion

As the study of behavioral genetics, ecology, physiology and evolution proceeds, particularly at the interfaces or areas of overlap between these disciplines (e.g., ecological physiology, evolutionary ecology) one can at least hope for a more coherent and universally accepted approach to behavioral problems. This does not mean a uniform approach, but rather a multifaceted one with workers in different areas clearly perceiving the relevance of each other's research in the construction of a broad, yet detailed picture of behavior. We have tried to argue here that evolutionary biology, a discipline so successful in uniting disparate zoological fields of study, a discipline that demands an integrated knowledge of genetics, physiology and ecology, is in the process of developing a broadly acceptable approach to questions about behavior.

The Goals of This Book

What should motivate you to read this book (other than threat of examination, a time-honored but otherwise least honorable motivator)? Of course, curiosity is a basic human attribute, albeit one that formal education seems singularly successful in suppressing. Nevertheless, should you possess some residual curiosity about an academic question you are to be encouraged to develop it. Behavior is a fascinating topic and much of the research reported in the chapters that follow is ingenious and insightful. A second motivator might be the search for self-understanding. All these chapters touch directly or indirectly on this subject and should provide you with new ideas about such questions and issues as the role genes play in the development of your behavior, how selected experiences alter the path of development, how nerve cells provide you with the sensation of touch, the physiology of sexual behavior, why humans spend so much time and energy on their children, the nature of human language, and why some things are easy to learn and others very difficult.

The Organization of the Chapters

The rest of this book is divided into three sections. The first of these has three chapters, which deal with the evolution, genetics and development of behavior. In his chapter on evolution Peter Klopfer examines the conceptual foundations of the notion that behavior evolves; the chapter is, among other things, a rewarding exercise in getting our ideas clear on this fundamental topic. The chapter on behavior-genetic analysis by Robert Murphey, gives a straightforward account of the principles of genetics as they apply to behavior. Murphey's treatment of the *components of behavioral variance* in a population will be particularly useful to the student. Gary Mitchell's chapter on development concentrates upon social and emotional behaviors and draws substantially from his own work and that of others on emotional development in nonhuman primates.

The next section, with two chapters, is decidedly physiological in emphasis. In Chapter 5 Robert Scobey and Thomas Harrington have presented the technically demanding, conceptually complex field of sensory physiology in such a way that the reader need not have extensive physiological background to understand it. A particularly original feature of this chapter is the extent to which it presents information on the cutaneous senses. Chapter 6, by Benjamin Hart, gives a comprehensive overview of our current understanding of spinal reflex mechanisms. Hart shows how the concept of inhibition, developed initially by Sherrington, is still extremely powerful in accounting for the control of spinal behavioral mechanisms.

The final section contains four chapters, each of which deals with a different functional behavioral category. In Chapter 7 Gordon Bermant and Benjamin Sachs describe the functions and causes of courtship and mating behaviors. Chapter 8 concerns parental behavior; in it Dale Lott presents some wonderful examples of the lengths to which parental behavior has evolved and the complexities of its environmental and physiological causes. Chapter 9 is by Robert Bolles and presents what may fairly be called "The New Look" in American learning theory. Bolles shows how the facts of species adaptedness influence the predispositions of animals to learn, or not to learn, certain responses or relationships. Completely generalized theories of reinforcement, so long dear to the hearts of American learning theorists, are now being modified to meet the demands of biological reality. And finally in Chapter 10, Jarvis Bastian and Gordon Bermant present an evolutionary view of communication, in which they argue that communication is not a part of social organization, but rather proof of it. They also present a discussion of the relation between nonhuman communication systems and human languages.

At this point it should be obvious why our book is entitled *Perspectives on Animal Behavior*. All the major components of an evolutionary approach are inspected here. Moreover, as you read subsequent chapters it will become quite clear that the various authors differ somewhat in regard to their assumptions and departure points in regard to studying behavior. We have not made an attempt to present a homogenized view. Uniformity has been willingly sacrificed for the greater benefits of diversity. We hope that you will read this book critically, prepared to challenge what does not seem reasonable or logical to you. In the references at the end of the book you will find both historical and contemporary sources in sufficient numbers to pursue your study for some time. The study of animal behavior is an exciting field, in part, just because we still know so little; there is a very great deal left to learn.

2

Evolution and Behavior

Peter H. Klopfer

What do behavioral scientists mean by the term "evolution"? Unlike physicists who have generally first discovered the nature of something and named it later, biologists and psychologists have all too often first named a phenomenon and then sought its properties. Hence the perceptions of these scientists have frequently been obscured by the constraints placed on their initial statements by linguistic requirements. This in turn has often led to the substitution of semantic for factual problems (Hardin, 1956), a reflection of the fact that "each language frames the world uniquely" (as Steiner, 1967, has eloquently put it). These problems have been especially persistent and important in the development of ideas about evolution.

For example, Lotka (1956) has argued that evolution implies the history of a system; but history itself is not evolution. That is, the description of the history of a swinging pendulum is not a description of a system that has evolved. By "evolution" we imply, as well, the history of a system that is not undergoing cyclic change, though some kind of change is involved. But not just any kind of changeful sequence is an instance of evolution. For example, three events occurring in the sequence A, B, C represent a changeful sequence, but then so does C, B, A. Evolution is not merely random change.

To speak of change which is not cyclic and not random is to point to an implication of "progress" when we speak of evolution. But what does "progress" mean? From lower to higher forms of life? This is of course a totally anthropomorphic conception, because there is no operational basis for determining what is lower or higher except according to very arbitrary criteria. Simple to complex? Unspecialized to specialized? These are criteria that are often used and they could be made operational. But while it may be argued that in any given instance evolution has proceeded from the simple to the complex, the reverse position can just as well be taken. There is no basis for arguing that a bacterium is inherently more simple than a protozoan, for example. And when it comes to deciding whether or not a camel, in terms of its morphology or physi-

---

Research supported by grants MH 04453 and HDO239 and an NIMH Research Scientist Award.

ology, is simpler or more complex than a man, the decision surely cannot be justified on objective criteria. The problem is this: characteristics such as simple or complex, specialized or unspecialized, are usually based on single components rather than on the entire system.

If we reexamine more closely the pendulum that was mentioned earlier, we will find it is not strictly cyclic. This is due to friction. The pendulum gradually slows, tending toward a state of maximum entropy or irreversibility. This is true of any natural system where there is a lag between the externally imposed stimulus and the response. Consequently, it might be reasonable to consider evolution to be the history of a system that undergoes irreversible change. Thereby we exclude cyclic mechanical systems, though not all mechanical systems. Indeed, if our knowledge of a given mechanical system is a statistical one, changes that appear to be irreversible may in fact not be so when the exact state of every molecule can be specified. So we have to then qualify our definition yet again by saying we mean by evolution the history of a system which we *regard* as changing irreversibly, though admitting that "irreversibility" may be a function of our ignorance. This is merely another statement of the second law of thermodynamics, which is perhaps as good a way to conceive of evolution as any other. (These details are considered more fully by Lotka in his *Elements of Mathematical Biology*, 1956.)

Suppose we do think of evolution as merely a statement or example of the second law of thermodynamics ("every system will run down"). The second law applies only to closed systems, i.e., systems which are not able to receive energy inputs. The earth is not a closed system, for the sun provides an energetic input. What this implies is that while the second law of thermodynamics may be inescapable for some systems contained within the earth, the occurrence of evolution may not be inescapable. Let us put the question in another form. Is evolution inevitable?

Suppose we have a machine that can reproduce itself. The machine consists of two parts: one for reproduction and the other carrying the blueprint for reproduction. Suppose further that the machine has a finite life, i.e., is mortal, and also that the parts from which new generations are composed are drawn from a finite store. And suppose, finally, that during reproduction there occur, infrequently but occasionally, errors in the copying process. Given sufficient time and a sufficient number of machines, there is a very high probability that a "mutant" will arise which is more efficient in reproducing itself than its predecessors were. It may be more efficient by being faster, or by having a prolonged capability of reproduction, or by requiring fewer parts. In any case, its progeny will come gradually to monopolize the finite store of parts. Thus, evolution will have occurred. In short, a system with the characteristics we have just stated *must* evolve even if the mutant that appears is only very slightly superior. A comparison with the biological situation reveals that important similarities do exist. Organisms are self-reproducing, they are mortal, and the resources on which they depend are finite. Errors in copying, i.e., genetic mutations which result in offspring that differ from their parents, do occur. And finally, the time that has been available since the earth was first formed is many times the reproductive span of any of its constituent organisms. From all this it follows that evolution on this planet has been inevitable.

It has also been argued, on the other hand, that the resources on the planet have not yet been exhausted; hence our assumption about the finite store of parts of which organisms are comprised has not been met. Therefore, how can one say that evolution is inevitable? Now it is obviously true that there has not been a planetary exhaustion of supplies. But this misses the fundamental point, which is that the reproductive potential of any organism that has lived on this earth has indeed exceeded the earth's carrying capacity. Any organism has a potential sufficient to produce a mass expanding outward with a speed greater than the speed of light! Consider just the elephant, with a generation time of about 16 years. In 160 years one pair could potentially double ten times; yield, 1024 living offspring; in 320 years, 20 times or 1,048,576 descendents in the 20th generation. Mice, with one pair producing ten young or five pair per year, could in a 20-year period produce a mass of 5^{20} mice. Obviously this potential has never been realized because there is mortality, a death rate, which is in some rough way proportional to the birth rate. Now since variations do occur and copy-errors are inherited, this mortality is likely to be selective, at least in part. By this is meant that all the variants are not equally likely to survive—some will be better suited to their environment than others. And, given selective mortality, the evolution of the organism in question once again becomes inevitable. None of this, of course, disproves other views of how changes have occurred, i.e., such views as "special creation" (by a creator) or "orthogenesis" (evolutionary changes whose outcome is inevitably set by mechanical causes), though these latter (and others) become superfluous explanations if the processes described above can be shown to have actually operated.

Is there ever an end to an evolutionary sequence, for example, the sequence which led to man? It has been argued that if mortality due to natural agencies were eliminated, then evolution would cease until the earth became totally saturated or depleted of its resources. But the relevant fact is that mortality cannot be eliminated. All one can hope to do is to alter the nature of the selective agency. For example, consider the death rate of humans as a result of automobile accidents. This shows a maximum at around age 24. The death rate gradually decreases thereafter, rising very slightly again somewhere beyond age 60. Since the mortality is highest in the preproductive period, there is a strong selection against any accident-prone genotype which might exist (which isn't to say that accident-proneness is necessarily heritable!). The substitution of death by auto for death by disease illustrates that civilization alters the nature of selection but doesn't eliminate selection. Or, take the hypothesis that the high population densities characteristic of human populations today produce physiological stress. This may favor genotypes that, in the past, under conditions of low density, were at a disadvantage. Another example is the blood disease *prophyria,* which is not harmful to primitives (who lack pills) but is fatal in conjunction with sleeping pills and certain analgesics. Here again, the nature of the selection has favored genotypes which in the past were not so favored, and *vice versa.* Finally, a genetic constitution leading to high efficiency of food utilization can, under conditions of food abundance, produce obesity and problems of high cholesterol levels. Under conditions of starvation or food shortage, this same genetic constitution is more favored. In short, it cannot be denied that selection and hence the evolution of man continues. The main difference between the present situation and that of the past is that we seemingly have now the capacity to control the selective agencies that favor or disfavor certain genotypes. Whether we exercise this control or not, the evolution of *Homo sapiens* (and other organisms, too) cannot cease.

The Evidence for Evolution

In the previous paragraphs we discussed the inevitability of evolution, having assumed that certain conditions obtain in the environment. However, it is at least equally desirable to have empirical support for a belief in evolution. Thus we must face the question, what constitutes valid evidence for the actual occurrence of evolution, and particularly for any specific evolutionary sequence?

The test of most theories lies in their predictive value, the fact that they can predict the outcome of specified contingencies. *Post hoc* theories, explanations after the fact, are of restricted utility. For example, assume that the numbers on automobile license plates are actually distributed randomly, without regard to the county in which the vehicle is registered. Nevertheless, if we were asked to collect a series of license plates from different localities and then try to find some order in the numbers assigned them, some code for different counties, we would in general be able to do it. A theory "describing" a pattern can almost always be generated, even from a table of random numbers. To make an accurate prediction of the pattern that will be found is another matter however.

Some have argued that theories can be explanatory without being predictive (e.g., Scriven, 1959). The argument takes the following form: Suppose condition A_0, A_1, or A_2 sometimes leads to condition X (for instance, smoking, sunshine, or mustard gas sometimes lead to cancer). Suppose, as well, that the incidence of X due to any single A is small. This means that if a particular individual has been exposed to condition A_0 (though not condition A_1 or A_2) our best prediction would be that he will not be subject to condition X. If the individual, however, does develop condition X, we do deduce A_0 as its cause.

It is possible, of course, that the nonpredictive explanation is not predictive only because of our ignorance. If our understanding of the mechanism of action of A_0, A_1 and A_2 (smoking, sunlight, and mustard gas) were more adequate we might well be able to predict what the outcome of the individuals being exposed to the different conditions would be. However, at a molecular level, ignorance of some conditions is inescapable. (The Heisenberg indeterminancy principle reminds us that we can never simultaneously determine the position and velocity of an atomic particle. To measure one of these attributes—velocity—requires altering the other—position.) Nevertheless, an evolution theory can sometimes allow predictions. For instance, given two populations with a different mortality due to adaptive differences, one can predict which will replace the other. Suppose there is a dark-colored species of *Peromyscus* (a mouse) living in a desert which consists of white sands interspersed by black lava beds. These mice are much preyed upon by predators that depend on vision. Protective coloration or camouflage is known to protect! If the pelage or fur of the offspring of one of the mice is light (due to albinism, or any of a number of possible mutations) we can predict (and will often be correct) that the dark variety will eventually dominate on the lava beds and a strain of the lighter ones on the white sand. Of course, when we deal with evolution which has already occurred, the problem is very different indeed. The rigorous tests to which predictive theories are put cannot then be applied.

If a predictive theory has limited utility in explaining events that have already occurred, the next best thing is a direct record of past events. Of course, records of past events are rarely direct; they always require interpretation. Interpretations, in turn, also depend on nonpredictive theories. But sometimes, at least, the gap which must be bridged by inference can be fairly small. For example, if one finds a series of fossils that are stratified so as to indicate the temporal sequence in which they were deposited, and if these fossils represent a graded series of types (one type merging into the other), and if one has evidence that there is a genetic relationship between the various types (or, if there are changes in the proportion of a given type within the populations of fossils at different levels), then it is reasonable to conclude that what one is seeing is a picture of progressive change (or evolution). The evidence of genetic relationship must be inferred on the basis of similarity in comparative anatomy, serology, or embryology, although this last criterion is only available if extant representatives of the animals represented by the fossils are available. The history of the horse as described by Simpson (1951) is based on this kind of evidence, that (a) there have been gradual changes in the predominance of particular types in the population, and (b) that the simplest possible way in which one can explain such a series is to assume evolution. Alternatively, instead of having a series of fossils, one may have a contemporary series which represents different geographic variants. For example, on the islands of the Galapagos Archipelago, six

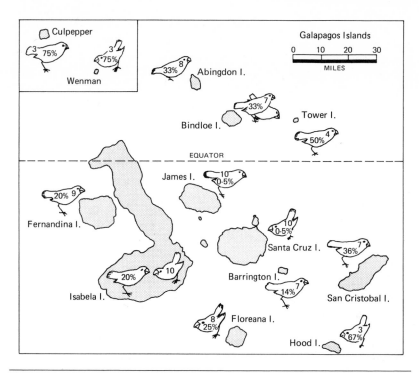

Figure 2-1 A map of Galapagos Archipelago. The total number of species of finches on each island is indicated by the number. Below each number is indicated the percentage of endemic forms of Darwin's finches. This percentage increases with distance from the central islands. Some of the island names shown here have subsequently been changed.

hundred miles west of the coast of Ecuador, Darwin noticed that the percentage of endemic birds increased with the degree of isolation of the islands, while the number of resident species decreased (Figure 2-1). He also noted that the most parsimonious explanation for this was a progressive evolutionary change that was the consequence of the isolation of the peripheral populations from the central. A geographic dimension has here replaced the temporal one. (Darwin's observations on the Galapagos Archipelago were a particularly potent factor in stimulating the development of his theory of evolution by natural selection.)

Establishing the genetic relationships alluded to earlier has depended, in the past, on the existence of anatomical, embryological, biochemical, or blood serum similarities though these are not independent. For example, the concept of homology (similarities stemming from evolutionary relatedness) is meaningless if it assumes the evolutionary relationship that it seeks to demonstrate. Second, embryological similarities are not themselves evidence for evolution. The most that embryological similarities can provide is support, not proof, of an evolutionary or genetic relationship. Nor do embryological similarities represent similarities that are independent of adult anatomic similarities. If adult structures are structurally similar, fetal structures are likely to be similar also. Hence if a relationship is deduced on the basis of adult anatomic homologies, embryologic similarities cannot provide independent support for the relationship. As for similarities in blood groups or serum antigens, they may or may not demonstrate affinities, depending on the ease with which new variations in an organism's proteins can arise. It is probably true that the antigenic properties of the entire organism are a relatively conservative trait. But it is not necessarily true that all antigens are so conservative as to be reliable indices of relationship.

In summary, empirical proof of the occurrence of evolution, in the strictest sense, is probably not possible. But the theory of evolution may nonetheless be considered valid because of (a) the arguments from logical necessity, (b) the fact that it does allow prediction as to the outcome of the competition between populations.

The Evolution of Behavior

It is no small irony that while one of the major thrusts of current ethology involves the deduction of the evolutionary history of particular behavioral mechanisms, the goal of early ethology was the converse: the deduction of the evolutionary history of species manifesting a relatively invariant behavior pattern. Heinroth (1910) and Lorenz (1941) began their studies of water fowl with the idea that since the behavioral displays involved in courtship represented very conservative characters in evolution, similarities and differences in the patterning of display movements would provide a more reliable index of the evolutionary relationships of different species than similarities or differences in the anatomic structures. Much the same reasoning underlay Petrunkevitch's studies on spiders (1926). But now the focus has shifted from ethology as a substitute for anatomic studies in the deduction of evolutionary sequences, to a direct study of the evolution of the behavioral sequences themselves. If this suggests a certain circularity, it is unfortunately a circularity that has in many instances plagued ethological studies.

Perhaps the clearest description of the methods for evolutionary studies of behavior has been made by Tinbergen (1962). He points out that to avoid circularity one must first independently establish the degree to which a pair or group of species are related to one another; this determination is to be based on conventional systemic procedures, exclusive of behavioral characteristics. Tinbergen lays particular stress upon anatomical homology for establishing degree of relationship. Thereupon, he argues, one can deduce the origin of the behavior in question. He assumes that the most aberrant species in the group under consideration, or the species with the most specialized niche, is least likely to be the most similar to the ancestral form and *vice versa*. Tinbergen admits these arguments "are all full of pitfalls," but goes on to say that "when the various criteria are applied together and all point in the same direction, they carry conviction."

I would argue that the addition of uncertainties does not increase the reliability of the sum, and, as we have indicated before, the various criteria that Tinbergen proposes for determining common ancestry are often not independent. Physical similarities in extant descendants of extinct ancestors are reliable indicators only if the genetic basis of the characters in question is known, as well as the susceptibility of the characters to environmental influences. For instance, it has been shown that the scutes ("scales") of snakes, which are frequently used as a taxonomic character, vary according to the temperature at which the

Figure 2-2 Footprints reveal the locomotory habits of their perpetrators, and footprints may be fossilized, thus becoming a permanent record of the species' evolutionary history.

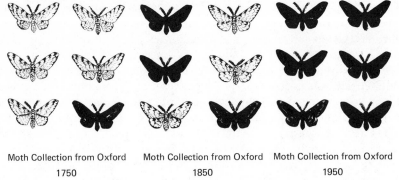

Moth Collection from Oxford   Moth Collection from Oxford   Moth Collection from Oxford

1750   1850   1950

Figure 2-3 Changes in the proportion of melanic forms in English collections of *Lepidoptera* chronicle evolution.

embryo develops (J. Bailey, personal communication). And as indicated earlier, serologic, embryologic, and other similarities are not independent of similarities in anatomy. A similar embryologic history implies similarity in both biochemistry and gross morphology. Finally, no satisfactory operational definition for "specialization" or "aberrance" has yet been advanced. It is unprofitable to toy with these terms until an *a priori* basis for defining them is available.

In actual practice, studies of the evolution of behavior have drawn on six different kinds of evidence. In the first instance, there have been behavioral "fossils." For instance, the body form of some insects has been preserved in amber. By the study of their morphology, which revealed the existence of different castes, it has sometimes been possible to deduce their social organization. Also, footprints preserved in mud may provide information on the gaits of those that left them (Figure 2-2).

Second, there have been instances where it has been possible to observe the direct effect of natural selection. One of the best examples of a direct observation of evolution in process has been the spread and then the demise of industrial melanism (abnormal darkening of pigment) in the *Lepidoptera* in many parts of England. Moth and butterfly collecting has been a favorite hobby of many in the English islands, and as a consequence it has been possible to document, for the past century, changes in the proportion of melanic and nonmelanic forms. These changes have been coincident with the prevalence of deposits of soot on the landscape. In the process of domestication, as well, it is possible to have a documented record of changes in characteristics, especially behavioral characteristics, of certain species (Figure 2-3).

Third, it is possible to make some deductions about the evolution of behavior by drawing inferences from the character of the environment in which the animal in question had to live. Pollen stratigraphy, for example, reveals the vegetation and climates of the past (Figure 2-4).

From these we can make some reasonable assumptions about the degree of dispersal of the ungulates that inhabited the area in question. (One further example of an inference from environmental conditions is provided in the section on the evolution of aggression, below.)

Fourth, in any population there may be much variation. It has been argued that these variants represent the past and future course of evolutionary changes. Of course, while this may be true in some instances, it is also true that polymorphism (or if we are talking about variation

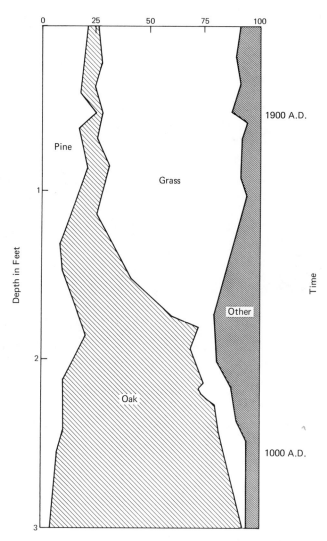

Figure 2-4 Temporal changes in the proportion of types of pollen reveal past changes in climate.

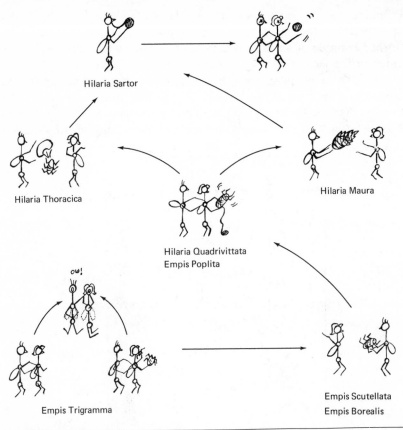

Figure 2-5 Differences in behavior among allied species often allow inferences about their evolution.

in behavior, polyethism) is itself adaptive. Consequently, the existence of variability may be a permanent feature and not a harbinger of evolution.

Fifth and sixth are comparative studies. These can be distinguished on the basis of whether they are "strong" or "weak." A strong comparative study is one in which an array of species exists which provides a spectrum of behavioral types. One nice example of this is to be found among the empeid flies (Schneirla, 1953) some of which have very elaborate courtship behavior, others of which have very simple courtship behavior, and many other species with courtship of intervening complexity (Figure 2-5).

This kind of comparison, however, does not allow one to make any deductions about the direction in which the evolutionary change has occurred, whether from the simple to the more complex courtship pattern, for example, or the reverse. And, finally, in the case of the weak comparative studies, we deal with the situation where an array of species has a variety of traits, some elements of which are held in common. In this case, the assumption is made that those elements that are common represent ancestral traits. The more species that hold the traits in common, the more primordial is the trait in question, and vice versa. We have already indicated some of the objections to conclusions based on arguments of this kind.

Much of the problem of deducing the evolution of behavior stems from a misconception of what we mean by behavior in the first place. We tend to treat behavioral phenomena as if they were something palpable, forgetting that "behavior" is not best regarded as a noun. Indeed, this is easy to forget because we are constrained to use a language that requires us to cast all our statements in the form of noun and verb. This is a requirement which often borders on the absurd, as when we have to invent the word "it" to express the idea that "it is thundering." Consider a very simple act which has often been the focus of discussions of behavior by ethologists. This is the response the young gull chick makes to the tip of the bill of its parents. It pecks at that bill tip. The act of pecking seems as simple and "instinctive" an act as any that has been described. But the analysis by Hailman (1967) has shown that while the final outcome, the contact of the chick's bill with the bill tip of the parent, does represent a relatively simple and stereotyped behavioral response, it is in fact a response composed of a multiplicity of alternative movements, many of which can vary without the final result being altered. The characteristics of the parental bill that elicit a response by the chick include such elements as figure-ground contrast, the orientation of the bill, its diameter, its rate of movement, its color; Hailman demonstrated that many of these attributes could be manipulated in a compensatory manner without the overall rate of pecking being influenced. Similarly, the motor response itself involves a step forward, a sideways twisting of the head, a lunge of the head, an opening of the bill. The various components of these movements, too, can vary independently. Analysis of the pecking could certainly be pushed further along the reductionist path. It is quite clear that in the forward movement of the chick's head, the same muscle bundles do not have to be involved each time. And even within one bundle different fibers doubtless fire at different times. Indeed, the more one approaches the molecular level in describing behavior, the more probabilistic and nondeterministic does the description become. Hence, we have to ask what is the behavior that is being genetically determined, that is capable of evolving? Must we postulate a separate gene for each motor element of pecking, suggesting that each can be independently inherited? And for the perceptual preference, is each attribute to be considered as having been separately coded by the genes? This is akin to the view that the second letter of every English word has a particular and unique function. The inheritability and evolution of behavior becomes more intelligible if we view the gene not merely as a repository of data or a blueprint from which an organism can be constructed, that is, as an inchoate homunculus, but rather as an information-generating device which exploits the predictable and ordered nature of its environment. This is a view that has been championed most cogently by Waddington (1966) and can be summarized as follows: a segment of the alpha helix (the chromosomal DNA) specifies a particular species of RNA which ultimately, and in an appropriate environment, and in concert with other proteins, leads to the synthesis of a particular enzyme, which in turn may repress or activate further synthetic activity by that portion of the helix, or repress or activate another segment. "Wheels within wheels," and all dependent as much on substances external to the helix as to the structure of the helix itself. Hormones, for instance, whose synthesis ultimately can be traced back to the action of particular segments of the alpha helix, are now known to activate genetic transcription at another portion of the helix. The transcription products may further feed back and regulate development. (Note Schneiderman and Gilbert, 1964.)

The implication that behavior is the outcome of feedback processes that are expressed in the context of a stable environment is profound. For one thing, great developmental stability is assured, for the epigenetic system (the system that acts with the gene in development) is buffered and self-correcting at many points. Many gene mutations will produce no phenotypic (or observable, measurable) effect. At the same time, the system is responsive to changes in environmental conditions. (Note particularly Waddington's 1966 reference to this flexible aspect of epigenetic systems.)

In short, behavior is not a thing, defined and determined by a discrete locus on a DNA molecule. It is a *process* that derives from a series of interactions, which at times can achieve a certain level of predictability and stereotypy. At some point this degree of inevitability and sameness becomes so great that we speak of "an instinct." The instinct is secreted by the process!

"Genetic" refers to inheritable differences. But, while I have argued that it is nonsense to talk about the inheritance of behavior as such, it is true that behavior may be more or less stereotyped. One can imagine a continuum with acts or perceptions or responses that range from highly plastic and variable at one end, to others that are highly constrained or stereotyped on the other. Further, it is of interest to ask how behavior which falls on one end or the other of this spectrum develops, or to make inquiries about its evolutionary history, or the mechanisms which underlie it, or the functions which it serves. The answers to such queries will reveal interesting differences between those kinds of behavior which we know to be highly plastic, those falling at one end of the continuum, and those kinds of behavior which we know to be more stable, less flexible, falling on the opposite end. The hope of finding an instinct in a chromosome, in any event, is illusory.

In the face of such disparagements, how does one study behavioral evolution? Is any kind of extrapolation from one kind of animal to another ever possible? Or does the notion of epigenesis preclude evolutionary studies? In the three sections that follow I want to use the subjects of aggression, territoriality, and altruism as examples, for all three have been much studied, have served to stimulate many speculations, and have provided particularly cogent instances of both useful and misleading approaches to the study of behavior and its evolution.

The Nature and Evolution of "Aggression"

"Aggression" is a wary tiger, and I suggest we not approach him directly. Instead, consider your reaction on a first visit to the rain forest of the tropics. It will most likely be one of puzzlement, for you will see far fewer animals, especially insects and the smaller invertebrates, than you expect. Only after careful scrutiny does one become aware that these animals are indeed present, but inconspicuously so. Many of them resemble their background to a degree that makes them all but invisible. Moths resemble bark or leaves, complete even to the inevitable imperfections; beetles mimic thorns, and caterpillars resemble bird feces or moss (see Cott, 1957). The extent and perfection of this cryptic coloration has excited naturalists for decades, though the reason for it was rather soon hit upon. The more perfect the resemblance, the greater is the individual's contribution to the gene pool of the next generation: in the parlance of the biologist, the greater its fitness. But explanations in terms of ultimate causes reveal nothing about the immediate cause or mechanism of the phenomenon. They tell us neither how cryptic coloration contributes to fitness nor on what mechanisms it is based. For instance, the benefit of camouflage may for one beast lie in its being able to escape predation; for another it may lie in being able to more effectively ensnare its prey; for yet another, the advantage may be in enhancing reproduction. Doubtless, yet other functions will be found. As to the mechanisms that permit camouflage, these include specialized appendages, pigments, refracting scales and hair, to list but a few. Even two animals of identical color may share this color by virtue of entirely distinct pigments. The point is that we find it entirely reasonable that the same ultimate purpose—camouflage—serves many different functions and is assured by a wide variety of mechanisms. How curious, then, that for a phenomenon no less widespread—aggression—we see no objection to assuming a unitarian point of view. The popularity of the simplifications of Ardrey (1966) and Lorenz (1966) attest to our penchant for explaining human aggression by the assumption that because man shares common elements with the hunting behavior of wolves, his behavior is dependent on similar mechanisms. Since aggressive behavior is widespread, it is reasonable to assume that it may enhance fitness, just as we assumed for cryptic coloration. But it is not reasonable, in the absence of evidence, to go beyond this and assume that a similarity in the ultimate cause establishes any basis for understanding the proximate cause and function or underlying mechanisms. Let us now approach the tiger.

Here is a man vigorously slapping a pretty girl. An aggressive act? By most standards, yes. But what about his emotional state? He could simply be in the state which he, on introspection, labels "angry." On the other hand, he could be quite dispassionate, a therapist who is coolly shocking a hysterical patient into calmness. He could even be a raging sex maniac whose libido will allow no suppression. The point, obviously, is that one and the same motor act can be associated with three different emotional states. In man, these states can be labeled and identified (at least subjectively), but it is as yet hardly possible to do this in nonlinguistic beasts. The reason for this lies in the somewhat surprising fact that the physiological parameters which had long been thought to provide an index of emotional state (as subjectively perceived) are themselves highly variable. Thus, an injection of epinephrine (adrenalin) may produce feelings of fear, elation, or no feelings at all, depending on the cognitive state or expectations of the subject (Scotch, 1967); and, in baby ducks, the presence of the mother-surrogate or an alien, presumably frightening, stimulus may be associated with the same, different, or no change in heart rate (Klopfer, 1968). Finally, one particular emotional state—extreme anger, for instance—may be expressed in a variety of ways: a violent, fatal attack; a coolly calculated shooting; or a long-term poison plot, where the victim is entirely removed from his aggressor.

What these examples are intended to indicate is that we can identify particular motor acts (some of which do violence to others and are therefore called "aggressive"), particular emotional states (some of which, at least in man, involve feelings of violence, and are called "aggressive"), and finally, certain cognitive states (which may include plans to commit acts of violence), and there is no dependency or reliable correlation between them. This independence has a neural counterpart as revealed by the work of Delgado (1967), who has shown that stimulation of a given site in the central nervous system may produce a highly stereotyped motor response whose aims appear variable, while at another site the stimulation produces variable sequences of motor responses with an invariable aim (and note the other evidence for independent or multifactor control cited by Washburn and Hamburg, 1968). The three features which we associate with an aggressive act, in short, the motor response, cognition (or the brain sites that are stimulated), and the emotional or affective state, are not necessarily tied to the activity of particular neurons or the concentration ratios of particular hormones (though it might be argued that the "relevant" hormones have simply not been identified).

What, then, do we mean by "aggression"? Is it the act, the cognition, the emotion? Can "aggression" only exist when we have all three? This, of course, would make it utterly impossible to discuss aggression in animals other than man, thus precluding an inquiry into the origins of affective and cognitive aggression.

Our dilemma stems, principally, I believe, from our tendency to turn aggression into a thing. Our language tends to compel this view, as noted above, for its grammar demands that every statement consist of a noun and a verb. The noun "aggression" is, I believe, responsible for the notions implicit in most theories of aggression, and made explicit by Lorenz (1966), of a specific drive which can be visualized as a fluid, accumulating in a reservoir, which must be periodically discharged. Lorenz has written, "Intraspecific aggression is in man just as much of a spontaneous instinctive drive as in most higher vertebrates" (Lorenz, 1966). He then goes on to equate the instinct "aggression" to the death instinct described by Freud.

The conclusion is that there is no such entity as aggression—it is a Cheshire cat. Acts may be performed that we label aggressive: some of these are highly stereotyped and may be elicited by the stimulation of specific brain centers; others are more plastic and may be elicited by stimulation in a variety of centers; further, there are affective states, at least in man, which we subjectively know as "aggressive," and these may be associated with a variety of hormonal states and motor acts.

The argument that aggression is inevitable in man because it exists in his forebears is thus nonsense. "Aggression" does not exist. The most we can hope is to be able to relate a particular way of responding to particular stimuli to an ancestral habit—i.e., to deduce the evolutionary history of the acts and their underlying mechanism from the selective forces which have shaped them.

Another approach to this problem is illustrated by Moyer (1968), who distinguishes several kinds of aggression, e.g., predatory, inter-male, fear-induced, etc., and proposes that each has a particular neural and endocrine basis. Though this is a more reasonable approach, it is still not immune to the previous criticism. And, again, there is scant justification for *a priori* assumptions about the similarity in mechanisms controlling predatory behavior in wolves and eagles. His scheme does have the merit of conceding that manipulations which facilitate one kind of aggressive act may suppress or have no influence on another.

We can consider another example, taken from J. P. Hailman's (1967) study of the pecking responses of gull chicks, the very model of a model "instinct." Newly hatched gulls (of many species, but particularly the Laughing gull, *Larus atricilla*) will peck at colored spots held before them; they prefer particular colors, so one can draw a curve relating wavelength to preference (as measured by peck frequency). The act of pecking is relatively stereotyped, though it consists of a number of component movements (head rotation, lunge, bill opening) that may vary independently.

Somehow, during development, links are established that assure a high probability of a particular color's perception being followed by a peck. The six questions we have to answer are: 1) What is the range of, and the constraints on, the variability of the perceptual preference mechanism (i.e., the width of the color preference curve)? 2) What is the range of variability of the components of that complex though recognizable act we call the peck? 3) To what degree is any given perception and the act linked? Always? Sometimes? Rarely? What factors in ontogeny affect the link? 4) How localized, if at all, are the neurons that subserve the link? How much variation in neural pathways (spatial and temporal) is possible without altering the motor outcome? Note particularly that if a particular neuron participates in many acts, and not always in any particular act, asking about its function may be akin to asking about the functions of the second letter in every English word (Bateson, 1972).

In the case of subjects capable of introspection we can inquire of them whether a given perception or act is associated with any particular emotion. If it is, we must further inquire as to whether 5) the emotion is always, sometimes, or only occasionally aroused in association with the perception or the act (or both), and what factors in ontogeny influenced its association. 6) And finally we need to ask whether a given emotion is dependent upon a particular combination of hormonal substances, and to what degree the latter may vary independently of the former. What these questions obviously suggest is the possibility that a system may achieve a particular and predictable state even while its component subordinate systems vary. This is almost a standard feature of many homeostatic (cybernetic) systems. The simplest example would be a volume of a gas, predictable enough, but whose component molecules may individually vary in velocity and position. The development of behavior involves applying constraints between the "molecules" so there are limits to their independence.

Consider this hypothesis on the causes of obesity developed by Schacter (1968). At infancy, the child is beset by a variety of stimuli that provide sensations of discomfort—it itches, is cold, wet, hungry. To all these, and sometimes spontaneously, it responds with a wail. Its unskilled mother responds to all wails as if they signaled hunger. The child may be warmed, dried, succored, etc., but only after it has first been fed. Feeding becomes an intervening signal: physical distress—feeding—then relief. These infants, then, as adults, are programmed to respond to all conditions of distress (including anxiety) by feeding, hence their obese condition. In them there has presumably been established a deviant, though stable and predictable, set of stimulus-response contingencies. Presumably many (any?) responses could be conditioned to many (any?) emotional states, just as we can be conditioned to subjectively either value or dislike particular emotions (i.e., a particular emotion can presumably function as a positive or as a negative reinforcer). Where variations of this kind do not occur, the contingencies have apparently been fixed at an earlier state in development—a reasonable evolutionary stratagem for responses that need not be variable and which occur to stimuli that do not vary.

Aggressive acts have the effect or the function of separating, repelling, or physically vanquishing another member of the same species. While too broad a statement for some purposes, and too narrow for others, this definition does cover many of Moyer's (1968) classes, and avoids the dilemma of confusing overt acts with their motivation. It allows for the possibility that the expression of acts called aggressive is dependent on a variety of sensory inputs, and a range of alternate neural pathways.

With this definition of aggressive behavior as "repelling," however, it is at least possible to ask specific, interesting questions about human aggression and even to hope that these could be answered. Why should "repelling" behavior evolve, for instance? Might a general explanation exist?

In animals that are highly mobile, requiring low-energy resources that are abundantly available, and requiring a high proportion of their time to consume a sufficient quantity of nutrients, "repelling" behavior is perhaps of limited value. Indeed, most ungulates tolerate conspecifics close to themselves, and, except for males during the rut, show little "aggressive" behavior. In contrast, among carnivores, which depend on large packets of food, needed but occasionally, and which are rendered less mobile by virtue of their helpless young, the proximity of many conspecifics may be less acceptable and the cost of aggression worth meeting.

Man behaves "aggressively," at least sometimes. Why? Consider the description by Birdsell (1953) of the probable composition of early human tribes. Assume that his picture is of general validity, that early man had to depend largely on gathering low-energy foods over a large region; that there were both lower and upper limits to the size of the band, the lower imposed by genetic considerations, the upper by economic (or ecologic) factors. To what extent would "aggressive" behavior be advantageous in such a society? If the reasoning as applied to ungulates is correct, repelling or separating behavior should actually be of limited value. And, since human females do not exhibit marked sexual seasons, and male libido waxes and wanes in daily fashion, there is possibly even a selective advantage to the avoidance of sexual dominance. In short, I suggest that, *contra* Lorenz (1966), it is at least equally plausible that early man showed very little "aggressive" behavior. I do *not* believe that the repelling behavior, dominance, and defense of property that characterizes many men today is a biological heritage carried from our primate ancestors. At most, it might be argued that later in man's history "aggressive" behavior did become adaptive, and hence more widespread, though the relatively brief time there has been for such a far-reaching evolutionary change to occur argues against it. The many different studies by such anthropologists as Benedict, Mead, and their students cast yet more doubt on the validity of "aggressive" behavior being an invariable component of man's makeup. Indeed, radical as it may be for a biologist to propose it, I suggest more serious consideration be given to the view that "aggressive" behavior is a cultural artifact dependent upon our learning to associate specific stimuli with particular emotional states and responses. The social (and practical) implications would be hopeful indeed, if this view proves the more nearly correct and we discard the picture of a fluid, aggression, slopping about in man's cranium and finding, inevitably, cracks through which to seep.

The Nature and Evolution of "Territoriality"

"Territorial" behavior has served as a springboard for evolutionary speculations even more often than has "aggression." This is at least partly due to the antiquity of studies of animal distribution and spacing, a factor of economic importance; and partly to the intrinsic appeal of a songbird driving intruders from its nest. No less important, territorial behavior is both ubiquitous and often obvious.

Since the distribution of organisms is rarely random, it is of interest to ask what factors influence their distribution. For example, are these factors related to their social behavior, or to the physical structure of the areas in which the animals dwell? In making reply, it is necessary to consider, first of all, the scale of the distribution with which we are concerned. In speaking of distributions, one must be explicit about whether it is the distribution of the entire species or of its particular representatives that is being considered. Second, when one talks about the distribution or occupation of space by animals it is necessary to specify the temporal span under consideration. The area occupied by an animal during the first year of life may be very different from the area occupied during adulthood. Both of these in turn may reasonably be expected to differ radically from the area inhabited by representatives of the given species over its entire history. The importance of these distinctions is all the greater when one recognizes the inverse relationship that appears to exist between the capacities for adaptability and adaptation (note Gause, 1942). That is, an organism is capable either of responding fairly quickly, within its own lifetime, to changes in the environment, (and in so doing sacrifices the ability to produce offspring that are preadapted to more stable, but extreme conditions) or produces preadapted offspring at the sacrifice of the ability to meet the stresses of short-term changes. One can be either a Jack-of-all-trades (adaptable) or a Master of One (have an adaptation)! A highly adaptable species might well, when its activities are measured over short periods of time, be found to have a much more varied geographic range than a species with a high degree of adaptation for a particularly harsh environment. But, quite the converse could become the case if the time scale were enlarged.

Another consideration that is relevant in discussing animal distributions and territoriality concerns the environment as perceived by the animal itself. Von Uexküll (1921) coined the term *Umwelt,* meaning the environment as perceived by the animal. It is essential to recognize that the sensory worlds of organisms are not identical. Two white blossoms may look equally bright to us and certainly the same color, i.e., white. But if one blossom absorbs and the other reflects ultraviolet radiation, to which our eyes are insensitive, then the honey bee, for example, whose eyes are responsive to ultraviolet radiation, will see the flowers as very different indeed. Similarly, the bat, whose range of auditory sensitivity is greater by almost an order of magnitude than our own, lives in a world of sounds very different from ours. The fact that the sensory and perceptual capacities of species differ one from the other introduces difficulties in planning analyses of the causes of particular distributional patterns. It is often helpful, therefore, to begin an analysis of dispersal and spacing mechanisms by first discovering the sensory capabilities of the animal under study. Having established its capabilities, one can then more easily determine to what (if anything) the animal preferentially directs its responses. For example, it is an observable fact that on many sandy beaches there is a close match between the color of the sand and the eggs which certain birds deposit in crude nests on the exposed beaches. Have the birds chosen the background that matches the color of their eggs? It could also be that predators, crows or foxes for example, have preferentially selected and removed from our gaze any eggs that did not match their background and which were therefore conspicuous. But if the shore birds did choose the appropriate beach, this would be evidence of a preference for, and selection of, a particular area—which could explain their spatial distribution. The answer to this particular question is not known. But, had we more information on color and contrast vision of shore birds, it might be possible to eliminate one of the alternatives.

A number of techniques have been developed for studies of habitat selection. Some involve rearing animals in simple, artificial environments in which elements of the natural environment can be successively introduced and varied. For example, Wecker (1963) has done this in studying the preferences of different races of deer mice, some of which habitually live in woods, some of which habitually live in pines. And Klopfer (1969) has done this for a variety of perching birds in an effort to find out what elements of the environment they select and respond to in distributing themselves. Correlational analyses have also been useful in this regard. I call attention in particular to the work of MacArthur (1965) and his students who have found that for many North American birds it is the diversity in the vertical distribution of foliage density that determines the presence or absence of particular species, not the particular species of plant.

Even within one particular kind of habitat and over short periods of time, one may find that some animals continue to space themselves in a decidedly nonrandom fashion. There may be a restriction of certain activities of one animal to a particular locus, from which some others, at least of the same species, are sometimes excluded. This constitutes territorial behavior. But, in considering the phenomenon of territoriality further, let us be forewarned; it is one thing to say that a rat has moved to a reward box, it is another to say that a rat has moved to chamber X (Hinde, 1966). In talking about territoriality, we must be certain that we see "the territory" as a "chamber X" expression, not as a "reward box."

It may be overly facile to say that there are as many kinds of territories as there are kinds of animals, but this statement is not very far from the truth. In some cases, the activities that are limited to particular places are feeding activities. In other species, it is breeding activities that are so limited. In some cases, it is only males of the same species that are excluded; in others it may be both sexes. In some, only the males seem to be doing the excluding, in others it may be the males and the females together, or even groups of animals together excluding members of other groups. The exclusion may operate year round or at only certain times of the year. Specific examples of these many and varied kinds of territorial behavior are listed in Klopfer (1969). But the question which is obviously of interest to us here concerns the origins or the evolutionary history of territorial behavior.

The variety of territorial types, and the uneven though widespread distribution of the territorial trait across the animal kingdom (with some species in any order showing the trait), argue against the notion that this behavior has arisen but once and represents a homologous phenomenon among different species. There is no reason at all to believe, for instance, that the kind of private property ownership exhibited by some nationalities of *Homo sapiens* is in any way derived from the defense of the nest-site shown by a few species of songbirds. Another argument supporting this statement draws heavily on the absence of territorial behavior in most primates other than man. Rather, the origins of territorial behavior are very likely as diverse as that behavior is itself. It has almost certainly originated many times in many different unrelated groups. Yet this is not to say that territorial behavior in all of its many forms may not have something in common. Brown (1964) has suggested that we consider territorial behavior as an example of site-dependent aggression. He then asks under what conditions it is economic for an animal to devote its time and energy, its resources, to aggressive encounters. The answer, of course, depends largely on the density, dependability, and the distribution of the resources which the animal needs, relative to its time and energy budget. For example, a bird such as the Great Tit of Europe nests in hollow cavities, which are rather scarce but of which a pair needs only one. It is quite economic for Tits to spend a certain amount of time defending this limited resource. On the other hand, during the winter, it has been estimated by Gibb (1960) that these animals must feed on one small insect every two-and-one-half seconds. The food supply is abundant but distributed in very small packages. The amount of the resource that is lost by diverting food-gathering energy and time to food defense makes this trade uneconomic. Food resources should not be defended and in fact a feeding territory has not evolved in this species.

Similarly, a bird that feeds on a fairly large parcel of food that has an even distribution might find it economic to defend a feeding territory, as, indeed, some owls do. But birds such as gulls, whose food is also available in patches but with these dispersed so that there are areas of great abundance and areas of paucity, would not find it economic to defend a feeding site. In short, the type of territory which evolves will depend on the economics of site-dependent aggression, rather than on the presence of innate territorial urges. Urges, too, have all too often been treated as simple physiological entities.

The Nature and Evolution of Altruism

Both territorial defense and aggressive behavior are implicitly involved in much altruistic behavior. Hence it may be of interest to conclude with a brief consideration of altruism's origins.

While altruism implies selfconsciousness, a deliberate sacrifice of self for the sake of others, this implication is unwarranted. For indeed, altruistic behavior may exist without any consciousness at all. (What we mean by "consciousness" and the story of its evolution is a separate question. See Klopfer, 1969.)

Altruism refers to a class of acts which increase the probability that the recipient of the act will survive and at the same time decrease the probability that the altruist will survive. It should be particularly noted that an act of assistance is not in itself altruistic if there is no cost attached to it. That is, an act, if it is to be considered altruistic, must in some way decrease the chances of survival of the performer. The question has often been asked, how can such behavior evolve? It appears so manifestly "unfit," and evolution favors "fitness." In order to resolve this paradox no small number of authors has resorted to inventions such as "group selection." By this is meant that, somehow, behavior that benefits a group will be maintained even while the individuals manifesting the behavior are removed. The genetic mechanism by which group selection might possibly occur has never been made clear. It would certainly need to be a mechanism that runs counter to the concept of natural selection as typically understood. But, while group selection seems an almost mystical and certainly unexplained concept, the fact of the matter is that drone bees do accept death once the queen has been fertilized, and lionesses do, in fact, fight unto death to protect their cubs. How can we explain the evolution of this self-destructive behavior?

In highly social species the helpfulness of strangers toward one another could possibly be explained by the fact that an individual who maximizes friendships and minimizes antagonisms is as a consequence more likely to survive, even where these friendships do occasionally involve him in acts which are individually more dangerous than avoiding them would be. The overall consequence of aiding others may still increase the chance of survival if receiving aid from others is also assured. It is not necessary that every such act of assistance has to lead to beneficial results to the doer, so long as on the average most such acts enhance survival. This kind of "altruism," then, is likely to depend on social organization and social forms, which make it profitable in the long run to maximize acts of charity. What about those cases where the overall consequence of the act does *not* enhance prospects for survival? This question, and the paradox of altruism, is resolved if we remember that fitness, in the evolutionary sense, does not refer to the physical survival of the individual or his phenotype. What it refers to is the proportion of the genes of some future generation that can be attributed to the individual in question. Every child carries half the genes of each of his parents. If one parent sacrifices himself for the sake of his children, all of his genes are not lost; half survive in his child. If the parent's self-sacrifice leads to the survival of two of his children, his fitness is not the least impaired, for then $2 \times \frac{1}{2}$ of his genes survive. This is genetically equivalent to the situation where the parent survives and the two children die. If, however, the sacrifice of the one parent's life leads to the survival (or to the increased probability of survival) of three of his children, then his death has in fact increased his fitness by a full 50% ($3 \times \frac{1}{2}$)! Where it is not children that are involved, but nephews or cousins or some more distant relative, the number of individuals that have to be saved before the altruist's fitness is increased by his death increases proportionately. The degree of altruism that natural selection will favor is directly proportional to the closeness of the genetic relationship of the individuals involved. Hamilton (1964) has developed this theorem in a fairly precise fashion and has demonstrated that if the gain to a relative of degree (R) is $(K) \times$ (the disadvantage to the altruist), selection will favor the altruistic gene, if K is greater than $1/R$. In short, it is almost inevitable that altruistic behavior will evolve. There is no necessity, therefore, to evoke such phenomena as group selection nor to require the existence of some form of consciousness in order to explain it.

If altruistic behavior is in a sense an evolutionary inevitability, what about its conscious concomitants? Man, at least, not only behaves altruistically at least sometimes, he has often elaborated ethical systems that encourage altruism. That is, he has articulated beliefs as to what constitutes right and wrong. Altruism is often identified as "right." Can one develop an evolutionary rationale to account for the emergence of ethical systems?

An ethical system represents a conscious belief as to what constitutes right and wrong. Waddington (1960) has pointed out that human communications systems have requirements which in themselves lead to the formation of ethics just as inevitably as natural selection leads to the evolution of altruism. Waddington writes, "Socio-genetic transmission of information from one generation to the next can, like any other system of passing on information, only operate successfully if the information is not only transmitted but is also received. The new born infant, in fact, has to be molded into an information acceptor. It has to be ready to believe (in some general sense of the word) what it is told. Unless this happens, the mechanism of information transfer cannot operate. Once it has happened and the mechanism becomes functional, then the socio-genetic system carries out a function analogous to that by which the formation and union of gametes transmits genetic information."

In summary, the evolution of a distinctive human animal involves the inheritance of traditions as well as genes, and these traditions can be transmitted only if the youngest generation is responsive to the teachings of older generations. From the very beginning, therefore, if communicative ability enhances fitness, selection must have favored those humans that accepted social structures and the teachings of their elders. The newborn infant, in other words, had from the beginning to be an authority acceptor. And, if ethical beliefs are merely strongly held beliefs that allow the individual to distinguish between "right" and "wrong," then their development, too, becomes inevitable. The consciousness of guilt that is present even in very young children may be relevant here. The fact that ethical beliefs once established are not impervious to change is irrelevant. The point is that given the existence of selection for social transmission, it was inevitable that man should evolve as an ethical animal, a beast willing to believe.

Finally, given the existence of belief systems, their content can also, within very broad limits, be partially dictated by natural selection. If, for instance, inbreeding reduces fitness, then advocates of beliefs that help prevent incest would be favored.

Most of this is speculative, including Waddington's portrayal of the infant as an authority acceptor. But the issues are open to experimentation. Authority acceptance in children is a subject that is being presently studied, and the universality of particular kinds of beliefs within the ethical systems of different cultures has also occupied many an anthropologist. At the very least, we now have a model which shows us that natural selection can shape such aspects of human behavior as altruism and its conscious articulation.

3 Genetic Correlates of Behavior

Robert M. Murphey

The purpose of this chapter is to point out some of the central problems involved in understanding relationships between genetic endowment and behavior. The first section presents a discussion of what inheritance means. Next there is a brief report and evaluation of some research tactics that have been used to increase our understanding of the genetic correlates of behavior. The chapter concludes with comments regarding the general efficacy of studying behavior-genetic relationships.

The Nature-Nurture Pseudoquestion

Suppose for the moment that you are a Stone-Age man. You are caught up in a bewildering world populated by men and other animals with whom you must cope if you are to survive. Successful coping requires that you avoid, exploit, and compete with the creatures around you; you must be able to assimilate and use information about them if you are to make your mark and leave your genes upon this earth. Of particular value will be information about their behavior: when will an animal fight, when will it hide, how fast can it run? Your ability to predict the behavior of those with whom you share the world will be an important factor in the duration of your survival.

You may arrive at an early generalization: animals that look alike behave alike. Animals with feathers and wings fly and sit on eggs; animals with four legs and fur nurse their young; animals that look like you make noises and use tools like you do; and so on. In time your ability to relate morphological features of other creatures to their habits and habitats may become quite acute, and obviously it will not depend on the experimental methods of modern science. Consider for example the achievement of the "primitive" hunting natives of the Arfak mountains in New Guinea, who have different names for 136 kinds of birds. When modern biologists applied scientific techniques to count the number of bird species in the area, they arrived at a number of 137 (Mayr, 1963).

Your generalizations about relationships between appearance and behavior need not prevent you from making another important generalization: there are differences between individuals that fall into any single category of creatures that you have devised. This conclusion should in fact be forced upon you by the others around who are, generally speaking, most like you. Some persons will be tall, others short; some cowardly, others fearless; some stupid, others, unpleasantly, more clever than yourself.

Armed with these two general insights, you may proceed to draw further conclusions and entertain additional speculations. Some of your reasoning, while adroit, may nonetheless lead you to hold incorrect beliefs. For example, you may speculate that the behavioral traits of another creature are inherent in its flesh. Thus you might eat fish in order to swim faster, or gazelle in order to run faster. Moving closer to home, you might eat the muscles of strong men, the brains of wise men, and the hearts of brave men (the word *courage* comes from the Old French word *cor* which in turn was derived from the Latin word for heart). Your cannibalism or other ritual eating habits constitute a gastronomic self-improvement program derived from your interpretations of biological variation.

Another speculation you might make, and one that sounds quite familiar to modern ears, is that the behavior of similar creatures is passed on from one generation to the next by way of whatever mechanism it is that makes them kin. If you are able to coin the words for it, you might postulate that behavior is inherited. However, as soon as you announce your stunning insight, you are likely to receive heated disagreement from some Stone-Age colleague. He might point out that much of what you do as an adult you did not do as a child. Someone taught you how to hunt, how to use your tools, how to build a fire, how to stay dry, and even how to talk. You were not born with all these bits of behavior, he goes on; you didn't inherit them, so you must have learned them. You acquired them by interacting with your environment.

You and your skeptical acquaintance might sit up many a night discussing these issues, but regardless of how reasonably you put your arguments, you might never get your friend to agree. Indeed we need not speculate on whether a general agreement was reached (see Breland and Breland, 1966, for further speculation on prehistoric, biological thought), for the controversy is still with us today, embellished of course by the arguments and counterarguments offered over hundreds of years by dozens of philosophers and scientists, but still fundamentally unresolved in the minds of many. The label that has been placed on the dilemma is "The Nature-Nurture Controversy."

The Nature-Nurture controversy can be traced backward at least to Greek times, and forward again through René Descartes, the German nativists, the English empiricists, the French mechanists, and finally (by way of a few other *ists*) to twentieth-century behavioral scientists. The older history will not be considered in any detail, but one point should not be overlooked. At least since the time of Descartes (1596–1650), a major theme or assumption of western philosophy has been the concept of *dualism:* the universe (reality) is composed of two different kinds of substance, the material (*res extensa*) and the immaterial (*res cogitans*).

In reference to the language of psychology, this distinction is equivalent to a division between physical events (the motions of our bodies) and mental events (the workings of our minds). The tacit acceptance of dualism has been tremendously important in helping to sustain a conceptual framework wherein the question of learned vs. inherited behavior could flourish.

By the time twentieth-century science took on the Nature-Nurture issue, the opposing forces were so well entrenched that there was little meaningful communication between them. Representing one side of the issue were the scientists who had been so highly influenced by Sir Francis Galton's *Hereditary Genius* (1869) that they became caught up in the notion of a direct relationship between physical heredity and almost any behavioral manifestation that one could mention. Galton's views, taken with the evolutionary theory of his cousin, Charles Darwin, gave rise to considerable emphasis upon the importance of inborn behavioral traits, which were typically called "instincts." During the first two decades of the century talk about instincts grew rampantly in regard to both animal and human behavior. When observed behavior could not be explained easily, its occurrence was simply attributed to instinct, as if labeling the behavior as "an instinct" explained its occurrence. The use of this circular reasoning became so pronounced as to become well-nigh intolerable; obviously (now, but perhaps not then), labeling a behavior as an instinct (or indeed labeling it as anything else) explains little or nothing about it. Perhaps inevitably, an opposing view came forward and became immensely popular. The name of the alternative approach was *behaviorism.*

Led by John B. Watson, the early behaviorists insisted that all behavior is acquired through learning; they were concerned to eliminate the word "instinct" from the vocabulary of American psychology. Watson's best-known statement regarding this issue appeared in his 1924 book *Behaviorism:*

Give me a dozen healthy infants, well-formed, and my own specified world to bring them up in and I'll guarantee to take any one at random and train him to become any type of specialist I might select—doctor, lawyer, artist, merchant chief, and yes, even beggar-man and thief, regardless of his talents, penchants, tendencies, abilities, vocations, and race of his ancestors. I am going beyond my facts and I admit it, but so have the advocates of the contrary and they have been doing it for many thousands of years.

Watson's denial of the importance of biological endowment in behavior was based in part upon a curious logical error that he had committed in interpreting the results of some important genetical research. Few of his readers bothered to verify his statements in this regard, and the erroneous interpretation remained virtually unchallenged until fairly recently (Hirsch, 1967a).[1] At any rate, in American experimental psychology at least, instinct theory and related views holding to the centrality of hereditary factors in behavior took rather a beating for some 30 or 40 years, until finally the impact of European ethology and modern genetics became felt and a truer perspective began to emerge.

The original Nature-Nurture question, asked: "Is behavior learned or is it innate?" By the middle of this century most textbooks of psychology that addressed the problem at all said something to the effect that the Nature-Nurture controversy was a dead issue. Because the environmental forces had to act upon *something* that was inherited, and because no inherited material ever developed without *some* environmental forces acting upon it, it followed that behavior must be some sort of joint consequence of the workings of Nature and Nurture. The argument went on to claim that the truth of this relation between Nature and Nurture was virtually self-evident; hence there was no need to discuss it further.

But in fact the issue is not dead at all; it has merely been abandoned by many because it is a difficult topic to discuss sensibly without dragging up all the old arguments from the past. No, Nature-Nurture thinking is alive and well and living in the twentieth century. Although few would ask, "Is behavior inherited or is it acquired?" we still speak of "inherited" versus "acquired" traits—a mode of conceptualizing which is not far removed from that of our Stone-Age intellectuals.

But it is possible to break out of the mental set that almost compels us to think in Nature-Nurture terms. Modern genetical theory provides the key for understanding the important issues hidden in the concept. An appreciation of at least four fundamental principles is required to comprehend what is really at stake when discussing the role of inheritance in behavior. The four principles will be considered one at a time.

The first principle is this: *behavior is ALWAYS a phenotype—it is NEVER a genotype.* A *phenotype* consists of any or all of the observable characteristics of an organism. Phenotypic characteristics can include almost anything: morphological traits (e.g., coloration, hair or bristle type, bone structure), physiological traits (e.g., metabolism rates, enzyme levels), or behavioral traits (e.g., reaction times, visual acuity). A *genotype* refers to the genetic material (DNA and a little bit of cytoplasm) that an organism receives from its parent(s). Genotypes are seldom if ever available for direct observation, given our present level of technology; rather, the genotype of an individual is typically inferred from the phenotype and, if possible, the antecedent conditions that preceded the production of that individual (e.g., who mated with whom). Thus behavior,

---

[1] Hirsch points out that Watson misinterpreted the significance of Johannsen's classic work on the determinants of variation in populations of bean plants. Watson maintained that the work proved that the "vast majority of variations in organisms is not inherited." In fact the work did not prove this at all, even for beans, let alone any other species of plant or animal. Watson's views on this matter appeared in his 1914 book *Behavior—An Introduction to Comparative Psychology*.

being by definition something that an organism does, is a phenotypic manifestation and is never a genotype.

The second of our four principles is this: *genotypes are inherited—phenotypes are not.* Your physical inheritance was the DNA you received from a sperm of your father and the DNA plus a bit of cytoplasm from the ovum of your mother; that is all. Thus you did not inherit your behavior; neither did you inherit your heart, your hair, or your fingernails. Each of these traits or characteristics (and it is not implied here that "behavior" is a single trait or characteristic) is part of your total phenotype.

The statement that you did not inherit your behavior does not warrant the conclusion that all of your behavior is learned; to draw that conclusion would be to repeat a central mistake in Nature-Nurture thinking. An appreciation of what conclusions are warranted depends upon understanding the third of the four principles. Comprehension of the third principle, in turn, is contingent upon familiarity with a few basic facts about modern genetic theory which we shall consider now.

Genetics must be studied by the examination of populations of animals rather than single individuals. Every biological species harbors a pool of genes, and no two species possess exactly the same gene pool although some of the same genes may be found in more than one pool. Within each species different individuals carry samples of some proportion of the total pool which is, in principle at least, available to the next generation. Because genotypes are causally related to phenotypes, a wide diversity of genotypes within a population is typically associated with a wide diversity of phenotypes. Phenotypes which are successful in the environment of the species will have a greater chance of passing the related genotypes on to the next generation. Particularly when the environment is undergoing rapid or extreme change, the greater the variety of phenotypes represented, the greater the likelihood that some will succeed in the new conditions. In any case, it is clear that the concept of fundamental importance here is the variability among individuals in the constitution of their individual genotypes.

We shall need a term or symbol to represent the idea of phenotypic variability within a population in regard to a particular trait; a symbol commonly accepted by geneticists is σ_p^2, which is to be read as "phenotypic variance."[2] Phenotypic variance is an index of observed diversity within the population for a particular trait; it is the diversity which we are "given" by observation and direct measurement. One task of biological science (including psychology) is to discover the determinants of this diversity.

---

[2]Variance is used here in its ordinary statistical sense. For example, if five animals are measured for a particular quantitative trait and yield values of 1, 2, 3, 4, and 5, then the mean value for the trait is 3.0 and the variance is $\Sigma (X - M)^2/N = 2.0$. See Roberts (1967) or Parsons (1967) for a fuller discussion. In many genetics texts the symbol V is employed rather than σ^2 as a designation for variance.

It was asserted above that genotypes are causally related to pheno-types. This means that some proportion of the observed phenotypic variance in a particular trait may be accounted for by reference to genetic diversity or variability within the population. The technical term used here to express the idea of genetic variability is σ_G^2, which is to be read as "genetic variance."

It should be obvious that genetic variance is not the only determinant of phenotypic variance; if it were, the Nature-Nurture controversy would never have arisen and survived. Some proportion of the phenotypic variance observed for a single trait is attributable to variability in the environments (both pre- and post-natal) in which animals grow, live, and are observed. The term used in reference to environmental variability is σ_E^2, which is read as "environmental variance."

There is yet a third factor that must be considered as a contributor to phenotypic variance; it is the variance that arises as the result of the interaction of particular combinations of genotypes and environments. In other words, some proportion of the total phenotypic variance is produced because a single genotype will respond differently in different environments and because different genotypes will respond differently in different environments. The technical name for this idea is the "gene-environment interaction variance," and it is symbolized as σ_{GE}^2.

We can now turn directly to the third of the four fundamental principles, which is that $\sigma_P^2 = \sigma_G^2 + \sigma_E^2 + \sigma_{GE}^2$. Put into words, this says that *the phenotypic variance observed for a particular trait in a given population is accounted for by the sum of the genetic, environmental, and gene-environment interaction variances related to that trait.*

Here we have a formula which provides a basis for a rational ap-proach to the question about the relative contributions of heredity and environment in phenotypic variation. Notice that the formula is expressed in terms of variance (variability) observed in a *population,* not in terms of the score or value of a trait in any one individual. This is what was meant when it was said above that genetics must be studied by observing populations rather than single individuals. The role of genetic influences on phenotypes can be viewed only in light of differences between indi-viduals, not by considering individuals in isolation. Marler and Hamilton (1966) have put it this way: ". . . at no point is the inference drawn that a particular trait in a given individual is inherited; rather a certain difference between the traits of two individuals is shown to be inherited." Or as Barnett (1967) has said, "One does not say that the possession of two eyes is inherited. An animal belonging to a two-eyed species may have one eye. . . . In this case we say that the difference from the normal type is genetically determined. Similarly, . . . we do not say that learning ability is inherited, though it is possible to find genetical differences in learning ability." Hirsch (1967b) has also given this topic a very thorough and thoughtful treatment. You may find it disturbing that Marler and Hamilton's and Barnett's usage of the word, "inheritance," is seemingly at odds with the second of our four principles. That is, if the second principle is correct, then "differences" cannot be inherited any more than behavior can. What is really meant by the two quotations is that the logic one must use regarding the effects of genes on phenotypes requires that the investigator compare one organism with another.

The fourth and final principle may be stated in two slightly different ways: 1) *for any population and any trait, σ_P^2 is always greater than zero;* or 2) *no two organisms are exactly alike.* There is of course no way to prove either of these statements deductively; they are simply very secure and credible biological generalizations.

But what about the components of σ_P^2 for a given trait and population; is σ_G^2 or σ_E^2 ever equal to zero? If σ_G^2 is zero then all the observed differences in a trait between individuals in a population must be due entirely to environmental variation σ_E^2. If σ_E^2 is zero then any differences must be due entirely to genetic differences between the individuals. This is at the heart of the Nature-Nurture controversy; more than that, it may be the Nature-Nurture controversy. Hence we need to look at the problems involved in trying to determine σ_G^2 and σ_E^2.

Consider σ_G^2 first; when may it equal zero? The most obvious case is a pair of monozygotic (identical) twins. Because the genotypes of the pair are identical, there is, by definition, no genetic variance. This fact simplifies the formula that relates phenotypic diversity to its components, and thus at first may seem to offer a tool for resolving the relative importance of genetic and environmental factors in producing phenotypic (e.g., behavioral) outcomes. However, it is very important to understand that studies of single pairs of identical twins do not allow statements to be made about the relative importance of heredity. Phenotypic differences may be attributed to genetic differences (i.e., "inherited") only if genetic differences exist; by definition they do not exist in a pair of identical twins. A failure to understand this obvious point can lead to serious misinterpretations of data collected on twins. In other words, all that can be said about identical twins is that any differences between a pair must be due to environmental variance.

Further, identical twins can, it seems, only be said to be identical genotypes, which does not mean they are physically identical. Recent research on monozygotic armadillo quadruplets has shown that the individuals are morphologically and physiologically far more distinct than had earlier been anticipated (Storrs and Williams, 1968). Similar results have been reported for twins in other species, and some possible causes of these marked variations have been suggested (Pollin and Strabenau, 1968). These studies were conducted under conditions in which there were no apparently large differences in the environments of the animals under study. The importance of these findings for psychologists is that we cannot assume that genetically identical individuals who have developed in the "same" prenatal environment will be morphologically and physiologically identical at any time after the exact moment of fertilization of the egg.

There are other conditions under which groups of animals are treated *as if* there were no individual differences in genotypes, and all observed phenotypic differences are treated as environmental in origin. Many generations of inbreeding can greatly reduce genetic variance (see for example Lush, 1945). Inbreeding, which will be discussed in more detail presently, refers to matings of biologically related individuals; the closer the relationship (e.g., parent-child, brother-sister), the greater the degree of inbreeding and consequent genetic homogeneity. On the basis of the available evidence, however, σ_G^2 does not appear to reach zero even under the most severe inbreeding regimens, at least when all of the genes that an inbred population carries are considered, but no one knows this for sure (see, e.g., McClearn, 1967).

To summarize, in those rare instances where σ_G^2 is presumably equal to zero, logic does not permit us to speak of what the organisms in question inherited, unless of course they can be compared with organisms with different genotypes, in which case σ_G^2 of the group of individuals under study ceases to be zero. Where σ_G^2 does not equal zero, then by definition, individual genotypic differences contribute to individual phenotypic differences. That is, even when we are dealing with individuals and not populations, individual genotypes are still important. After all, every organism has some kind of genetic endowment, and every organism has observable (though not necessarily observed) characteristics which are correlated with its genetic endowment. The rather obvious point that "environmental influence must be an influence on something" (Hirsch and Tryon, 1956) is well taken here. However, when we study a single individual we simply cannot determine the components of phenotypic variance; this can only be done by comparing differences between individuals.

Not only is σ_G^2 almost never zero, but the same is true of σ_E^2. We have seen that even when (presumably) genetically identical individuals are produced, they turn out to be different in at least some (if not all) respects. Available evidence (e.g., Pollin and Strabenau, 1968) suggests that the environments of all individuals are different from the moment of conception. Insofar as one can determine, there has been no convincingly reported example of two or more animals living in identical environments. Moreover, because behavior is by definition a response of an organism to some set of environmental circumstances, a response which indeed is likely to alter those circumstances, it becomes reasonable to assume that σ_E^2 is never equal to zero. And finally, given that σ_G^2 and σ_E^2 will seldom if ever be equal to zero in situations that are both genetically and behaviorally interesting, and that genotypes tend to respond differentially to a range of environments, it is quite likely that σ_{GE}^2 will not be equal to zero either.

The significance of the discussion so far for behavioral research may perhaps be clarified by the following example. Suppose that two organisms (*A* and *B*) are selected "at random" from a population which, for reasons of taxonomic and linguistic convenience, we shall call "Norway rats." Each of them has a unique genotype and each has undergone a unique developmental history. Each is tested, individually, in the same apparatus (environment), which we shall call a "T-maze." The experimenter has decided that he will present each rat with a food pellet every time the rat turns left when it passes the choice point of the maze. After a number of training trials in the maze, both rats turn left far more often than they turn right.

How are we to interpret this set of events in the light of the earlier discussion? To begin with, have the rats learned to turn left in a T-maze? Recall that the experimenter reinforces the rats for what they already "know" how to do, so in that sense they did not *have* to learn to turn left in the maze. Why did the rats turn left in the first place? Obviously the experimenter did not create the ability and willingness to turn left in the subjects' response repertoires. He waited until the rats turned left (for whatever reason) and then presented a food pellet when the rat turned left. What was learned? What was inherited? $\sigma_P^2 = \sigma_G^2 + \sigma_E^2 + \sigma_{GE}^2$. Without more information we can say very little. Suppose further that the performance of rat A in the T-maze, i.e., the probability of his turning left after a given amount of practice, is superior to that of rat B, but when the animals are tested in a different task, say avoidance conditioning, rat B masters the task more rapidly than rat A. Now, with regard to "learning ability" in these two contexts, we have a very crude indication of a genetic-environmental interaction as well as an illustration of the effects of the other components. That is, the two rats differed from one another (σ_P^2) in terms of both their genetic endowments (σ_G^2) and their environmental histories (σ_E^2). Moreover their behavioral phenotypes seemed to vary with the particular combination of genotype and environment that was studied (σ_{GE}^2).

If one is willing to accept the four principles that have been presented here and the discussion that has surrounded them, it follows that behavior is neither inherited nor acquired, neither innate nor learned (we shall take up the probable biological meaning of these words later in the chapter). We have argued for the position that behavioral differences vary with genetic and environmental differences. This may seem to be just a small change in terminology, but accepting it leads to a more realistic understanding of behavior and its correlates. And the Nature-Nurture question, in either the classical or modern form, ceases to be meaningful.

We hasten to add, however, that nowhere in this discussion has it been either stated or implied that genetic and environmental variances contribute in equal proportion to the production of phenotypic variance. The proportions of σ_P^2 attributable to σ_G^2, σ_E^2, and σ_{GE}^2 will vary with the phenotype under study. Now we may turn our attention to the kinds of questions that have been asked, should have been asked, and can be asked with respect to the matter of genetic involvement in behavioral variation.

Behavior-Genetic Analysis

Recall that the first half of the twentieth century in American psychology was characterized by the ascendency and dominance of behaviorism; that talk of the importance of hereditary factors in behavior was not well thought of, not fashionable, and not encouraged. In that context it is not surprising that those who were still interested in the influence of genetic variables on behavior were more or less obliged to spend much of their time demonstrating their own intellectual and scientific legitimacy. This meant that, until very recently, investigators in the area[3] found themselves assigned to the rather unsatisfying position of having to advertise the relevance, and even the existence, of basic biology rather than pursuing their work solely for the purpose of generating knowledge. Under the heading of "A False Start," Hirsch (1967a) has provided a concise summary of the period as he sees it:

Because of the stubborn and persistent opposition to the study of heredity and behavior, far too much effort has been spent proving the trivial points that this or that behavior shows a genetic component or a genotype-environment interaction, or in chasing single genes in order to have a more clear-cut case. Behavior geneticists have been so preoccupied with the defeat of environmentalist opposition that they have had little time for the more important task of a critical analysis of their own work. . . . I shall assume that the battle to overcome ignorance and the behavioristic opposition to according heredity its proper place in the behavioral sciences has been won effectively and decisively.

Apparently the work on behavior-genetics (or whatever you wish to call it) has had some effect, as evidenced by the recent influx of books and papers on the subject and by the playful title of Richard J. Rose's 1970 American Psychological Association Convention paper: *Revised perspectives on clinical research and training now that we know heredity exists.* Future historians of science are apt to look back upon this facet of our perplexed century with considerable amusement. Even so, much of the experimentation and observation that has been carried out so far was well done, intriguing, and imaginative. The next few pages contain some highly selected examples from the behavior-genetics literature. The studies that are cited here were chosen because they were particularly influential, interesting, representative of the state of the art, and/or otherwise supported this author's biases. Space considerations, regrettably, do not permit the inclusion of many other worthy investigations.

---

[3] There is not as yet a universally acceptable label for the field of inquiry that takes the role of genetics in behavior as its subject matter. The area has been termed "Behavior Genetics," "Behavioral Genetics," "Behavior-Genetic Analysis," and (particularly in England) "Psychogenetics." See Hirsch (1967b) for a discussion of the merits and demerits of these terms.

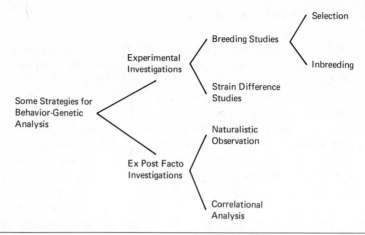

Figure 3-1 Relations among various strategies for Behavior-Genetic Analysis.

Before plunging into specific studies, however, it must be pointed out that behavior-genetics (like the other biological sciences) has made use of both *experimental* and *observational* approaches in collecting data. Ordinarily, the former has restricted itself to the laboratory, where breeding and environmental factors could be controlled in some fashion. The latter approach tends to focus its attention upon events that take place in Nature—phenomena over which the investigator has limited if any control. Two common approaches to data-gathering under these conditions are naturalistic observation and correlational analysis. These kinds of studies (*ex post facto* investigations) attempt to establish some relationship between a behavior pattern and a possible genetic basis. They tend to be imprecise and subjective but do deal directly with the real world whereas experimental studies are fairly objective and provide data which can be rigorously analyzed but tend to be artificial. In this chapter, due to space limitations and personal preference, only experimental studies of behavior genetics will be discussed in detail.

Generally, the experimental investigations that will be described here can be divided in terms of what can be called *breeding studies* and *strain difference studies.* Breeding studies can be (albeit somewhat artificially) further categorized as involving *selective breeding* or *inbreeding.* For purposes of clarity, a diagram of the approaches just mentioned is presented in Figure 3-1.

Now is a good time to define some terms. What is meant by *strain difference studies* is put off until later in the chapter. *Breeding studies* will refer to situations wherein animals are mated with one another in some systematic way as an intentional experimental procedure. If the animals are bred solely on the basis of some shared phenotypic characteristic(s) of "unrelated"[4] individuals or without regard to how closely the animals are related, we are dealing with *selective breeding;* if the subjects are mated mainly or in part because they are close kin, then *inbreeding* is said to be taking place. In the real world of laboratory experimentation it is usually the case that a breeding program will encompass both selection and inbreeding. Often an investigator's inbreeding tactics (whether by design or by accident) will include culling out animals that do not meet certain phenotypic criteria as well as breeding only closely related individuals. A typical selection experiment, especially when the foundation population is small, will involve some amount of inbreeding regardless of whether or not inbreeding is an intentional feature of the investigation. Both inbreeding and selection techniques have advantages and disadvantages from a practical and theoretical standpoint. But a discussion of that can wait for the time being. At the moment, some examples are in order.

Selective Breeding Studies

Among the selective breeding studies that have been reported, at least two series of investigations stand out clearly as being classical in the behavior-genetics literature. The earlier of the two was carried out by the late R. C. Tryon, a student of E. C. Tolman at Berkeley. Tryon's studies are important historically as well as scientifically because of their influence on the acceptance of heredity as a proper interest for American psychology and because they led directly to an impressive amount of research by other workers. Even so, Tryon's experiments were very much a continuation of some work initiated by Tolman (1924), who had attempted, in some respects unsuccessfully, to breed selectively for "maze-learning ability" in rats.[5]

Tryon (1930, 1931, 1939, 1940a, b, 1942) introduced improved genetical and measurement techniques to the same problem, and was able to report considerably more definitive and positive results. Employing multiple-unit mazes, which were much more complicated than those generally in use today, Tryon tested a random sample of 142 unselected rats for "maze-learning ability," as defined by the number of blind-alley entrances (number of errors). The "brightest" animals (i.e., those at the extreme left end of the scale in Figure 3-2) were bred with one another, as were the "dullest" animals (those on the right-hand side of the distribution). Notice that in the first generation (F_1) there was not much difference between the "brights" (B) and the "dulls" (D) but by F_8 there was very little overlap between the two distributions, and the difference between the two strains persisted thereafter.

---

[4] Degree of relationship as employed here refers to individual-familial kinship rather than to taxonomic and presumed evolutionary relationships.

[5] Tolman gave credit for the original idea to Warner Brown.

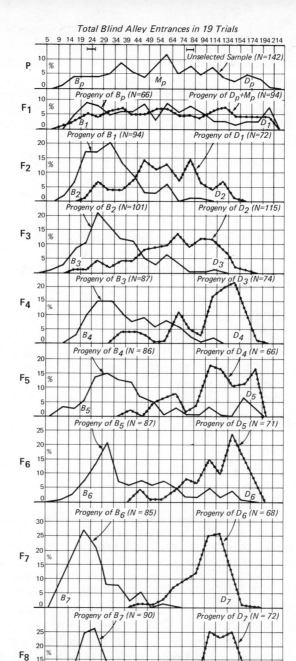

Total Blind Alley Entrances in 19 Trials

From "Individual Differences" by R. C. Tryon, in F. A. Moss (Ed.), *Comparative Psychology*. New York: Prentice-Hall, 1942.

Figure 3-2 Effects of selective breeding for eight generations on maze performance. Generation P was the parental, unselected stock; Group B in generations F_1 through F_8 was selected for "maze brightness," and group D for "maze dullness."

Aside from the obvious lesson that Tryon's experiments contributed, a very straightforward demonstration that heredity does have something to do with behavior, he also devised a conceptual framework for distinguishing components of large units of behavior ("behavior domains") and analyzing individual differences between animals as those differences relate to the components ("subdomains"). That is, Tryon's interests went far beyond merely establishing the existence of relationships between inheritance and behavior. He wanted to find out something about the nature of those relationships, and he provided some means for doing so. This aspect of Tryon's work has yet to be universally appreciated by behavior-genetics.

A second illustration of selective breeding methods in animal behavior is furnished by an unusually creative series of studies reported by Hirsch (a student of Tryon) and his students (Hirsch, 1959; Erlenmeyer-Kimling, Hirsch and Weiss, 1962; Hirsch and Erlenmeyer-Kimling, 1962; Hostetter and Hirsch, 1967), with fruit flies (*Drosophila melanogaster*) as experimental subjects. The fruit fly or vinegar fly (or "drunkard" if you are from the Deep South) is a tiny insect that you have no doubt seen swarming about garbage cans and overripe fruit in the summertime. You may have thought that they were gnats, but they are not. It is often claimed that *Drosophila* have been the most useful of any bio-instrument in genetic research: "It would not be an exaggeration to say that we have

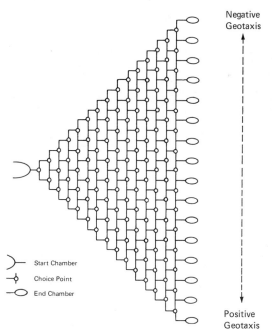

From "Some correlates of negative geotaxis in *Drosphila melanogaster*" by R. M. Murphey and C. F. Hall, in *Animal Behavior* **17**. © Animal Behavior, 1969.

Figure 3-3 Schematic diagram of mass screening for geotaxis.

learned more about the basic laws of heredity from the study of this fly than from work on all other organisms combined" (Demerec and Kaufmann, 1962). As Fuller and Thompson (1960) have pointed out, genetic investigations require that the species under study be highly variable, prolific, easily maintained, and have a small number of large-sized chromosomes. An additional advantage would be a short generation time (i.e., a brief period of time between birth and reproduction). Fruit flies excel in all of these criteria.

Exploiting the biological advantages of *Drosophila*, Hirsch and his collaborators set out to see how far they could pursue the genetic analysis of a behavioral trait, and the trait they chose was *geotaxis*. Geotaxis refers to an animal's movement with respect to gravity; *negative* geotaxis de-

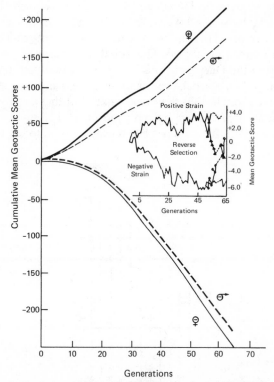

Erlenmeyer-Kimling, L., Hirsch, J., and Weiss, J. M., "Studies in Experimental Behavior Genetics: III. Selection and hybridization analyses of individual differences in the sign of geotaxis," *Journal of Comparative and Physiological Psychology*, **55**, 1962, 722–731. © (1962) by the American Psychological Association, and reproduced by permission.

Figure 3-4 Cumulative curves showing the difference between male (♂) and female (♀) mean response to selection for negative and positive geotaxis. The insert shows the effects of continued selection and reverse selection (where negatively geotactic flies are selected for positive geotaxis and vice versa); successful reverse selection demonstrates that even after two strains become statistically separate, there is still overlap between the two—that there is genetic material associated with positive geotaxis in the negatively geotactic strain and genetic material associated with negative geotaxis in the positively geotactic strain.

notes locomotion away from the pull of gravity, while *positive* geotaxis is movement toward the earth. After having used some relatively crude but ingenious measurement techniques for estimating the reliability of individual differences in geotactic behavior (Hirsch and Tryon, 1956), a sophisticated multiple-unit maze was constructed which provided a fairly easy means for testing many subjects simultaneously (the method is called "mass screening"). The apparatus (see Figure 3-3) was constructed in such a way that the flies assorted themselves into groups which differed in the strength of their geotactic tendencies. This was accomplished by presenting the animals with a series of vertical choice points, where a fly could walk either toward or away from the pull of gravity when he came to the end of a horizontal tube. If, for example, a subject turned upward 15 times (see Figure 3-3), he achieved the highest possible negative geotaxis score and was trapped in the topmost end chamber of the maze; if he turned downward consistently he was trapped in the bottommost end chamber and was said to be positively geotactic. The location of the terminal chamber in which the fly ended his sojourn through the maze told the experimenters the number of responses toward or away from gravity that the subject had made. Altogether, there were 16 end chambers or 16 possible geotaxis scores. As expected, the animals distributed themselves over all of the possible outcomes (the distribution was not symmetrical, however).

Flies were then sampled from the two extremes of the distribution. Negatively geotactic males were bred with negatively geotactic females, and likewise for positive geotaxis. After several generations of selective breeding, the lines began to stabilize, such that negative and positive strains were developed (Figure 3-4).

Next, the investigators introduced flies which had certain visible, morphological characteristics ("genetic markers") that were known through past research to be associated with particular chromosomes. Such markers are mutations which differ substantially from the "normal" or "wild type" fly. In essence, and this is something of an oversimplification, the idea was to determine whether any of the morphological traits were correlated with particular patterns of maze behavior. If so, it would imply that the maze behavior was associated with the genetic material being carried on the chromosome that was associated with the morphological trait. Perhaps this will become a bit more clear if some visual representations of what the investigators did are presented. Parenthetically, it should be noted that *Drosophila melanogaster* have four sets of chromosomes—one pair of sex chromosomes (XX in females, XY in males) and three pairs of autosomes, labeled II, III, and IV.[6] The markers, their phenotypes and related chromosomes, were as shown in Table 3-1.

By way of some rather complicated breeding procedures, eight combinations of phenotypes were generated (remember that each animal

---

[6]An autosome is a nonsex chromosome. Chromosomes Y and IV were not marked in the Hirsch *et al.* series.

Table 3-1

Marker	Phenotypes[7]	Chromosome
B	Bar (eye is shaped like a narrow bar; wild type eye is round)	X
Cy	Curly (wings are curled upward; wild type wings are straight)	II
Pm	Plum (eye is brownish and flecked with darker spots; wild type eye is red)	II*
Sb	Stubble (bristles behind head are shorter and thicker than wild type bristles)	III
Ubx	Ultrabithorax (balancing organs or halteres, located behind the wings, are swollen)	III*

*Not used in testing

receives a full set of chromosomes from both parents, and ends up with two sets). The eight combinations are shown in Table 3-2 (p. 89), where empty parentheses indicate unmarked chromosomes.

Hirsch and his group then sought to estimate the extent to which the morphological phenotypes (and by implication, the chromosomal material) were associated with behavioral phenotypes. This is obviously a complicated matter which will not be explored in depth here. Very briefly, some of the conclusions of this research were: in an unselected population, genes on chromosomes X and II were significantly associated with positive geotaxis, and selection acts strongly on genes located on chromosome III to contribute to negative geotaxis.

The details of these findings are not so interesting as their implications, the most important of which for our purposes is that behavioral phenotypes can be amenable to sophisticated genetic analyses just as morphological phenotypes can be. Moreover, and once again, it can be seen that successful genetic selection involving behavioral traits has furnished persuasive evidence for the importance of genetic considerations in the analysis and understanding of behavior. Let us turn now to some other sources of data relating to the same point.

Inbreeding and Selection:
The Good, the Bad, and the Ugly

A few pages ago a discussion of the relative merits of inbreeding and selection was promised. You will recall that inbreeding refers to matings between related individuals. Roberts (1967) gives a practical definition of inbreeding in a population as ". . . the mating of individuals that are more closely related than the average relationship between all the individuals of that population." Typical inbreeding procedures involve mating progeny with parents or siblings with siblings, perhaps in deliberate

---

[7]See Lindsley and Grell (1968) for more complete descriptions of these and other *Drosophila* phenotypes.

Table 3-2

Chromosome Types		X	II	III
		Chromosome Numbers		
	1)	( )	( )	( )
		( )	( )	( )
	2)	( )	( )	Sb
		( )	( )	( )
	3)	( )	Cy	( )
		( )	( )	( )
	4)	B	( )	( )
		( )	( )	( )
	5)	( )	Cy	Sb
		( )	( )	( )
	6)	B	Cy	( )
		( )	( )	( )
	7)	B	( )	Sb
		( )	( )	( )
	8)	B	Cy	Sb
		( )	( )	( )

conjunction with some sort of selection criterion, even though Broadhurst (1960) discourages such an approach in behavior-genetics. Selective breeding in its purest form, on the other hand, attempts to breed animals that share some phenotypic characteristic(s), where the scientist doing the selecting ordinarily makes some effort to minimize inbreeding or, at least, promotes "random" mating *vis-à-vis* biological kinship. The major arguments as to which is more efficacious—inbreeding or selection—have been articulated clearly by McClearn (1967) and Roberts (1967), whose ideas on the topic will be borrowed rather liberally during the course of this discussion. In their own work, McClearn makes use of inbred strains almost exclusively, and Roberts almost never. A general resolution of the issue becomes in some ways about as much an aesthetic as a scientific judgment, once the theoretical arguments are fully understood. As an arbitrary beginning, we shall take up inbreeding first.

Suppose that you were to examine a sample of highly inbred subjects and found that they were homogeneous (alike, though not identical) in their behavior on some task, say T-maze performance. It is most likely that, should you measure several other phenotypic characteristics of the same sample, the animals would be homogeneous with respect to those also. The reason for the tendency to be alike on all traits is in the nature of inbreeding itself. That is, by virtue of mating closely related individuals, the members of successive generations will become more and more similar in all of the genetic material that they inherit. In contrast, the products of selective breeding (taking Tryon's studies as an example) should become more alike "only" with regard to the genetic material that is involved in maze performance. An explanation of the quotation marks around "only" should become evident after a while.

The homogeneity that characterized inbred strains has a number of important implications for behavior-genetic research. On the positive side, the property of alikeness enables the researcher to maintain reasonably good control over genetic variables (in the sense of reducing chaos), while manipulating environmental variables. Exhuming the components of variance model for the moment, if σ_G^2 is small, then the greatest proportion of σ_P^2 can be attributed to σ_E^2, and environmental effects can be varied systematically (examples of this kind of research are presented in the next section). Furthermore, if the animals really are alike, then the strain can be used as a sort of behavioral reference point or standard; McClearn (1967) has said it quite eloquently: "One of the prime advantages of employing an inbred strain is the relative reproducibility of the key feature of an experiment: the living subjects whose behavior is being examined." This is a consequential advantage indeed, for it means that experimental results are likely to be reliable, and that investigators who make use of the same strains have a fair chance of replicating findings from other laboratories.

To recapitulate, the greatest overall advantage in the use of inbred strains is precision and reliability. Ordinarily, reliability is a necessary condition (but not a sufficient one) for validity. Reliable and valid results are not enough, however. Eventually the biological scientist ought to be able to make statements about how his findings relate to other realms of the biological world. It is at this point that inbreeding can be monumentally weak and for precisely the same reason that it is reliable—the homogeneity of the subjects.

It is tautological but worthwhile to point out that, in inbreeding, homogeneity is brought about at the expense of heterogeneity. Unless some selection procedure is employed in connection with the inbreeding program, the genetic material that becomes "fixed" (homogeneous) gets that way by a more or less (but not quite) random process. Suppose that a population contains genetic material associated with both black and brown coat coloration, and that the relevant genes in the population have particular proportions, e.g., 25% black and 75% brown. If the specific individuals who are sampled "at random" from the population for purposes of founding an inbred strain happen to produce, say 75% black-related genes, then in one generation the frequency of black shifts from 25% to 75%—it triples. Consider now that thousands of genes might be involved in the heredity material received by most organisms that are of interest to us. This means that, in many ways, the products of inbreeding are highly atypical in that they will resemble only a fraction of the original population. It is not hard to see that any findings generated through the use of an inbred strain may well be specific to that strain and not generalizable to the original population—not to mention the species at large, other species, other genera and so on. Related problems arise in the statistical analysis of data (see Roberts, 1967).

There are, interestingly enough, mechanisms which at the same time counter homogeneity and foster biological uniqueness of inbred strains. One of the most influential of these is the action of deleterious (disadvantageous in the reproductive sense) genes. Statistically speaking, most harmful genes are recessive. They tend not to be manifested phenotypically in heterozygous individuals—those who inherited one kind of genetic material from one parent and its counterparts (alleles) from the other parent. So, if an individual receives a deleterious gene from one parent and a dominant, nondeleterious allele from the other, it is generally unlikely that the phenotypic effects associated with the former will be expressed. We have already noted that inbreeding promotes homogeneity at the expense of heterogeneity; at the level of the individual, this means that homozygosity is favored over heterozygosity. Inbreeding, then, is more apt to bring about the phenotypic expression of deleterious genes that would be found in selected strains or in the original population. Among the more obvious outcomes of this state of affairs are the following:

1) Inbred strains have a tendency to *become inviable;* this is called "inbreeding depression." It is commonly difficult to keep an inbred strain alive for more than a few generations.

2) Given intense selection pressure against deleterious phenotypes, inbred strains that do not succumb to inbreeding depression tend to be "loaded" with nondeleterious genes (i.e., they will have unusually low deleterious gene frequencies) so that they become *highly atypical* of the population from which they were derived.

3) Almost paradoxically, there tends to be selection pressure to preserve heterogeneity, which of course would tend to suppress the phenotypic manifestation of deleterious genes.[8] Natural selection in favor of heterogeneity appears to be a powerful factor in maintaining σ_G^2 at some value above zero in inbred strains (see Lerner, 1958, for further thoughts and data relating to this).

At this point it would be easy to say that what inbreeding lacks in generality, selective breeding accomplishes at a cost of a reduction in precision. But the matter is not quite so straightforward. Selection has its own shortcomings in generality as well as in practicality. Bear in mind

---

[8] What is and is not harmful must ultimately be assessed in light of environmental demands. The "sickle-cell trait" in red blood cells, for instance, which is closely associated with genetic endowment, can lead to severe anemia and death—especially in the homozygous condition. As one might expect, the relevant gene frequencies are low in most human populations. Some exceptions are found in certain African and other warm climate groups, where the trait has an inordinately high probability of occurrence. It turns out that the sickle-cell trait seems to have an association with resistance to malaria and perhaps some other diseases. This is an example of natural selection in favor of heterogeneity (e.g., Stern, 1960; Lerner, 1968; and especially Wiesenfeld, 1967).

that the constitution of gene pools of unselected inbred strains is arrived at through something of a random process. By and large, whatever genetic endowment that was carried by the parental stock is the genetic endowment of succeeding generations because alternative genetic material from unrelated individuals is not introduced. There are, of course, some modifications of the parental material by way of biological accidents (e.g., mutations) and incidental selection.

Selective breeding, on the other hand, represents a deliberate attempt on the part of an experimenter to bring about strains that suit his genetic fancy. Almost inevitably, the investigator strives for homogeneity with respect to a particular phenotype (e.g., maze brightness or negative geotaxis) and heterogeneity of all other traits (e.g., coat coloration or scutellar bristle number), something which is difficult to achieve for several reasons. Apart from the fact that selective breeding is commonly bothered with such mundane problems as difficulty in maintaining a large enough subject sample to minimize inbreeding, unnerving phenotype fluctuations (particularly during the early stages of the program) and certain perplexities in the statistical analysis of data, there are serious genetical considerations that act at cross purposes with the goals of the investigation.

In the first place, as you are already aware, genes are carried on chromosomes. Genes which are carried on the same chromosome are said to be "linked," and they are known collectively as "linkage groups" inasmuch as they are transmitted together.[9] Indeed, this knowledge was the basis of the chromosome studies reported by Hirsch and his associates which were mentioned earlier. Recall that if a behavioral trait (A) was associated with a morphological trait (B) which was in turn known to be associated with a particular chromosome (C), then a relationship between (A) and (C) was implied. This means that when one is breeding selectively for a particular trait, one would expect the successive generations to become more and more homogeneous not only with respect to the gene or genes that are related to the trait of interest but also with regard to all of the other genes that are carried on the same chromosome or chromosomes. Suppose now that the genetic correlates of the trait of interest are *polygenic:* i.e., at least two and possibly many genes contribute to the development of the trait (which is probably true insofar as all behavioral traits are concerned—e.g., Parsons, 1967). Successful selective breeding should bring out homogeneity of all chromosomes (and hence their related genes) that are connected with the trait of interest. Under these conditions, to the extent that the experimental subjects are endowed with a small number of chromosomes (*Drosophila* have 4 sets, laboratory rats 21), the strains will be atypical of the original population and the

---

[9]Accidents do happen on occasion, and the linkage groups do at times become fractionated. Space does not permit a discussion of the topic here, but see Srb, Owen and Edgar (1965).

general applicability of findings will be reduced. There are other genetic complications, which we shall not go into here, but which nonetheless detract from clarity in the analysis of selective breeding studies.

Another key problem is that the trait of interest generally consists of more parts than is immediately evident. Tryon made it clear, in speaking of domains and subdomains, that an observed behavioral phenotype consists of many constituent parts, or to quote Murphey and Hall (1969): "A behavioral act, even though simple in appearance to the human observer, may in reality be a cluster of traits consisting of three components: the behavior of interest, genetic correlates, and task correlates." Genetic correlates of the trait refer to all of the genetic material (and correlated phenotypes) that are directly related to the trait, including the kinds of things mentioned in the previous paragraph and discussed elsewhere by Fuller and Thompson (1960) and Roberts (1967). Task correlates are traits which, though not "directly" a part of the behavior of interest, are intimately involved in the activities of an animal when he is performing whatever behavioral task that the experimenter is measuring.

Selective breeding is likely to bring about homogeneity of genetic material related to both the behavior of interest and task correlates. For example, Murphey (1965) reported that when negatively geotactic males were tested in a different kind of maze, which consisted of a series of sequential choice points but did not measure geotaxis, the subjects showed a tendency toward repeating responses rather than alternating them. If you will look again at Figure 3-3, you should see that the geometry of the Hirsch *et al.* geotaxis maze requires that the flies repeat responses at choice points in order to receive high negative geotaxis scores because every consecutive response away from the pull of gravity is a repetition of the previous response. Subsequent research by Murphey and Hall showed that selection for negative geotaxis can also encompass selection for response stereotypy (a tendency to repeat responses at sequential choice points), "maze hardiness" (the ability to survive in a dry, plastic environment), and reduced locomotor activity while negotiating a maze (high activity levels under those circumstances were shown to be mal-adaptive, leading to mortality). The investigators contended that, "The total environment in which the behaviour of interest is measured imposes an array of biological requirements on the organism. The gene pool will be modified in response to these extraneous demands, regardless of their conceptual relevance to the behaviour of interest." Broadhurst and Bignami, in 1965, and Broadhurst and Eysenck in 1964, have studied similar problems having to do with emotionality in rats. Now if the task correlates have their own genetic correlates, which of course they must, then we would expect selected strains to become progressively more homogeneous with respect to those also, with the result that the subject pool becomes less representative of the original population as the components of the behavior of interest, task correlates, and their genetic correlates all increase.

There is one further problem common to both selective breeding and inbreeding that should be mentioned before leaving this subject. It is well-known in today's genetic theory and practice that a particular phenotype can be brought about in any number of ways. Three hypothetical strains (derived from inbreeding or selection) might be "equally maze dull," where one of the strains is characterized by a metabolism defect, another by poor vision, and the third tends to have peculiar ratios of neural transmitter substances. Furthermore, it is not uncommon to find instances where environmental influences are such that a phenotype which is normally associated with one genotype is derived from a totally different genotype—that is, several different genotypes may lead to the development of the same phenotypes. This kind of mimicry is termed "phenocopying." The points made in this paragraph, and covered in much more detail by Fuller and Thompson (1960), have tremendous implications for what behavior-genetics has and does not have to offer in the way of general biological knowledge. We shall return to the issue toward the end of the chapter.

The question as to which is preferable—inbreeding or selection—has not been settled by virtue of this discussion. Maybe it is just as well to end it with some opinions of McClearn (1967) and Roberts (1967). According to McClearn, "Perhaps the most straightforward evidence on the role of heredity in a behavioral trait can be obtained from examination of differences between inbred strains." Now Roberts: "Any work on inbred lines refers . . . very strictly to those inbred lines only. Any conclusions from such work should not be deemed to apply to the species at large without supplementary evidence." To give McClearn another turn: "An apparent disadvantage of an inbred strain is the lack of generality obtained from research upon it. This cannot be construed as an argument for, or a defense of, studies using nonspecified or nonstandard groups, however. There the generality is no greater, but many are deceived into thinking that it is. In fact, the narrowness of applicability of results from inbred animals can be regarded as a sterling virtue. A researcher familiar with various strains is never tempted to overgeneralize his findings. If data are available for only one particular strain, the issue is regarded as completely open with respect to any other strain." Both McClearn and Roberts do agree that there are circumstances under which the use of inbred strains is more appropriate and others that call for selected strains.

Strain Differences and Heritability

Strain difference studies make use of comparisons between two or more preexisting groups of animals that are known to differ genetically from one another. In the behavior-genetics literature, a majority of the strain difference studies have employed inbred strains, although selected strains have been compared as well. Probably the best-known examples of the latter have come from examining differences between descendents of Tryon's maze-bright and maze-dull rats. In one of the earlier investigations in this connection, Searle (1949) demonstrated several differences between the two strains, including superior "learning ability" by the "dulls" when they were tested in certain situations other than the one which Tryon used. Krech, Rosenzweig, Bennett and their associates have carried out a comprehensive research program over a number of years on biochemical and environmental correlates of descendents of Tryon's strains.[10] They have reported numerous physiological and behavioral differences (e.g., Krech, Rosenzweig and Bennett, 1962; Rosenzweig, Krech and Bennett, 1960). The reader may wish to relate these findings to the discussion presented in the preceding section.

There is an enormous literature on inbred strain differences reviewed by Fuller (1960) and McClearn & Meredith (1966). A representative list of reports of behavioral differences between different strains of mice is presented below:

Investigators.	*Demonstrated Strain Differences*
Carran, Yeudall and Royce (1964)	Fear of electric shock.
Collins (1964)	Number of correct responses during avoidance training.
Hall (1947)	Audiogenic (sound-induced) seizures.
Henderson (1967)	Effects of prior experience on open field activities.
Maas (1962)	Emotionality and brain chemistry.
McClearn (1961)	Activity levels.
McClearn and Rodgers (1959)	Alcohol preferences.
McGill (1962)	Male mating behavior.
Weir and DeFries (1964)	Effects of trauma administered to pregnant females on exploratory behavior of offspring.
Winston (1964)	Effects of infantile trauma on water escape performance.

Most practitioners of behavior-genetics now recognize that if two or more strains are measured on almost any behavioral task they will in all likelihood differ from one another (e.g., DeFries, 1967). Hirsch (1963) has been delightfully blunt about this:

When different strains within a species are compared, it actually becomes a challenge not to find differences in one or more behaviors. When strain comparisons are followed by appropriate genetic crosses, genetic correlates of behavioral differences are demonstrated.

---

[10]After discontinuance of selective breeding in 1940 and prolonged maintenance of the animals after relaxed selection pressure, descendants of the original maze-bright and maze-dull lines were renamed S_1 and S_2, respectively, after 1950.

. . . with the wisdom of hindsight, we can ask why so many demonstrations were necessary. Should it not have been common knowledge that within each population the variation pattern for most traits will be conditioned by the nature of the gene pool, and that this will differ among populations? The answer lies in one phrase: the heredity-environment controversy.

Inbred strain studies do have a variety of other possible goals besides demonstrating the obvious. One such purpose is to compute a statistic called "heritability" which is related to σ_G^2 and purports to say "how much" phenotypic variance is due to genetic variance.

Bear in mind that "heritability," as geneticists use the term, is a statistic—nothing more. Even within that limited statistical context it has a double meaning. Heritability "in the broader sense" (h_B^2) or "the degree of genetic determination" can be estimated by finding σ_G^2 (this requires some genetical, logical and statistical manipulations that are covered by Falconer, 1960, and Roberts, 1967) and dividing it by the total variance; thus $h_B^2 = \sigma_G^2 / \sigma_P^2$, which is the ratio of genetic variance to total phenotypic variance. It turns out the h_B^2 is not a very useful statistic for most purposes (e.g., Parsons, 1967) because σ_G^2 includes a number of genetic goings-on that are not strictly connected with the transmission of hereditary material from one generation to another. By way of further experimental and mathematical gyrations, σ_G^2 can be partitioned into smaller components, one of which is additive genetic variance (σ_A^2). This term is supposed to provide an estimate of variance attributable to the actual genes that are passed from one generation to the next (i.e., additive genetic effects), and σ_A^2 / σ_P^2 gives us the more useful estimate of heritability "in the narrower sense" (h_N^2) or, more commonly, h^2.

Generally, h^2 will vary from about 0.00 (where none of σ_P^2 is attributable to additive genetic determinants) to 1.00 (where all of the observed variation is due to additive genetic effects and none to other sources). For reasons brought out by Roberts (1967) among others, under most circumstances it is much easier to estimate h^2 by using inbred strains, although the statistic has been calculated for selected strains as well. At any rate it is of some comfort to know that it is within the realm of possibility to estimate *how much*—not *whether*—variation in our observations of behavior is directly attributable to inheritance.

Returning for the moment to the first section of this chapter, the probable biological meaning of the words "innate" and "inborn" is that h^2 is high, which is to say that for a particular set of behavioral observations, much of the observed variation is attributable to additive genetic effects. "Learned" and "acquired" are terms suggesting that h^2 is low.[11]

---

[11] In order to savor the complexity of this notion, you might begin by thinking of a selective breeding experiment involving, say, "maze learning ability" where h^2 for "learning" should be low and h^2 for "ability" might be high.

But is that what we want to know about behavior? Is h^2 an explanation of a phenomenon or is it just a description? And for that matter, what have the selective-breeding and strain-difference studies to offer in the way of increasing our understanding of behavior and its variation? Before pursuing these questions further, a few words about *ex post facto* investigations are in order.

Ex Post Facto Investigations

As was mentioned earlier, *ex post facto* investigations are studies which take place in situations where the investigator exercises limited or no control over the variables which he is undertaking to investigate. In the strictest sense, he looks at Nature *as is;* he cannot or does not manipulate breeding regimens; he cannot or does not manipulate environmental contingencies. To the extent that such variables are controlled (rather than sampled), the study becomes an experiment.

Of course an *ex post facto* investigation of the kind just described is not apt to occur in the real world. The mere intrusion of an observer into some ongoing eco-behavioral process is likely to bring about some discrepancy between observed events and what would have happened had he not appeared upon the scene. The marking of animals that are under scrutiny and the introduction of extraneous noises, objects, and scents all conspire to impose a modicum of fictitiousness upon the situation. Some *ex post facto* investigations require that subjects be brought into the laboratory for observation. Even though no attempt may be made to structure the variables that affect the animals' behavior, some amount of it is bound to take place anyway.

Ex post facto investigations are the very basis of what meager knowledge exists about human behavior-genetics, but the approach is an historical outgrowth of naturalistic animal studies. The usual means by which such investigations are carried out is no more specifiable than those of painting a picture. That is to say, there are some basic techniques to be learned and used, but the actual practice of this kind of science is primarily an art form requiring an intelligent, sensitive master who is capable of converting observed external events into a language that can be understood by those who did not observe the events. In doing this, he may record his observations verbally, or he may quantify them so as to make whatever he saw amenable to correlational and other mathematical adjuncts. In terms of behavior-genetic analysis, the events that the naturalist observes may have to do with differences between strains, species, genera, and the like within one or more naturally occurring environmental contexts. Sometimes this is combined with experimentation, which is another kind of art form, but an art form nonetheless.

Enough of this book is concerned with *ex post facto* data gathering and analysis that no specific examples of the approach need be given here. There is, however, one additional point that ought to be made before leaving the topic. *Ex post facto* investigations tend to be imprecise while experimental studies tend to be artificial; moreover, the former tend to be more subjective than the latter. For those reasons, data derived from experimentation are apt to be more tractable to rigorous analysis; *ex post facto* data are more prone to have something to do with the real world as human perceptual systems know it. On that perhaps redundant note, we turn now to some comments on what behavior-genetics has offered and has to offer in the way of helping us to understand behavior and its variation.

Behavior-Genetics and Scientific Efficacy

From the material that has been presented in this chapter and elsewhere in the book, it is clear that common sense, naturalistic observation, and laboratory experimentation all converge upon the inescapable (and by this time trivial) conclusion that genetical and environmental events have something to do with behavioral variation. Somewhat less obviously, it is nevertheless plain that the components which combine to bring about behavioral variation are themselves inherently variable. They are so variable, in fact, that their ubiquitous presence might at first appear to be an almost insurmountable and unfortunate nuisance that interferes with studying the real issues—but variation *is* the real issue. Variation is the subject matter of biological science, of which behavioral science is a part. Our domain of study is variation at any or all levels of biosocial organization, from the arrangements of atoms to how whole organisms treat one another, singly and in groups.

What is to be learned about behavioral variation from the source of data that we have called "behavior-genetics," at least as it has been interpreted and represented here? In this writer's opinion, behavior-genetics can be credited with two important accomplishments:[12] 1) by way of overwhelming evidence and rhetoric that is occasionally revivalistic in character, it has been instrumental in forcing American psychology to recognize the relevance of biological individuality as it relates to behavioral variation; 2) it has demonstrated that behavior is not so ephemeral and disjunctive as to be inaccessible to biological analysis. Other approaches to the study of behavior can also be credited with these two achievements.

Behavior-genetics, by itself, does not appear to have gone very far in furnishing general, explanatory principles with respect to how behavior comes to vary as it does. Rather, its role has been largely that of description and definition. To be more specific, selective-breeding and strain-

---

[12] If you will read the theoretical articles by Hirsch that have been cited throughout the chapter, you should recognize that his influence upon this discussion has been substantial. This does not imply that he would agree with everything that is said here.

difference studies can almost always provide concrete testimony in behalf of the importance of genetic endowment as a factor in the observed variation, but in and of themselves, they tell us little about the intricate webs (not chains) of events that intervene between the fertilization of an egg and, say, fear of a strange object.

Even in those cases where sophisticated experimental and mathematical procedures have been used to estimate heritability, a more than superficial understanding of what the data mean for the bio-behavioral economy of the organisms involved is not immediately evident. For one thing, it may be that those behavioral traits that are most important to the animals in terms of coping with the world about them also happen to be those with the lowest heritabilities (i.e., the proportion of phenotypic variance attributable to additive genetic variance is small). As Falconer (1960) has pointed out, "On the whole, the characters with the lowest heritabilities are those most closely connected with reproductive fitness, while the characters with the highest heritabilities are those that might be judged on biological grounds to be the least important as determinants of natural fitness."

Be that as it may, there is no *a priori* way of interpreting a given value of h^2 with respect to the latitude of possible phenotypic expression under a wide variety of environmental conditions. Suppose, for instance, that h^2 (and, hence, σ_A^2) was high in Tryon's first generations of maze-bright rats. Remember that when Searle (1949) tested the animals in other learning situations, performance of the "brights" was inferior to that of the "dulls." Genes are known to have what has been termed a "norm of reaction," which refers to the amount of phenotypic plasticity that can be expressed in connection with the genotype. For example, the "bar-eye" condition (to which reference was made in discussing the work of Hirsch *et al.* in correlating chromosomes with *Drosophila* geotaxis; see also Figure 3-5) is a phenotype characterized by a great reduction of the number of facets (ommatidia) in the compound eye of fruit flies. The shape of the eye is restricted to a narrow, vertical bar of about 90 and 70 ommatidia in males and females, respectively, as opposed to some 740 and 780 facets for "wild-type" males and females. The condition ordinarily occurs when a particular, single gene is present. Thus, it would be safe to say that

Figure 3-5 Approximate eye shapes of wild type (left) and bar-eye (right) conditions in fruit flies.

the genetic contribution to the trait is quite high. Krafka (1920) raised bar-eyed flies under a range of temperatures: at 30°C the average ommatidia number for males and females combined was about 57; it was near 101 at 25°C; close to 141 at 20°C; under rearing conditions of 15°C, the average number was increased to approximately 242. Moreover, Lindsley and Grell (1968) cite studies indicating that the number of facets can be increased by introducing certain chemicals into the flies' food medium.

As a general rule, heritabilities are estimated under a very restricted range of environmental conditions, and unless a great deal is known about the phenotypic effects of various environmental influences, it is anybody's guess as to what h^2 would have been under other circumstances. Thus, when one reads of "racial differences" in intelligence, granting that the basic data on IQ scores, for example, are reliable, we cannot say what the results would have been had environmental contingencies, test instruments, and the like, been different. In the introductory paragraphs of many textbooks on behavior (particularly those in psychology), it is common to read a platitude to the effect that: "genes set the limits of the development of behavior, but environment determines the degree of development within those limits." As we have seen, the statement may be just as true in the reverse: "environment sets the limits of the development of behavior, but genes determine the degree of development within those limits." Certainly the "norm of reaction" notion is very deserving of much further study.

But what about the associations between genetic, physiological and behavioral correlates? Surely they have something to teach us about the nature of behavioral variation. Suppose two inbred or selected strains are known to differ physiologically and behaviorally as well as genotypically. Relationships of that kind do not necessarily offer much in the way of generally applicable principles either. You will remember that ostensibly similar phenotypes can be brought about in any number of ways, including environmental mimicry, in which case the phenotype is termed a "phenocopy." Now add to this the problems of generality that were brought out in the discussion of selective breeding and inbred stocks as research tools. Whether selected or inbred strains are studied, the observations made on those subjects may well be unique to them and the unique biological and environmental combinations that they represent.

If findings associated with phenotypic manifestations of particular genotypes are tenuous within species, it is even more difficult to imagine how direct behavior-genotype relationships can be generalized to other species. For example, in one of the most interesting and straightforward demonstrations of a single gene-behavior relationship, Kaplan and Trout (1969) found strains of mutant *Drosophila* that are prone to seizures when subjected to certain kinds of environmental stimulation. Moreover, the researchers were able to relate the phenotypic expression of the seizures to specific, single genes and to a particular neurological lesion of the thoracic ganglion. Although fascinating in their own right, it is not clear how these findings can shed light upon the nature of seizures *qua* seizures in other species—mice, for instance. Although some mice have seizures they most certainly lack anything like a thoracic ganglion, nor do they share gene pools with fruit flies.

The matters just brought up should not be construed to mean that genetically oriented studies of behavior have no more to offer the behavioral scientist than the two major accomplishments listed at the beginning of this section. On the contrary, such investigations can generate precise, testable hypotheses about the nature of a wide variety of behavioral phenomena. Suppose that a selected or inbred strain of rats is shown to be both highly intelligent and characterized by elevated ribose nucleic acid (RNA) content of the brain cells when compared with another strain that tends to be stupid and deficient in brain cell RNA. Knowledge of the association does not permit us to assert that intelligence is correlated with RNA in brain cells of rats or any other organism. We can, however, put the statement in the form of a testable hypothesis—a hypothesis that might not have occurred to us had those particular strains of rats not been studied. Testing the hypothesis can then lead to further findings, more hypotheses, and, with any luck and intellectual opportunism, knowledge regarding individual differences in intelligence. Some of these hypotheses can be tested through the use of behavior-genetic methods.

In fact, current trends indicate that behavior-genetics is becoming more and more an important adjunct to other scientific strategies in the study of behavioral variation rather than being content to demonstrate still another strain difference or estimate another h^2. Interdisciplinary combinations of behavior-genetics and techniques growing out of such fields as biochemistry, physiology, psychology, and ethology are potentially formidable. It is within the realm of possibility, for instance, that in the future (given enough technology and creativity) a team of behavioral biologists will be able to perform gene transplants, taking tiny units of genetic information from the germ cells of some individuals and placing them in the germ cells of others. The joint effects of small, systematic genetic and environmental modifications may then be traced from zygote formation all the way through individual differences in the subjective feelings of sadness and joy. At the moment, though, we are still some distance from that point, which is probably for the better given that the stewardship of the biological knowledge we have right now has not necessarily been in the best interests of ourselves or those with whom we share the world.

4 Comparative Development of Social and Emotional Behavior

G. Mitchell

Animals do not miraculously appear on the earth full grown and immediately set about behaving in adult ways. Instead they start life as a tiny fleck of protoplasm and gradually develop into specific kinds of creatures frequently composed of billions and billions of flecks of protoplasm. This process takes time, and while it is happening the behavior of the developing animal is changing. Developmental psychologists are interested in these behavioral changes and what causes them. *Comparative* developmental psychologists believe that by studying what happens as a variety of animal species mature, they can hope to make some generalizations about the process of behavioral development.

To illustrate the approach of comparative developmental psychology, I have chosen to explore a single topic in this field—the development of social behavior and social emotions. Comparative studies take advantage of the fact that many animal species live in groups and appear to possess quite similar emotions related to this way of life: for example, love, fear, and anger. Perhaps by studying the way in which social behavior develops in other species, as well as our own, we can better understand how and why humans form emotional attachments to their mothers, come to love or hate the people around them, and gradually acquire the spectrum of abilities which make up being a social creature. This chapter is intended to give the student a selected view of contemporary ideas and research on this topic.

Research supported by NIH grants MH-17425, MH-19760, and MH-22253 to G. Mitchell; HD-04335 to L. Chapman and RR-00169 to the California Primate Research Center.

The Organization of Development

The development of an animal is not random. As the human embryo grows it follows a pattern characteristic of the human species, with arm buds appearing at one time, what will become the legs at another, the rudiments of a nervous system at another, and so on. That there is such an organizational scheme is apparent when one considers the obvious fact that all human beings are recognizable as human, different from all chimpanzees or all wolves. An equally clearcut piece of evidence for the overall organization of growth is the occurrence in many different animals of *critical periods* for the development of particular structures and behaviors, most particularly social behavior patterns.

Critical Periods and Imprinting

During behavioral as well as morphological growth there are often special times during which specific behaviors (such as following one's mother when she moves) and in some cases quite general ones (such as a general tendency to explore one's surroundings) may be most responsive to both organizing and disruptive influences. Any period in life when a major new relationship is being formed is a critical one for determining the nature of that relationship, and any period when very rapid organization is taking place is a critical period (Scott, 1968). Scott notes that critical periods are characterized by great sensitivity to special kinds of changes in conditions. For example, a small amount of contact between an infant dog, other pups (not necessarily its brothers and sisters), and a lactating female (not necessarily its biological mother), will determine which individuals will be the close companions of the animal during its infancy and perhaps for as long as it lives. A similar amount of contact between two strange adult dogs will often have no lasting effect at all.

Just how critical these important times are has been a topic of some discussion. Since they are usually not sharply defined in time (and are thus optimal rather than critical in nature) the term "sensitive period" is preferred by some researchers (Hinde, 1962; Sluckin, 1965). A sensitive period does not necessarily "switch on" and "switch off" at a certain point in the development of an animal; environmental variables will often affect the timing and duration of this interval. Nevertheless the important point is that development may be organized in such a way that an animal will be most sensitive to certain kinds of events during special restricted periods in its lifetime. The classic example of the role of a critical period in behavioral development is provided by *imprinting*.

Imprinting

In 1873, D. A. Spalding wrote an article entitled "Instinct, with Original Observations on Young Animals" in which he said: "Chickens as soon as they are able to walk will follow any moving object. And, when guided by sight alone, they seem to have no more disposition to follow a hen than to follow a duck, or a human being" (p. 282).

Spalding also noted that development of the "following" response occurred only at a certain period very early in the chick's life (beginning a few hours after emerging from the egg). During this sensitive period the chick will follow almost any moderately large object which it sees moving nearby. Naturally this is normally the mother hen. Having followed something, the baby chick will form a lasting attachment to it even if the moving object happens to be a beach ball or a human being rather than its mother. When the chick forms a lasting attachment to something it has followed it has become *imprinted* on that object.

While Spalding was the first to describe the phenomenon, the famous German ethologist Konrad Lorenz was the first to subject it to a systematic analysis (Lorenz, 1937). Lorenz pointed out that imprinting (which he called "Prägung" or "stamping" just as Spalding had used the phrase "the stamp of experience") was fairly widespread among birds and appeared to be a very special sort of learning that produced a number of striking effects on the behavioral development of the imprinted bird. The key features of imprinting appeared to be 1) that it had to occur during a very brief sensitive period in the bird's life as a result of following a moving object. Failure to accompany the moving object during this period of time meant that social attachments would not be acquired at

Figure 4-1 Konrad Lorenz, followed by goslings.

all. 2) However, once imprinting had taken place it had effects on the bird's behavior as an adult. 3) The stability of these effects were most dramatically expressed in the sexual behavior of imprinted birds. For example, male mallard ducklings which had followed a male instead of a female mallard shortly after hatching, would as adults court males but not females. Similarly, as the result of being the first moving object in the lives of a substantial number of greylag geese (Figure 4-1), Lorenz himself had been courted by a good many male greylags; these geese readily chose him (or any other available human being) in preference to female greylags. Thus during imprinting young birds normally learn to recognize their species, and male birds learn to recognize females— information which they will not employ in sexual behavior for many, many months but which is absolutely critical for normal reproductive behavior.

As with many important concepts, things have turned out to be somewhat more complex and less clearcut than perhaps Lorenz originally suspected. For example, the uniqueness of imprinting as a form of learning has been questioned (Sluckin, 1965); the sensitive period may be more open-ended than Lorenz thought it was (Hinde, 1962); and under certain circumstances it is possible to show that imprinting is not necessarily rigid and irreversible. Salzen (1968) reared chicks with either a ball or a sponge, and at three days a clear preference could be demonstrated for the familiar object. However, after exchanging the objects so that the chicks were exposed to the opposite stimulus, a subsequent test showed that the birds had switched their preferences. This shows that imprinting does not occur instantaneously nor is it entirely irreversible (see also Sluckin, 1965). Moreover, it appears that animals sexually imprinted on a biologically unsuitable object will sometimes alter their sexual behavior in maturity and eventually court a proper mate.

Finally, although the term "imprinting" was originally applied to attachments based on the following responses of goslings, ducklings, and chicks, imprinting-like behavior may be much more widespread. For example, Sluckin (1965) and others have pointed out that very similar phenomena occur in mammals. Human infants deprived of an opportunity to have close contact with a loving adult during the first six months of life (as sometimes happens in orphanages and elsewhere) may, if they survive, have great difficulty forming close social attachments later in life (Spitz and Kobliner, 1965; Stone and Church, 1968, p. 209). Because of similar kinds of reports for various mammalian species, it seems reasonable to use the term *imprinting* more freely. Although approach and following are important components of imprinting by young precocial (mobile-at-birth) birds, they are not crucial for imprinting-like phenomena in other animals. Attachment to any configuration of sensory stimuli may occur as a result of simple *exposure* to these stimuli.

However it comes to be defined and modified, the experiments on imprinting stand as an important illustration of special effects of special experiences during a critical period in an animal's life.

From "Mother-infant relations in langurs" by Phyllis Jay, in H. L. Rheingold (Ed.), *Maternal Behavior in Mammals.*
New York: John Wiley & Sons, 1963. Photo courtesy of Dr. Phyllis J. Dolhinow.

Figure 4-2 Langur monkeys in play groups.

Sensory Stimuli, Optimal Arousal and Development

We have seen that the behavioral development of the chick, other precocial birds, and other animal species is organized to be especially receptive to certain events during restricted periods in the animals' lives.
The same principle of developmental organization is illustrated in a more
subtle way by reference to the relationships among sensory stimulation,
the theory of optimal arousal, and behavioral development.

Many animals prefer moderate levels of sensory stimulation; they
will try to avoid both sensory deprivation and massive stimulation. Using
infant monkeys as an example, Mason (1967) has shown that there is
a relationship between physiological measures of arousal level, such as
heartbeat rate and neural activity in the cortex of the brain, and the
monkey's tendency to engage in play behavior. Social play appears to
increase arousal. Monkeys will seek to play with one another when their
arousal levels are low but will avoid play if it becomes too exciting.
Clinging, huddling, or embracing another individual (particularly the
infant's mother) occur most frequently at high levels of arousal and serve
to reduce it. A change, either an increase or a decrease, in physiological
activity which moves toward an average optimal amount is rewarding
to the individual.

Other external stimuli besides those offered by playful companions
provide sensory input affecting the arousal level of an animal. Dember
and Earl (1957) and Dember (1965) have developed a general theory

of how animals respond to different sorts of external stimuli in the course of their development. Although their results have been presented in the form of a "complexity-dissonance preference theory" we can translate their main ideas into English without great difficulty. Essentially they argue that objects and events differ in their novelty and complexity. Individual animals tend to ignore objects which are highly familiar and very simple; however, stimuli which are somewhat familiar but slightly different arouse interest and curiosity; objects and events which are extremely strange and highly complex are avoided and will elicit fear, anger, or rage in an animal forced to maintain contact with them.

What is interesting from a developmental standpoint is that what is "novel and complex" clearly changes as the animal matures. Very young animals prefer to interact with simple objects which older animals ignore altogether. As the young creature grows older it manipulates and plays with objects which are somewhat more complex and slightly unfamiliar. The developing animal interacts with increasingly complex stimuli in increasingly complex ways. For example, a very young baboon at first initiates very brief, very simple contacts with other juveniles (it may just touch a companion and then run back to cling to its mother). As the infant matures these contacts grow longer and more complex; baby baboons engage in short grappling bouts with each other. Later still the juveniles seek each other out and form a play group in which wrestling, running, jumping, chasing, climbing, and tumbling last practically all day long.

As we shall see later, the process of behavioral development can be severely and permanently disrupted by artificially depriving the animal of the possibility of engaging in simple interactions early in life. Such an animal has nothing to build upon and therefore cannot go on to explore more and more complex stimuli. On the other hand, development can be somewhat accelerated if early in life the young animal has access to more complex objects than it would otherwise not encounter until later (Dember, 1965; Sackett, 1965a; 1965b).

In a normal environment, the developing animal comes to derive progressively more rewards from arousal induction and fewer from arousal reduction (Mason, 1967). Once a human baby has become familiar with a certain pattern of sensory stimulation, large changes in stimulation first provoke struggling and distress behavior; later the baby simply orients away from strange toward familiar stimuli. As the baby begins to perceive familiar objects it is reassured and experiences pleasure. It is possible that the sensation of pleasure experienced as the baby returns to a familiar pattern of stimulation actually comes to motivate the child to *leave* familiar objects and seek out mildly novel ones (Salzen, 1968). In any event, it is well known to all human parents that human infants become increasingly curious and investigatory as they grow older. This is also true of the young of many other animal species. Instead of using the familiar arousal-reducing mother as a source of reward, the maturing individual utilizes exploration and increasing independence to relieve "boredom" and achieve optimal stimulation. As the animal becomes an adult the sources of motivation appear to change; adult animals usually engage in only a fraction as much play and exploration as juveniles.

Mason's arousal theory of behavior development is based on the premise that young primates are "designed" to be motivated in certain ways as a young animal and in different ways as an adult. As it grows, the young monkey tolerates more and more stimulation from exploration and social play and is motivated to seek levels of arousal which would have been aversive earlier in development. In adulthood, the activities so rewarding to a juvenile no longer have similar effects, probably because of intrinsic changes in motivational mechanisms (Mason, 1967).

In summary to this brief section on the organization of social development, I wish merely to note that behavioral development appears to follow a characteristic pattern for each animal species. Animals at certain stages of maturation are "ready" for rather specific kinds of information. The chick shortly after hatching is sensitive to fairly large moving objects and will respond to them in a distinctive way with special long-term behavioral results. Juvenile primates are sensitive to the stimuli associated with their playmates and are strongly motivated to seek out high levels of social stimulation via play as they grow older. Later, as adult rhesus monkeys for example, they will respond to their age mates in totally different ways; they are only weakly motivated to engage in play behavior. It is worth emphasizing these "rules" for development here, because most of the rest of this chapter is dedicated to discussing the environmental factors which influence the development of social behavior and social emotions.[1]

Sensory Stimuli and Behavioral Development

The Importance of Touching

How does the young animal respond to the things to which it is first exposed, and what sensory modalities mediate these first exposures? In the development of animal species, the sense of touch is an extremely conservative sensory modality. Responsiveness to touch exists in all forms of animal life; and, in the ontogeny of the human being, the sense of touch is the first modality to mature *in utero;* thus the first early exposures are exposures to tactile stimuli. Many immature mammals depend upon touch almost exclusively in the early days of life. As we shall see, normal development of social and emotional behavior in higher mammals is initially correlated with appropriate exposure to skin and muscle senses early in life (e.g., touch, warmth, movement, etc.). Basic individual autonomy and independence cannot develop appropriately without ade-

---

[1]See Chapter 9 for a discussion of species-specific learning capacities [Ed.].

From *Primate Behavior: Field Studies of Monkeys and Apes,* edited by Irven DeVore. Copyright © 1965 by Holt, Rinehart and Winston, Inc. Reproduced by permission of Holt, Rinehart and Winston, Inc.

Figure 4-3 Mother baboon grooming her infant.

quate early contact exposure, an example of which is shown in Figure 4-3.

Touch, however basic, is not a single simple sensory modality, but rather utilizes a complex assortment of receptors (see Chapter 5 for a discussion of the physiology of touch receptors). One can, for instance, distinguish between active and passive touch, both objectively and subjectively. When the animal itself does the touching, there is often a different response from the response which may occur when something touches the animal. For example, some coelenterates (jellyfish, etc.) will not release their nematocysts (stinging, dart-like weapons) when they do the touching, yet will release them reflexively when they are touched by something else. A difference between active and passive touch is also apparent in humans. For example, as we move about, the hairs on the surface of our arms may be repeatedly brushed against our clothing without our noticing the contact. The touch receptors in our skin apparently adapt specifically in response to this kind of repeated stimulation. However, let one small fly alight on a tiny group of exposed hairs, thus providing a qualitatively different kind of stimulus, and we become very aware of *being* touched (Milne and Milne, 1962).

Tactile contact during rearing is important to the development of most animals, even to arthropods. There were, it was once believed, two *different species* of large African grasshoppers, a migratory species found in large social groups and a solitary species found living alone. Later, however, it was found that the solitary species was the same as the migratory species. The solitary variety could be changed into the migratory type by rearing in a group. Conversely, the migratory became solitary when reared in isolation (Faure, 1932). Still later it was discovered that tactile contact during rearing was the crucial factor in the development of the migratory behavior. The social aggregation depended on being touched. Ellis (1953) stroked grasshoppers reared in isolation with a fine wire and the isolates aggregated. Before stroking, isolate-reared animals were less willing to interact with a line-tethered decoy than were group-reared animals (Ellis, 1963a). However, the effects of isolation could be reversed by as little as four hours of "social experience" or stroking (Ellis, 1959; 1963b) (see Figure 4-4).

In most vertebrates, stroking leads to behavioral and physiological relaxation such as decreased heart rate and lowered activity (Gantt *et al.*, 1966). Evidently the stroking is similar to tactile stimuli received during early rearing and is pleasurable to the animal. Even young fish receive tactile contact during early behavioral development; in some

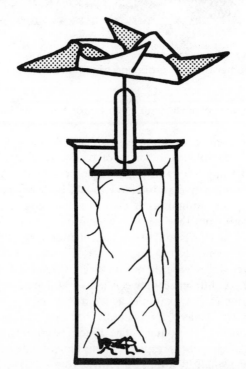

From the book *Life on a Little-Known Planet* by Howard E. Evans. Illus. by Arnold Clapman. Copyright © 1968 by Howard E. Evans. Published by E. P. Dutton & Co., Inc. and used with their permission.

Figure 4-4 Peggy Ellis' locust-tickling machine.

species (e.g., orange chromides) this contact involves primarily tactile stimulation on the ventral surface of an adult, just as it does in mammals and birds. In orange chromides such a contact is called "glancing" (Ward and Barlow, 1967).

The work of Harlow and Zimmerman (1959) on behavioral development of the rhesus monkey demonstrates the importance of early touch in primates. New-born infant rhesus monkeys prefer soft, cloth-covered artificial mothers to wire-covered surrogates even when they receive milk only from the wire dummy. (Puppies also prefer cloth to wire mothers, Scott, 1962.) If a cloth diaper rather than a surrogate mother is provided, the infant clings to the diaper; and if nothing is provided, the infant clings to himself. All of this is also true of chimpanzees, humans, and other primates. This tendency to orient toward tactile comfort is so strong that monkeys will continue to approach the mother for touch even though the mother repeatedly punishes them. Artificial "rejecting mothers" for infant rhesus monkeys were set up by Harlow. Some of these rejecting surrogates repeatedly push the infant away by shaking them off or by pushing them away with air blasts. But, as Harlow puts it, "the blasted babies" kept returning for contact (Harlow, 1958b). The same thing occurs with those real mothers who are brutal. The infant monkey keeps "returning for contact" despite repeated rebuffs (Harlow, Harlow, Dodsworth, and Arling, 1966). Being able to touch and be touched is apparently rewarding and important to many young primates (Figure 4-5).

Reprinted by permission, *American Scientist,* journal of the Society of The Sigma Xi.

Figure 4-5   Young rhesus monkey with cloth surrogate mother.

Movement

Another aspect of passively received physical contact which is of tremendous importance to early behavioral development is the movement of muscles, tendons, and balance receptors (i.e., kinesthesis or proprioception, and equilibration). The young isolate-reared animal will provide its own movement (just as it provides its own touch when touch is not supplied by a mother). Parrots and canaries, for example, develop head bobbing and rocking (Sargent and Keiper, 1967). So do young dogs, monkeys, apes, and humans. Isolate-reared dogs develop unusual postures and stereotyped pacing (Melzack and Scott, 1957). Socially deprived monkeys, apes, and humans also show stereotyped behaviors such as rocking, pacing, and bizarre postures and movements (Berkson, 1967). Harlow and Suomi (1970) have demonstrated that movement stimulation is important in the early attachments of rhesus monkeys, and Mason (1968b) has pointed out that:

. . . under natural conditions the mother ordinarily provides a great deal of passive movement stimulation to her infant in the course of her routine activities. . . . We might conjecture that the infant monkey or ape that is deprived of such passive movement stimulation may supply it for himself through self-rocking or similar repetitive activities. (p. 83)

To test this hypothesis, Mason compared two groups of rhesus monkeys; both groups had been separated from their mothers at birth. One group was reared on a moving cloth-covered dummy and the other on a stationary dummy. All of those reared on the stationary dummy developed persistent stereotyped rocking, while none of the monkeys reared on moving surrogates rocked. In addition, the monkeys reared on moving dummies were less fearful and more active than those reared with stationary surrogates. The robot-reared monkeys spent less time contacting their surrogate, they moved about the cage more often, and they were quicker to approach and interact with humans than were the monkeys reared on stationary dummies (Mason, 1968b).

Primates which have been reared alone are very withdrawn animals. When they do interact with their environment, which is rarely, their movements are more rigid than normal animals and they prefer only the simplest of stimuli. For example, Sackett (1965b) has shown that monkeys reared in isolation will choose to touch or manipulate an immovable bar rather than a moving bar or a loose hanging chain. Wild-reared rhesus prefer the object which will produce the most complex sort of movement and stimulation: the loose hanging chain. The simpler the rearing environment, in terms of movement and contact, the less complex the preferred level of input from moving and tactile stimuli when the monkey matures (Sackett, 1965b).

Temperature Changes

Early stimulation via temperature changes produces effects very similar to those produced by passive movement and passive touch. For example, baby rats which are picked up and handled and babies which are simply subjected to temperature changes both mature more rapidly than non-stimulated animals (Schaefer, 1968). Harlow (1958b) and more recently Harlow and Suomi (1970) found that warmth enhanced the degree of attachment of infant monkeys to cloth surrogates.

Handling

Early handling and being moved about by one's mother or a substitute handler of another species usually involve the combined stimulation of tactile, balance, and temperature receptors. The effects of these stimuli in very early life are usually beneficial to structural and behavioral growth. For example, baby rats picked up and handled by humans show earlier sexual maturity, increased activity, and increased viability under stress. They are also heavier, more dominant in social interactions, more social, and have a heavier cerebral cortex than nonhandled rats. Handled mice apparently respond similarly, and are more active and social (Denenberg, 1962; Levine, 1962). Likewise, handled kittens show increased maturation. They open their eyes earlier, emerge from their nest box sooner, show characteristic changes in brain activity sooner, and develop adult coat coloration earlier than nonhandled kittens (Meier, 1961).

Rheingold and Bayley (1959) provided extra tactile, kinesthetic, and warmth stimulation by handling institutionalized human infants; the handled infants' developmental quotients were much higher than those of nonhandled infants. Landauer and Whiting (1964) have suggested that extra stimulation through physical contact may also increase the rate of growth in humans. They compared the heights of adults in societies in which extra skin stimulation was given (such as piercing the ears or noses of infants early in life) to heights in societies in which such practices were absent. Those societies providing the extra sensory stimulation had significantly taller adults.

There is no doubt that an absence or near-absence of tactile and kinesthetic stimulation in the first year of a human infant's life leads to physical deterioration. Growth slows down, the skin becomes wrinkled and brittle, and muscles atrophy in extremely deprived institutionalized human infants. This syndrome, resulting from a lack of physical contact and/or movement, is known as marasmus (Ribble, 1944).

These and other data involving the absence of physical contact early in life have led several investigators to theorize that the disorders of emotional and behavioral development associated with maternal-social deprivation are primarily attributable to somato-sensory deprivation (cf. Prescott, 1967). These theorists maintain that it is the *kind* and not just the *amount* of sensory stimulation that is crucial for normal emotional development, and that the important *kind* of stimulation is somato-sensory: touch, kinesthesis, warmth and pain.

One more comment should be made regarding early physical contact in humans. There are some data which suggest that female infants receive and prefer more passive stimulation (Bell and Costello, 1964). While it may be argued that this is a function of the expectations and behavior of parents, it is certainly clear that there are differences between individual infants in their response to contact. Schaffer and Emerson (1964) observed that some human infants actively resist and protest when certain types of physical contact are administered to them. Some babies, the "cuddlers," prefer being hugged and held close, while others, the "noncuddlers," prefer being romped with, bounced, and swung. Prescott (1967) suggests that cuddlers and noncuddlers be referred to as somesthetics (touchers) and kinesthetics (movers). It may be that female infants on the average are primarily somesthetics and male infants are kinesthetics, although one would of course expect great overlap between the two and both sexes usually "enjoy" both sources of physical contact in early life.

Some recent reports on rhesus monkeys suggest that such a sex difference may be present in nonhuman primates (Mitchell, 1968a). Male monkeys receive more punishment from and play more frequently with their mothers than do female infant monkeys. Female infant monkeys, on the other hand, cuddle with their mothers more frequently and for a greater amount of time than do males.

Although early contact stimulation is probably of primary importance for the normal development of social-emotional behavior, the data discussed in the following section suggest that other forms of stimulation are also necessary for proper development.

The Importance of Hearing and Seeing

Recently Walters and Parke (1965) have discussed the role of distance receptors in human infancy and early childhood. Although they admit that physical contact is of great importance, they believe that the formation of psychological attachments is primarily fostered by distance-receptor experience. Their argument rests primarily on the idea that, in the human, the majority of social attachments in adult life are developed and maintained in the context of visual and auditory interactions and are accompanied by relatively little physical contact stimulation. As adults living in western cultures we simply do not go around touching and moving most of our acquaintances. However, in other cultures much more intense and prolonged physical contact is characteristic of adult social interactions. And even in our own society, touching remains important in adult life, serving critical functions of greeting others, providing comfort, and as a signal of an intimate relationship. Recently, touching as a form of communication that fosters feelings of comfort and intimacy has become increasingly popular in American culture, as evidenced in everything from "love-ins" to rather more earnest self-improvement sensitivity sessions and other currently popular forms of "group therapy" (Montagu, 1969).

Be that as it may, it is nevertheless true that at least the human and probably also other animals apparently must move gradually from a process which is based primarily upon contact to one which can be developed and maintained through distance reception alone. Sometime early in the life of the human being, distance reception (primarily audition and vision) apparently must play a role in social-emotional attachment.

Auditory Stimuli—The Heartbeat

Salk (1962) has argued that the human baby imprints to its mother's heartbeat *in utero*. He found that newborn babies cried less and gained weight faster than control babies if they were exposed for four days to heart-beat-like noise. Furthermore, he found that older children were more easily lulled to sleep by seventy-two paired beats per minute than by any other sound. Salk even believes that music and dance are natural outcomes of heartbeat imprinting that are created so that humans might "remain in proximity with imprinted stimuli." These are fascinating but speculative ideas which have been questioned by other writers (e.g., Sluckin, 1965).

Auditory Stimuli—Vocalizations

Just as the preferred complexity levels of kinesthesis and touch are decreased by early isolation, so are the preferred complexity levels of auditory stimuli. The frequencies and varied utilizations of "coo" vocalizations in monkeys and "yelping" in dogs are decreased in animals which have been raised in social isolation (Mitchell *et al.*, 1966; Fox and Stelzner, 1966), and institutionalized human beings are often reported to show delayed or retarded speech development (Goldfarb, 1945).

When an immature animal is separated from an attached object (e.g., its mother), it will display distress calls. In the rhesus monkey it is a "coo" (Mitchell, 1968b), in the dog, a "yelp," and in the baby chick, a "peep" (Bermant, 1963). When the mother is returned, these distress calls diminish or disappear and may be replaced by contentment calls (e.g., a "twitter" in the chick, or a "girning" sound in the rhesus monkey). The mother responds to distress calls by approaching the infant animal and, especially in the case of mammals, by contacting the infant. In the case of the turkey hen and her chicks, vocalizations by the chick are extremely important to the development and maintenance of attachment for the chick by the mother. If a turkey hen is deafened, she will peck her chicks to death (Schleidt, Schleidt, and Magg, 1960).

Sounds emitted by the mother are also of some significance to early social and emotional attachments. In chickens, the sound of a mother hen's "clucking" increases the likelihood of imprinting (Sluckin, 1965). There are sensitive periods for auditory as well as for visual imprinting. The finding that a bird's song, which is to some extent characteristic of its species, can be sometimes learned early in life in an imprinting-type fashion (Sluckin, 1965) supports the general idea that there may be sensitive periods for auditory imprinting.

Visual Stimuli—Sign Stimuli

It has been found that certain rather simple visual cues (sign stimuli) may serve to trigger specific kinds of behavior patterns (fixed action patterns) in entire populations of animals. For example, the elaborate courtship zig-zag dance of adult male stickleback fish is quite mechanically released by the sight of the swollen abdomen and inclined posture of an egg-laden female. A wooden model, simply a stick with a swollen "belly," held at the proper angle will also prompt a mature male to perform its zig-zag dance! The male fish is *not* responding to all the features of the female but only to a few key characters (sign stimuli). An example of the importance of such sign stimuli in early development is provided by the thrush. Baby thrushes gape when they are fed by their parents. This gaping is released by sign stimuli provided by certain shapes as they move horizontally above the babies' eyes. These blackbirds have evolved a capacity to respond to certain visual cues in a specific way at a particular point in development. Another example of this sort of thing involves the response of the infant rhesus monkey to a threat gesture. If rhesus monkeys are raised from birth in an isolation chamber where they are provided with pictures or slides of other monkeys but nothing else, the responses of the isolates vary with age and according to the slides shown to them. For example, between two and one half and four months of age, threat pictures yield a high frequency of disturbance behaviors which include a smile-like gesture known to primatologists as a fear grimace. Lever pressing to see threat pictures declines dramatically during this age interval. In other words, only when rhesus babies reach a particular age will the sign stimuli present in the face of a threatening adult release stereotyped species-characteristic fear responses (Sackett, 1966) (see Figure 4-6).

The fear grimaces of the rhesus monkey and of apes or chimpanzees look very much like a human smile (Figure 4-7). In fact the human smile may be fear motivated. At six or eight weeks of age most human babies spontaneously respond with a smile to a moving human face seen in front view, apparently as a fixed action pattern (a species attachment). Also at six or eight weeks of age the baby sheds tears when he cries and his crying begins to occur in response to social phenomena. By the age of three or four months the baby appears to enjoy what Stone and Church (1968) call playful fright. A parent says "peek-a-boo" and the child smiles. By four months of age many babies laugh in response to such stimulation. There is at first a mild fear response or startled response to "peek-a-boo" (accompanied by heightened arousal). This is almost immediately followed by smiling or pleasure (accompanied by arousal reduction to an optimum). The child is first startled by the strange, but then he recognizes that the strange is actually familiar. Many fear-inducing strange things become fear-reducing loved things through repeated exposure and repeated changes in arousal.

From *Primate Behavior: Field Studies of Monkeys and Apes,* edited by Irven DeVore, Copyright © 1965 by Holt, Rinehart and Winston, Inc. Reproduced by permission of Holt, Rinehart and Winston, Inc.

Figure 4-6 Threat behavior of an adult rhesus monkey.

From *Primate Behavior: Field Studies of Monkeys and Apes,* edited by Irven DeVore. Copyright © 1965 by Holt, Rinehart and Winston, Inc. Reproduced by permission of Holt, Rinehart and Winston, Inc.

Figure 4-7 An adult rhesus grins.

Individual Attachment

Baby rhesus monkeys (and many other animals, including humans) learn to recognize the face of their mother and become attached to it. If a baby rhesus is raised on a surrogate cloth mother it becomes devoted to whatever the surrogate's face happens to be.

Harlow and Suomi (1970) recently described the behavior of a female rhesus infant who was reared on a surrogate having a plain round wooden ball as a face. When the infant was 90 days of age, the plain face was removed and replaced by a face having eyes, a nose, and a mouth. Harlow and Suomi (1970) described the infant's response to the new face as follows:

The baby took one look and screamed. She fled to the back of the cage and cringed in autistic-type posturing. After some days she revolved the face 180° so that she always faced a bare round ball! We could rotate the maternal face dozens of times and the infant would turn it around 180°. Within a week the baby resolved her unfaceable problem once and for all. She lifted the maternal head from the body, rolled it into the corner and abandoned it. No one can blame the baby. She had lived with and loved a faceless mother, but she could not love a two-faced mother. (p. 164)

Security at a Distance

It is easy to demonstrate that a rhesus monkey becomes attached to a soft, cuddly object and will run to such an object (surrogate mother) when frightened (Harlow, 1958b). It is also very easy to demonstrate that such an infant monkey will play much more frequently and display much less fear-related behavior in the presence of its surrogate mother than it will in its absence. In other words, the rhesus infant seems to be more secure when he can see his surrogate mother than he is when he cannot. Security at a physical distance from the mother-figure, whether we are speaking of monkeys or humans, depends upon distance reception, in this case vision and probably audition. Mother monkeys and their infants certainly respond positively to each other's calls (Simons, Bobbitt, and Jensen, 1968). We have all heard a human baby babble and have gurgled back.

Security in the presence of the mother but not in physical contact with her does not spontaneously develop out of security based upon physical contact alone. An infant may have received an adequate amount and proper pacing with regard to touch, temperature, and kinesthesis, yet remain insecure in the presence of the mother when physical contact is not provided. Many female monkeys reared under different degrees of early social deprivation become indifferent mothers when they mature and are impregnated (Arling and Harlow, 1967). Such a mother may allow her infant to contact her but she will not retrieve or protect her infant and will even withdraw from it when the infant is threatened. The infant, by all objective vocal, postural, and gestural indices, is not disturbed when in contact with the mother, even when a threatening object appears.

However, when it is away from its mother, and thus dependent upon distance reception alone for its security, the infant screeches, rocks, fear-grimaces and crouches in the face of danger. It has not acquired security at a distance, because its mother has either continuously withdrawn from it or has protected it in an arbitrary or inconsistent manner (Arling and Harlow, 1967).

A similar pathology probably prevails in some human homes. The mother may nurse her infant child, tell the child of her love, and hold and cuddle it when there is nothing strange or potentially dangerous (e.g., strangers and visitors, whether children or adults) in the vicinity. But as soon as something or someone strange or novel appears in the home, the mother may ignore her own infant or even criticize or punish it rather than provide comfort. The child certainly does not develop security at a distance in such a situation, despite the fact that he may have been adequately nurtured as far as physical contact is concerned. Observations such as these in monkeys and man make theories which emphasize physical contact to the exclusion of distance reception untenable (cf. Prescott, 1967).

While on the subject of insecurity at a distance from the attached object, we might reflect upon the implications which such pathological predicaments might have for later behavior. A rhesus monkey reared by such arbitrary mothering has some difficulty in dominance relationships and develops some abnormal stereotyped behaviors (Arling and Harlow, 1967; Møller, Harlow, and Mitchell, 1968). We can only speculate on what might be the result in the human situation; but let us speculate anyway.

If one judges on the basis of the incidence of mental or emotional illness alone, there must surely be many American adults living in social situations which are too complex for them emotionally. It is difficult to believe that the high state of arousal in such people has developed solely from insufficient early exposure to physical contact. Most American mothers cuddle and rock their babies; American babies *are* soothed physically. Most American marriages involve physical intimacies, most American husbands and wives, at least gradually, come to provide contact security for one another. However, security at a physical distance may be missing both early in life and later in marriage. This may be reflected in a variety of social upsets, more or less serious in nature. For example, at a cocktail party a husband may be threatened and a wife may not come to his defense (in fact, all of us know that many a wife will aid the offense against her husband). The same thing applies to a wife who may feel threatened in a complex social situation. Reassurance, affection, noncontact comfort is often not supplied by either the husband or wife. It may be that this failure can in some cases be traced to early rearing conditions and failure to develop a feeling of security when separated physically from one's mother.

But even with security based upon distance reception, there is still visual or auditory contact between parent and infant, between husband

and wife. But the healthy, self-sufficient, autonomous primate infant or adult, whether man or monkey, must also be capable of functioning (without overwhelming aggressive or fear responses) in complete absence of the mother or of the spouse. Most primates are capable of behaving in a socially responsive and appropriate way to others of their kind. It is as if the sense of security first developed with one's mother can be generalized to embrace a wide variety of social situations.

Preferred Levels of Visual Stimulation

We have already seen that the less complex the rearing environment in terms of physical contact, the less complex the preferred levels of physical contact when the animal matures (Sackett, 1965b). The same thing is true for exposure via the distance receptors. Visual stimulus deprivation during rearing produces monkeys that prefer visual stimuli of low complexity. Sackett (1965a) compared monkeys which were reared under different levels of overall visual input and found that the most visually enriched group of monkeys preferred the most complex visual stimulus patterns. One-year social isolates, who were visually exposed only to the inside of their isolation chambers, exhibited a very low amount of interest in complex visual stimuli when tested at two and one half years of age. Since other animals provide complex visual stimulation as well as complex physical stimulation, it is not surprising that social isolates do not prefer to interact with other monkeys.

It is clear that under normal conditions early interactions with the mother have a great impact upon social emotional development. The importance of exposure to varying social and nonsocial stimuli in early life is best emphasized by removing many or most social and nonsocial exposures and observing the consequences. If an animal is raised in social isolation, it does not develop social-emotional security at a distance from a maternal object. In fact, it does not even develop a secure attachment at the pysical contact level. Animals raised in isolation are *not* affiliative but are extremely fearful, extremely aggressive, sexually inadequate, and are almost complete failures as parents. Some examples of comparative-developmental research on affiliation, fear, and hostility follow in the next sections.

The Development of Affiliation, Fear, and Hostility

Affiliation

Most animal groups are held together by social bonds, and we have seen how such attachments may develop. Early physical contacts with species-mates are usually of primary importance for bonding, although exposures through other sensory modalities are also important. Attachments to (or affection for) fellow members of one's own species (conspecifics) as adults can be called *affiliation*.

One way to determine the effects of exposure to conspecifics is to raise an animal in social isolation and determine, after emergence from isolation, the effects of *no* exposure to conspecifics. Such socially isolated animals, in general, show very little affiliation. As we have already noted, locusts raised in isolation do not approach a tethered decoy animal as frequently as do locusts reared together (Ellis, 1963b). Several species of fish reared in isolation do not school (Shaw, 1964; Breder and Halpern, 1946). Many birds raised in isolation do not seem to recognize their own kind and remain alone rather than join the flock (cf. Sluckin, 1965). Rats reared in social isolation also display low affiliation relative to socially reared rats. Such rats make few social approaches (Ashida, 1964). Isolated mice are also less affiliative than socially reared mice (Terman, 1963).

Scott isolated a ewe for ten days after birth; it subsequently failed to become part of the flock, even after several years (Scott, *et al.*, 1951). Isolate-reared kittens are very slow in orienting toward the mother (Rosenblatt, *et al.*, 1962). Dogs raised in isolation never reach the levels of positive social responses toward other dogs or toward humans that are reached by socially reared dogs (Fuller, 1961). Even puppies which have received nothing but punishment from human beings are more affiliative toward humans than are isolates (Fisher, 1955). The yelp, a call for social affiliation, occurs much less frequently in the isolated than in the socialized dog (Fox and Stelzner, 1966). Neonatal social experience up to three-and-one-half weeks, at the onset of Scott's critical period, facilitates the development of normal social affiliation in the dog (Fox and Stelzner, 1967).

Rhesus monkeys raised in social isolation show very little affiliation as adults. One excellent measure of what a layman might call monkey friendliness, or what we have been calling monkey affiliation, is grooming behavior (Figure 4-8). Isolate-reared monkeys groom very infrequently, if at all (Senko, 1966). Not only grooming but also huddling and clinging to other monkeys are almost completely absent (Mitchell, 1968b). The coo, the monkey's call for social affiliation, is markedly reduced in frequency in isolate-raised monkeys (Mitchell et al., 1966).

In tests of gregariousness, near-adult isolate-raised rhesus monkeys make fewer *social* choices than do pairs of adequately socialized monkeys (Mason, 1961). Monkeys reared from birth away from other monkeys but handled by humans during early life prefer humans to monkeys when tested at the age of two to three years. Rhesus monkeys reared in complete social isolation spend much less time with either humans or monkeys (Sackett, Porter and Holmes, 1965).

Chimpanzees raised in social isolation also avoid physical and other kinds of social contact with their species-mates. Grooming in the chimpanzee, for example, is infrequent following a history of social isolation. Isolated chimpanzees are not as sociable with humans as are normally reared chimpanzees, and this difference persists over several years (cf. Mason, 1967).

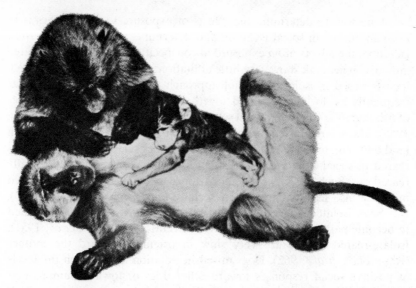

From *Primate Behavior: Field Studies of Monkeys and Apes,* edited by Irven DeVore. Copyright © 1965 by Holt, Rinehart and Winston, Inc. Reproduced by permission of Holt, Rinehart and Winston, Inc.

Figure 4-8 Adult baboons grooming.

The conditions necessary for the development of motivation for affiliative contact in the human include contact with an adult(s) who provides sensory stimulation to the child through talking to him, touching him, and cuddling him. In short, the adult encourages affiliation through protection and controlling the rate of stimulation of the child (cf. Report of National Institute of Child Health and Human Development, 1968). Thus, there is a massive amount of evidence documenting the importance of early social contacts for the development of the ability and desire to form social attachments.

Fear

In many vertebrate species, a period of fear appears to follow the initial attachment. The end of the so-called sensitive period for imprinting has always seemed to correlate with the beginnings of fear of the strange (cf. Sluckin, 1965). If an animal is reared in social isolation through this sensitive period for attachment the fear responses develop in an accentuated form.

Rats reared in social isolation are nervous or emotional. They often evince marked startle responses upon being touched. They are more easily frightened by novelty than are socially-reared rats. Adding extra stimulation in early life, in contrast to deprivation, makes rats much more viable under stress as adults (Hatch *et al.,* 1963). Isolate-reared mice, upon first emerging from isolation, show deficits in assertiveness (Rosen and Hart, 1963). Isolated cats show exaggerated disturbance in a novel environment and are fearful toward other cats (Schneirla, *et al.,* 1963).

According to Fuller and Clark (1966), who have studied isolate-reared dogs, the stress of emergence from isolation is as important as the isolation sojourn itself. A sudden bombardment by very complex stimuli upon emergence from isolation provides the complete opposite of what is required for adequate social-emotional development, i.e., gradual, continuous, paced increments in stimulation (see also Melzack, 1954). Fuller and Clark also found that the immediate post-isolation syndrome of freezing and crouching could be ameliorated by gradual adaptation to the post-isolation test environment during social isolation and by the administration of chlorpromazine (a tranquilizer) at emergence. Thus, fear could be controlled by controlling either the amount of stimulation or the physiological level of arousal.

Rhesus monkeys raised in enclosed isolation chambers also show extreme fear upon emergence (Mitchell and Clark, 1968). Animals which were allowed to adapt gradually to the post-isolation test environment did not show exaggerated fear upon leaving isolation. These animals did however behave very fearfully in later social situations which were obviously more complex than the environment the animals faced upon emerging from isolation.

Isolate-reared chimpanzees show high arousal and high fear upon emergence and for some time later. They also avoid physical and other social contacts. In the normal development of the chimpanzee, early clinging, associated with contact with a familiar mother, reduces arousal under conditions of stress or in the presence of fear-inducing objects. Interaction with mildly frightening objects in the normally reared chimp

Reprinted by permission, *American Scientist,* Journal of The Society of the Sigma Xi.

Figure 4-9 Infant monkey returning to surrogate mother for reassurance.

is gradually reinforced through the arousal reduction associated with continual retreat to the mother (Figure 4-9). The isolate-reared chimp, on the other hand, got no stroking comfort for such behavior (cf. Mason, 1967). It is not surprising therefore that isolate chimps and many other animals fail to cope with excess external stimulation (Melzack and Scott, 1957; Riesen, 1950). They often cower or crouch in a corner when confronted with anything new while clinging to themselves and rocking.

At the same time, these animals appear to attempt to raise levels of physiological arousal sometimes by self-biting and other forms of self-mutilation. Like other animals, optimal arousal is what they prefer; but unlike normal animals the only deviations from optimal arousal they are able to tolerate are deviations which they themselves can produce and control. Their own body is their only object of attachment. Social isolates fail miserably in dealing with external changes in stimulation. These apparently cause uncontrollable internal changes which are reflected in such outward signs as excessive fear, bizarre postures and movements, and excessive and inappropriate aggression.

Hostility

As we have seen, rearing animals in isolation interferes with the development of fear and arousal control. If an animal is kept isolated long enough, its ability to behave aggressively in an appropriate way may be just as dramatically upset. The older an animal gets, regardless of how he is reared, the more hostility he displays (cf. Cross and Harlow, 1965).

The objects toward which hostility is directed vary with rearing experience (Møller, Harlow, and Mitchell, 1968). Animals rarely attack familiar social objects; individuals which know each other may behave aggressively toward each other but usually in a variety of rather indirect ways instead of attacking each other directly.

The developmental precursor of aggressive behavior is play. Mock aggressive play is very widespread among animal species. This behavior (often some kind of wrestling with weak biting) is energetic but rarely produces fear or hostility initially. However, as animals grow older and strong bonds are formed between peers, incidents in which individuals are hurt and become angry grow increasingly common. Gradually, play decreases and true aggression (with the intent to dominate another) increases between older juveniles and young adults. In primates, at least, the prolonged period of physical development permits individuals to learn their social roles and aggressive capabilities. In addition, during this time lasting attachments are formed which will be the basis of friendly adult relationships.

An isolate-raised monkey has no friends, unless one wishes to consider his own body as his friend. At an early age isolates cling to themselves, groom themselves, and play with themselves. At a later age, in

Reprinted by permission, *American Scientist*, journal of The Society of the Sigma Xi.

Figure 4-10 An isolate-reared infant monkey "freezes" in fear.

the isolation chamber, assertive play begins to appear as the animals chew on themselves and play both parts in a wrestling match. Arousal in this context remains low and is apparently easily controlled by the isolate. Upon emerging from isolation, the isolate's self-directed play gradually changes to self-directed aggression. Stimulation from the external environment is not adequately handled by an isolate-reared animal. The greater the arousal (or the more complex or novel the external exposure), the more intense the self-directed aggression, unless the arousal is so extreme as to produce freezing in fear (Figure 4-10). If another animal is introduced (obviously not a friend) the adult isolate usually attacks both himself and the other animal.

Some fish and some birds reared in isolation from their own kind have been reported to show at first extreme fear and then (with time) excessive and abnormally directed hostility (Kuo, 1960a; 1960b). When isolate-reared mice are put together at three to four months of age they initially do not fight as much as normally reared animals would. However with time they come to fight much more than control animals (Davis, 1964). Cats raised in social isolation for ten months are extremely aggressive and hostile. In one experiment isolate cats actually killed a newly-introduced animal (Kuo, 1960c). Even though the isolated cat is more hostile than the socialized cat, he is much less dominant socially. Dominance and aggressive behavior do not always go together.

Kuo (1960c) also reported that isolate-raised dogs were hostile. Isolate dogs, like cats, mice, and fish, after emerging from isolation are not hostile and if attacked, fight back ineffectually (Fisher, 1955). Later they become excessively aggressive and may kill strange dogs. Although the dog post-isolation syndrome eventually changes from snarling and crouching to attack (Fuller, 1964), isolates remain inferior to controls in dominance (Fox and Stelzner, 1967).

When isolate-reared monkeys first emerge from the isolation chamber they are submissive, fearful animals which rarely if ever display hostility. By the time the isolate-raised monkey is two or three years old, however, he has become extremely hostile, both toward himself and toward other animals. This abnormal hostility apparently continues to increase up to eight years of age (Mitchell, 1968b; Cross and Harlow, 1965; Mitchell, 1970). Despite or perhaps because of socially-directed hostility, isolate monkeys, particularly males, are socially subordinate to controls (Mitchell, 1968b). Social isolation which extends through the ages of six to twelve months, the period when dominance relations normally develop in play groups, is particularly important with regard to producing excess hostility. The crouching, rocking, self-clinging syndrome, on the other hand, is produced primarily by isolation in the first three or four months of life (Clark, 1968).

Exposure to early playful biting and wrestling is evidently necessary as far as the appropriate control of hostility is concerned. Mitchell and Clark (1968) observed several eighteen-month-old monkeys which had been reared in isolation between months three and nine. These isolates did not seem to show excess fear but would approach strange stimulus animals in a manner which reminded one of a three-month-old infant attempting to wrestle. The normal stranger reacted with apparent surprise and responded with mock-aggressive play. The isolate became highly aroused in this situation and hostility was often the result (Mitchell and Clark, 1968).

Isolate-reared chimpanzees do not begin to develop hostility until at least the end of the second year of life. Those objects which elicit the maximum amount of chimpanzee aggressive behavior are ranked inter-mediate in degree of novelty (McCulloch and Haslerud, 1939). Perhaps objects which are slightly unusual arouse curiosity and play, those which are more strange arouse hostility, and those which are intensely novel arouse fear.

Other Emotional-Motivational States

Abnormal emotional development has a wide range of behavioral conse-quences other than those already discussed. For example, the develop-ment of sexual behavior is intimately linked to emotional maturation. Isolate animals fail to interpret the sexual advances of other animals correctly and may respond with excess arousal, fear, or rage (Mitchell, 1970). Not only is sexual behavior likely to be abnormal in these animals, but maternal behavior is also absent or deviant.

In summary, it is evident that early social experiences are critical for the normal development of affection, fear, and hostility. All these emotions develop gradually, but there are sensitive periods during which their development proceeds at a more rapid pace. Affectionate behavior appears first even though its continued maturation is overlapped by the initial development of fear and then hostility. The extent to which an animal displays these emotions, and the manner in which it controls them, depend on the degree to which it has become familiar with and attached to objects and other individuals around it. Isolated animals have no opportunity to explore things and interact with others. As a result, as adults they do not demonstrate normal affiliation, fear, and hostility, nor do they control their emotions in appropriate ways.

Afterthoughts Regarding Man

Many beginning students of animal behavior bemoan the lack of relevance of animal studies to the present plights of people. This is particularly true of psychology students who may be tempted to make categorical distinctions between cultural and biological phenomena. Cultural phenomena *are* biological and technoligical advances are not unnatural, they are a part of cultural evolution. The beginning psychology student admits, somewhat parenthetically, that the human being is an *animal,* yet often does not really understand that the human being *is* an animal. We have continuously developed, and we *will* continuously develop as a species and as individuals.

If there was ever a sensitive or critical period in the phylogeny of man it is now. There are facets of our behavior that are developing very rapidly and which are therefore extremely sensitive to deleterious agents. Inappropriate or inadequate early exposure, or a nonnurturing and poorly paced type of child rearing, could have consequences which could spiral our species into extinction in a generation or so. Our society is now exceedingly complex and, therefore, if incremental pacing toward greater and greater psychological complexity has been important to social-emotional behavior in the past, it is more important than ever now. But what or who will be responsible for the human condition?

Generations present and past have spoken of social *responsibility.* Somehow it has always been felt that man should know himself and take care of himself, i.e., that he should be responsible for himself and for his species. But what is responsibility? Can such an attribute be studied in some other animal in the laboratory so that we can apply the findings to man? At least one example comes to mind.

A female rhesus monkey raised in social isolation becomes behaviorally and emotionally abnormal. She becomes sexually inadequate. She may run away from a socialized male in fear, aggress against him when he approaches her, or collapse passively as he attempts to mount. To be impregnated, she must be raped or inseminated artificially. When impregnated, she carries the infant through her 150-day gestation period and usually delivers a healthy-looking baby. A normal female monkey would show *responsibility* for her baby in the following sense: If a human observer were to reach into the cage to retrieve the infant of an adequately socialized mother monkey, the mother would threaten the human and aggress against him. An isolate-raised mother, on the other hand, will allow the human to take the infant away; she does not act to protect her infant. In fact the isolate-raised mother often hurts the infant herself (Arling and Harlow, 1967). It seems reasonable to say that the mother is socially irresponsible.

This demonstration at the monkey level is not very different from naturally occurring events one frequently sees in human societies. A man stands on top of a bridge threatening suicide, and people below scream, "Jump!" A young girl is knifed and raped on a city street and people continue to pass by. In the monkey's situation and in the human condition, there has apparently been an inadequate development of affection for species-mates, an absence of imprinting (if you will) to one's own species. If what *Homo sapiens* needs today is social responsibility, then what he needs is early attachment to, or affection for, mankind. Responsibility results from attachment and attachment results from the proper development of affection early in life.

In a recent review of human and animal research on social deprivation (National Institute of Child Health and Human Development, 1968) it was stated that:

The absence of motivation for affiliative contact during early childhood is a more severe sign of social deviance in children than is the absence of a desire to be self-reliant and socially assertive. Just as dependence and social attachments form the "glue" for human personality development, these motives also supply the cohesive elements for all orderly social institutions. Unless contact with, and recognition by, people are important to the individual, organized society cannot exist. (p. 14)

Early affection muffles later fear and aggression. "Love thy neighbor" therefore applies as a universal truth in *all* animal societies. The idea is not new but the evidence is, and it has come from the laboratory data and field notes of the animal behaviorist. Love seems simple but is actually much more complex than most of mankind will admit. Most people can be warmed, touched, or moved by it, but few have claim to that autonomy which develops through the distance receptors. Few people have developed security at a distance from, and security in the absence of, their objects of attachment. Our objects of attachment have not generalized to all of mankind. It seems that under many of the conditions of our everyday lives, the complexity of exposure often arouses us beyond our limited levels of love.

5   Sensory Systems

Robert Scobey
Thomas Harrington

The Human Vantage Point—
Peeking Through a Sensory System

Every human activity and every human perception involves "the real world," a complex of objects and relationships that we each see existing "out there" with unquestionable reality. We may see things differently at different times, but we continue to believe wholeheartedly in one "real" world full of "real" things and relationships.

Actually, there is no single sacred real world, but rather, an infinity of possible ones. Which one of these "realities" one subscribes to is dictated by the kind of sensory system one is looking through. If you had different sensory equipment you would see a different reality. This is true at two levels. First, because our sensory systems miss many things, and second, because of the logical impossibility of bypassing one's sensory system, of cheating and peeking around it to infer what is *really* out there.

I Hear There's Going to Be a Storm

The senses of men and animals are limited. They suffer either a relative or a complete lack of sensitivity depending on the form of energy involved. These sensory limitations are necessarily imposed upon us by evolution, because an efficient sensory system *should* only accomplish certain limited goals for its owner to survive. We *must* be sensitive to certain things, while responding to others would be a needless burden.

A sensory system should warn of danger. Inanimate hazards like cliffs and sharp rocks are one kind of threat, and predators are another. The sensory system must allow its owner to find appropriate food, to reproduce, and so forth. The requirements of survival depend on a complex interaction of the animal's metabolic appetites, its sensory and motor capabilities, and the appetites and capabilities of its natural enemies. One animal's food is another's poison. The complex issue of who eats whom in the jungle is very much linked to the type of sensory equipment the process of evolution develops for each kind of animal to complement his motor equipment. Thus, the limitations of sensory systems are quite functional, filtering out unimportant things that would be so numerous

129

as to overwhelm an animal, rendering it unable to pay proper attention to things that matter. A male butterfly can detect the presence of a female several miles away by only a few molecules of her scent. For a human, smelling butterflies from miles away would be quite unnecessary at best.

Cats hear sounds up to 50,000 Hz, bats and dolphins hear as high as 150,000 Hz, while the human is limited to hearing sounds between 20 and 20,000 Hz. The sounds important to our survival are within this range. We talk there, lions roar there, and noises of animals moving are adequately represented there. Above this range we miss the high frequency subtleties of birds singing and bats echolocating, but if we could hear these sounds they would drown out the important ones. Similarly, if we could hear well at the lower frequencies, we would hear things like blood surging through our capillaries with each heart beat and be deafened by the breathing of other creatures. At the extreme of low-frequency sound are barometric pressure changes. If sensitive to these you would be able to say in a literal sense, "I *hear* there's going to be a storm."

In the Realm of Vision

Certain insects are sensitive in a different way to high-frequency electromagnetic energy (light) than we are. A honey bee would tell you that some flowers are colored "ultraviolet," but the human system cannot detect ultraviolet light. Because of this limitation, the sharpness and clarity of our visual world is better, since ultraviolet light wouldn't focus well with other light, but we can't gather "ultraviolet" flowers or put on "ultraviolet" socks. At the other end of the spectrum of "visible" electromagnetic radiation, pit vipers, such as rattlesnakes, exceed us. Their "pit" organ is extraordinarily sensitive to radiant heat and can detect changes in temperature of a few tenths of a degree centigrade in the surroundings.

Many animals can even sense forms of energy to which humans are completely insensitive. We know electricity only as an aberrant sensation of excess energy, but the African knife fish, living in murky waters of the Amazon River, produces and senses small gradients of electrical current. Using these abilities, this fish navigates around obstacles and detects the presence of intruders. We cannot sense carbon dioxide in the environment (although we unconsciously detect it in our own blood streams), but the mosquito searches out its human victims by sensing the small quanties of carbon dioxide and heat emanating from them. Indeed, water does not uniquely stimulate human or rat taste receptors, while dogs, pigs, cats and rhesus monkeys have receptors in the mucous membrane which signal the presence of this ubiquitous molecule.

Each animal peers out at the world through its own peculiar little complex of sensory capabilities and weaknesses, and these completely structure what it finds "out there." Humans, of course, are no exception; thus one of our first cautions in studying sensory systems is never to consider the framework of human reality to be universal.

Alone with Your Senses

Even so, by considering our sensory limitations and making allowances for them, it would seem that we could easily *infer* what is *REALLY* "out there." We cannot. As Ernst Mach said many years ago, "The world begins and ends inside the head." We are forever confined within the limits of our sensory awareness.

This is the second level of sensory limitation. We can never peek beyond our senses, inductively or any other way, and infer what absolute reality is like. The concept of an absolute reality that we approximate with our limited senses is comforting, but false. Isn't the world of physics "real," apart from any considerations of human sensory capacity? Certainly not. The fundamental units of physics, i.e., mass, length, and time, are not sacred; these are products of the human sensory system and of human imagination. Isn't this book you're holding a real entity with at least some real aspects that you are approximating with your limited sensory system? To gain a feeling for the relationship between perceptions and reality, let us consider an elementary sensory system.

An Elementary Sensory System's Reality

Imagine an organism with only one incoming communication channel. This channel is a single neuron, a cell with specialized properties for transferring information. A neuron together with its ancillary structure is a *receptor*. Say that the peripheral or outside end of the neuron is sensitive to what we humans call mechanical stress. Mechanical stress results in a single brief response of the neuron, called an action potential, which can propagate from one end of the neuron to the other, thus informing the organism that a mechanical stress occurred at the peripheral end of the neuron.

Allow this creature as much "IQ" as you wish, perhaps 2000, with perfect memory and perfect ability to correlate events in time. Now, give him our environment and see how he does. Stretch his single receptor across the intersection of Broadway & 52nd Streets. What kind of reality can our creature expect? Pitifully impoverished? It can not appreciate forms of energy which are not adequate to stimulate its receptor, such as light; still, with its wealth of intelligence, it can discover things we could never know. Neural signals will arrive at its central nervous system (CNS) in patterns signifying rush hours, periods of traffic lights, holidays, short cars and long cars, rainfall, pedestrians and dust. Each of these things will have different, interacting daily and yearly cycles. Millions of subtle phenomena, obvious to our creature, would be hopelessly beyond analysis by a human.

Furthermore, these phenomena would seem as perceptually real to the creature as this page does to you. Yet, it is only responding to neural signals coming up its sole sensory line and *making up "real entities;"* it can't appreciate its neural signals as being artificial any more than you can appreciate this book as being only a torrent of neural signals from your sensory surfaces. This creature is only hypothesizing its reality. Our brains receive neural signals, and from them we fashion entities, books and such, which we then *seem* to perceive directly.

How a Sensory System Works

General Principles of Sensory Neural Coding

Subjectively, there is a compelling trichotomy of sensory attributes that one experiences; first there are different *kinds* of sensation, often loosely called the "senses," such as visual, auditory, vestibular, gustatory, olfactory, temperature, touch, pain, and limb position. Tastes are rarely confused with sounds, lights are rarely confused with smells or touches. These different senses have relatively exclusive connections to the brain. Within each sense there are different qualities or modalities. Sounds have varying pitches and timbres; lights have different hues and saturations; and there are different qualities of touch, taste and smell. There is a neural reason for this.

To the Brain, Certain Specific Fibers Mean Specific Qualities from Specific Locations

In the mid eighteen hundreds, Johannes Mueller formulated the doctrine of specific nerve energies or qualities. He noted that no matter how a particular nerve is stimulated it will produce the same sensation. If you push on the side of the eye, light will be seen, even though pressure is the stimulus, because the brain always interprets activity in sensory nerve fibers of the optic nerve as signifying light. Light is the adequate stimulus for the eye. The brain does not know how a particular neuron was stimulated. It can only adhere to the anatomical labels it has for each incoming sensory nerve fiber. Even when the cortex (the outside rind of the brain where sensory information arrives) is stimulated electrically, a human subject will report lights, sounds, etc., depending upon where the stimulation was applied. One of Mueller's students, Hermann von Helmholtz, extended the doctrine to one of specific fiber energies, saying that each fiber in a nerve bundle would yield a specific sensation, no matter what the stimulus. Thus we have the concept of *labeled lines:* the brain views each incoming sensory line as implying only one quality of sensation. There is some question as to how far the doctrine of specific fiber energies can be generalized; however, it is a good first-approximation organizing principle.

Localization is also labeled. By analogy, all telephones (receptors) are connected through channels (nerve fibers) to main telephone centers. A telephone center can trace the location or origin of any telephone call. A brain "knows" the location of a touch because the anatomical connections of channels from the skin to the brain are fixed in our embryological development just as each house on the block has a separate telephone wire to it, and the location of the originating call can be assigned to that spatial location at the central "brain."

The reader may have difficulty reconciling stimulation on the sensory surface with perceptual phenomena because the sensory surface can be represented as a flat, two-dimensional array of receptor endings. Yet, perception seldom seems to be two dimensional. The image on the retina of the eyeball is perceived in depth outside of the body. It is difficult to appreciate the visual world as only a pattern of electromagnetic energy on a two-dimensional retina. Sounds also are perceived as outside the body, not as the bending of receptor hairs within the coiled, but conceptually two-dimensional cochlea. Obviously, the localization of visual objects and sounds beyond the sensory surface is essential to survival.

In contrast, the skin surface has not developed as a system of external projection of sensation. A vibrating point on the skin is located on the surface of the body. The localization of tactile sensation in free space is not uncommon, but is apparently a learned abstraction, and not "given" to us as are our visual and auditory localizations.

Von Békésy (1967) has shown that cutaneous sensation can be projected beyond the body by learning. For example, the application on adjacent fingers of two vibrators with equal amplitude of vibration but different time sequence of application can result in the projection of a sensation of vibration in the space between the fingers: the trained observer "feels" the vibration move in this space. Similarly, when vibration is applied to the thighs, a trained observer can localize the sensation in the space between his knees. People sense the pressure of a screw at the tip of a screwdriver, and a trained medical practitioner can project the vibratory and pressure sensations of palpation of tumors into the patient.

Intensity of Stimuli

At each location on the sensory surface a particular modality has the attribute of *intensity*. In order to signal the attribute of intensity, the quantity of energy at the receptor is usually transcribed into a temporal code to be transmitted to the brain. For example, the frequency of firing of a neuron often increases as stimulus intensity does.

When telegraphers send messages over the wires they employ Morse Code. The telegrapher encodes the English language into sequential bursts of electrical energy. In a similar manner, receptors and other neurons must communicate about intensity according to various kinds of code. A code, then, is a language. The act of pressing the telegraph key releases stored electrical energy for communication on the telegraph line. An important part of the study of receptor physiology has been to discover what elements in the body release stored cellular energy and thus transduce and encode impinging energy.

Transduction and Coding in All Sensory Systems

At each telephone in a system, sound in the air is changed into electrical signals in the telephone wires. This is an example of the process of transduction: converting one form of energy to another. Analogously, the skin contains specialized structures called receptors, which selectively encode different forms of energy into neural signals for transmission to the brain. These structures are relatively but not absolutely, more sensitive to one form of energy. For instance, the eye is sensitive to light, but a pressure on the eye will also encode a neural signal in the visual receptors and neurons, resulting in a sensation of light and "seeing stars." Thus transduction is a kind of selective filtering of stimulus information. Considering the telephone analogy once more, one notes that a familiar musical selection heard on the telephone is surely different from that sound when heard through a high fidelity reproducer. The telephone company has constructed the telephone to accept only the frequencies of sound important to voice communication. In order for the present telephone lines to transmit with full fidelity, additional channels would be needed to encode low- and high-frequency sound waves.

The nervous system is also organized this way. Different wavelengths of light are seen as different hues because three different kinds of photoreceptors called *cones* each respond differently to any specific wavelength. The sense of touch has at least three "lines": one for steady or low frequency skin displacements, one for medium frequency, and one for high.

There is often more than one kind of "line" for a modality at one spatial site. For example, according to one theory, there are two receptors in the skin for temperature. One receptor encodes decreases in surface temperature (cool or cold) while another encodes temperature increase (warm or hot).

Receptors

On a sensory surface, energy does not directly induce neural signals in the receptors. The energy of the stimulus simply "releases" stored electrochemical energy; thus the receptor is not a passive device. Stimuli to receptors are more effective because the impinging energy *controls* or modulates stored electrochemical energy of the receptor. This is "active transduction," a principle also used in vacuum tubes and transistors.

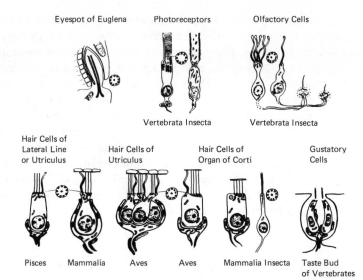

Eyespot of Euglena   Photoreceptors   Olfactory Cells

Vertebrata Insecta   Vertebrata Insecta

Hair Cells of
Lateral Line   Hair Cells of   Hair Cells of   Gustatory
or Utriculus   Utriculus   Organ of Corti   Cells

Pisces   Mammalia   Aves   Aves   Mammalia Insecta   Taste Bud
                                                           of Vertebrates

From "Principles of structural, chemical, and functional organization of sensory receptors" by J. A. Vinnikov, in *Cold Spring Harbor Symposia on Quantitative Biology* **30,** *Sensory Receptors,* 1965.

Figure 5-1   The structural organization of sensory receptor cells. The figure illustrates photoreceptors, olfactory and gustatory receptors, and hair cells of various sorts in different animal classes. Basic organizational similarities may be seen. Photoreceptors evolved through the infolding of membranes which subsequently become photosensitive. The hairlike protrusions of the olfactory and gustatory cells developed a chemical sensitivity, while the hair cells in the lateral line organ of the fish and in the inner ears of mammals and birds specialized in the transduction of mechanical displacements. In general, these cells do not have action potentials; rather, they communicate to the nervous system by release of a chemical transmitter at a synapse.

As noted, the transduction of energy into a neural signal occurring at a receptor is more or less selective. In order to provide a feeling for how this happens and to lay the basic groundwork for what follows, we will consider some receptor systems in detail.

The rods and cones of the eye are maximally sensitive to light; the hair cells of the auditory and vestibular parts of the inner ear are maximally sensitive to movements; taste buds and olfactory cells are maximally sensitive to chemical stimuli. The wide variety of anatomical, chemical, and functional features of receptor cells attest to specialization, but there is a common denominator. Figure 5-1 illustrates the structural organization of a variety of sensory receptor cells. The figure shows the eyespot of Euglena, a Protozoan, compared with the photoreceptors of the vertebrates and insects, and with the olfactory, gustatory, and hair cells of insects, birds, fishes, and mammals.

All sense organ receptor cells have hairlike protrusions and have a common ultrastructure, seemingly derived from a primitive prototype with motile cilia or flagella. The particular form of energy that a cell responds to appears to result from its evolutionary and ontogenetic lines of development from this primitive cell type. Likewise, the cells have a common biochemical basis. For example, Vitamin A is always associated with cells that have hairs or flagella; as a result, Vitamin A deficiency can cause sensory impairment. Diets deficient in Vitamin A can cause the rods of the eye to swell and break apart and lose the outer segment of the cell (derived from the ciliated section of the primitive cell) which is sensitive to light. The auditory and vestibular hair cells of the inner ear are similarly dependent upon Vitamin A. Thus, the specialized receptor cells in both vertebrates, and invertebrates, seem to have evolved by a process of natural selection from a primitive cell with a cilia or flagellum. These motile hairs represent the ancient organ for both mobility and the interaction between the primitive organism and the environment.

Electrical Signs of Nervous Activity in Receptors and Other Neurons

The change in permeability of the membranes of receptors and other nerve cells is what most researchers depend upon for the measurement of neural activity. This change itself cannot be easily quantified, but the electrical events which stem from this change can. When a nerve is stimulated in an excitatory way the membrane which surrounds and contains the cell's contents usually becomes more porous than previously. Because the fluid surrounding the nerve has a much different chemical makeup than the contents of the cell, there is a tendency for small particles to move through the membrane until the concentrations of different chemicals are equal on both sides. This tendency is reinforced by the fact that the distribution of ions (positively or negatively charged particles) is such that the interior of the cell is substantially negative (about -70 millivolts) with respect to the exterior. Positively charged ions on the outside (basically sodium ions) are therefore strongly attracted to the interior of the cell. The movement of positive ions into the cell naturally changes the electric charge difference across the membrane (from -70 mv to -60 mv or -50 mv depending on the nature of the stimulation). This change in membrane potential can be detected and recorded by a strategically placed electrode. Although one can do this by placing the electrode through or near the membrane of a single cell, it is also possible to record the electrical signs of nervous activity at great distances from active neurons. For example, nervous activity in the brain can be detected by electrodes taped to the skin surface of the head. This is called the electro-encephalogram or EEG—an electrical reflection of permeability changes at the membranes of millions of neurons in the brain. We will now consider the basic properties of electrical signs of neural action: the *local potential* and the *action potential.*

Local Potentials

These are simply changes in membrane permeability and ion flow restricted to a relatively small area on the nerve cell. The amount of change is roughly proportional to the amount of stimulation, and the membrane quickly restores its *resting potential* of −70 mv upon cessation of the stimulus. Because of the localized nature of these changes they are not suitable as signals over a long distance; however, they are important in a variety of ways. For example, in some receptor cells local potentials cause the cell to release a chemical transmitter which will excite another neuron. In all neurons local potentials are necessary if an *action potential* is to be generated.

Action Potentials

The action potential is the electrical sign of long distance neural communication. It is not proportional to the intensity of stimulation nor is it restricted to a local area. It is an all-or-nothing event of relatively constant magnitude that can be propagated for any distance along the nerve fiber. Very simply, what happens in the generation of an action potential is that in response to some stimulus—environmental energy or a chemical messenger transmitted from another neuron—the membrane at one end of the nerve *depolarizes*. That is, it becomes permeable to positively charged sodium ions which enter the cell changing its charge differential. This starts a local potential sweeping over the surface of the cell to the start of the axon or nerve fiber. This is the long process which carries neural messages toward the central nervous system (*afferent* fibers) or away from the central nervous system (*efferent* fibers). If the local or generator potential only slightly depolarizes the membrane at the start of the axon, little or nothing happens and the membrane rather quickly resumes its original resting potential of −70 mv. However, if the generator potential is sufficient, the threshold (about −50 mv) for an action potential may be reached. When this point is reached, the membrane undergoes a brief but massive change in permeability; there is a huge inrush of sodium ions and the interior of the membrane actually becomes positive briefly before an outflux of positively charged potassium ions restores the membrane's original condition. The change in membrane permeability at the beginning of the axon affects a neighboring section of membrane leading to similar events at that point and so on, in a chain reaction, down the length of the axon. It is these events which make up the neural message, the action potential.

If the axon of the nerve cell is wrapped in a fatty chain of *myelin,* the action potential jumps from gap to gap, in this chain, from one *node of Ranvier* to the next. See Figure 5-3 for a diagram of a myelinated nerve. These kinds of nerves enjoy faster conduction of messages than nonmyelinated neurons.

Once begun, the action potential continues to the end of the axon, reaching there in the same form as it began its journey. Action potentials do not vary in magnitude according to the strength of the stimulus. Once the threshold has been reached a standard message unit is generated. In other words, the action potential caused by a bright light is the same as that caused by a weak glow, and an action potential in the visual system is basically the same as an action potential from a nerve in the ear. Although these neural messages cannot move as fast as electricity, i.e., with the speed of light, they can, in man, achieve speeds up to 208 meters (about 600 feet) per second (von Békésy, 1967).

We will now turn our attention to two key receptor types, one in the visual system and one in our tactile system.

A Receptor for Light

In order to explore the events occurring at the receptor cell after the physical stimulus reaches it, let us use the photoreceptor as an example. Photoreceptors are beautifully arranged in a mosaic at the back of the eyeball. The more orderly anatomical arrangement of photoreceptors and the easy surgical approach into the eye have led us to a better understanding of the membrane events at the photoreceptor than at the hair

Photoreceptor (First Order Neuron)

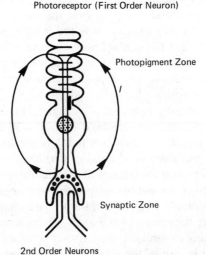

Photopigment Zone

Synaptic Zone

2nd Order Neurons

Figure 5-2 A schematic diagram of a photoreceptor. The photoreceptor is a cell specialized for the transduction of light into neural signals. The photopigment zone is the part of the cell sensitive to the impinging light. The presence of light causes the permeability of the membrane about the photopigment zone to *decrease,* producing a decrement in the extracellular current *I.* This decrease in the current leaving the membrane in the synaptic zone causes the membrane of the zone to become hyperpolarized. A subsequent release of transmitter agent from the photoreceptor at the synaptic zone carries the visual information to the second order neurons.

cell. Since both cells are presumed to have a common ancestor, the sequence of membrane events may be the same for the hair cells, taste and olfactory cells, and the photoreceptors, but this has not been proved.

Figure 5-2 illustrates schematically a generalized photoreceptor, either a rod or a cone, of a vertebrate. The rod-type photoreceptor can detect the presence of one photon, the smallest possible amount of light, and indeed the fully dark-adapted human can sense light when only a few photons are present. Therefore, the rod photoreceptor is optimally sensitive to light energy. In order to transduce light energy to an encoded neural signal, the light must strike a membranous disc located in the receptor's outer segment. These discs, shown in a cross section of the photopigment zone in Figure 5-2, are derived from infoldings of the cell membrane originally covering the cilium. The discs contain photopigments which are molecules sensitive to light. The absorption of a photon by the photopigment is thought to initiate a process of biochemical amplification leading to a decrease in the permeability of the outer segment membrane of the photoreceptor. Around the vertebrate photoreceptor there is a steady electrical current flowing extracellularly from the synaptic zone to the outer segment. This current is reduced by the permeability decrease at the membrane of the photopigment zone and hence the membrane of the photoreceptor is then less influenced by the current flow, and the membrane voltage in the synaptic zone increases. The molecular mechanism leading to the permeability decrease, as well as the process producing hyperpolarization and chemical transmitter release at the synaptic zone to excite the second-order cell, remains unknown.

The Pacinian Corpuscle

We now compare the basic things which happen in photoreceptors with those which occur in mechanoreceptors, for example the Pacinian corpuscles (Figure 5-3) which are located well below the surface of the skin in the dermis and also in connective tissue often quite deep within the body. These receptors are especially sensitive to quick changes in pressure. Unlike a rod or cone, the Pacinian corpuscle is an example of a receptor which generates and transmits its own action potentials. Anatomically, the Pacinian corpuscle is a single neuron associated with a body (the corpuscle) of concentric layers which surrounds the terminus of the innervating nerve fiber (Figure 5-3). The corpuscle itself is insensitive to stimulation. Its alternating layers of solid and fluid adapt mechanically so that no steady pressure, but only rapid changes in pressure, reach the wrapped nerve fiber. These layers slip relative to one another and absorb stimulus energy in their elastic components. As a result, after the corpuscle and the end of the nerve fiber have been moved by a change in pressure, the nerve will fire only briefly because the corpuscle quickly blocks further stimulation of the neuron.

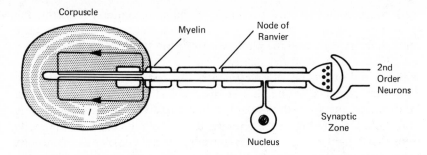

Figure 5-3 A schematic diagram of a Pacinian corpuscle. The transducer of a Pacinian corpuscle is located far from the nucleus of the cell (the "bullseye" hanging from the neuron). The receptor membrane is part of the neuron and is inserted into a laminar corpuscle. The corpuscle is layered; it absorbs the steady pressure but not the onset or removal of a stimulus. The stimulus energy, upon reaching the neural membrane, causes an *increase* in the permeability of the neural membrane. The current, *I,* then depolarizes the first node of Ranvier and initiates an action potential. The action potential sequentially stimulates the next adjacent node and the action potential skips down the axon to the synaptic zone where a transmitter (small dots) is released to excite the second order neuron.

The terminal end of the nerve fiber within the body of the corpuscle is sensitive to strain on the neural membrane, responding to about one-tenth of a micron (approximately four one-millionths of an inch) movement. The threshold for firing is reached when skin is displaced a few microns at the optimal movement rate. When pressure changes occur, causing these kinds of displacements, the neural membrane becomes more permeable to ions. The resulting current (I in Figure 5-3) then depolarizes the first node of Ranvier at the start of the axon and may initiate an action potential. As previously discussed, action potentials then skip down the myelinated axon to the synaptic zone (or area of near contact with another cell) where a transmitter chemical is released which will excite the next neuron. Information is on its way toward the brain where it may ultimately contribute to the perception of pressure change in a particular spot in the body.

In the following sections we shall attempt to show in more detail what happens to information on its way from a receptor to the brain.

The Properties of Sensory Systems

The Relationship Between Perceived Stimuli and Neural Activity

A stimulus may be perceived to have certain qualities such as intensity, movement, location in space, and so on. The question is how sensory receptors provide the necessary information which is ultimately translated by the brain into these characteristics. A major difficulty facing anyone who wishes to tackle this problem is that animals other than humans cannot provide us with direct reports of their sensations, and therefore there is no easy way to determine what they perceive. Humans can make verbal reports about their perceptions, but it is usually difficult to study the activity of sensory neurons in human subjects. However, by studying the neural responses to specific stimuli in a primate with a nervous system similar to our own, it may be possible to relate neural "behavior" to perceptual reports of humans. This approach is currently being employed with great success by a number of neurophysiologists. First, human sensory measurements through subjective reports are made. Then measures of nervous system activity are recorded electrophysiologically from single neurons of monkeys. For example, the skin may be vibrated precisely at one point. The subject is asked to judge the magnitude of the vibratory stimulus or to make a forced choice between presence or absence of the vibration when the stimulus is very weak. The same stimulus may then be applied to a monkey's skin. The electrical signals from the stimulus measured in the nervous system of the monkey may be studied to find the relations between human sensations and neural signals. Obviously, it is necessary to assume that the nervous system of the monkey under consideration is a good model of the corresponding system in man. In this manner, and given these major assumptions, human sensory phenomena can be related to neural data.

In the following sections specialized techniques of comparing neural with psychophysical data in sensory systems will be outlined. The nature of peripheral and cerebral neural codes and the relations between coded signals, sensations, and perceptions will be discussed.

Perceptual and Neural Thresholds

The threshold of a sensory system is often the first of its properties to be studied. We need to distinguish between the *psychophysical* threshold and the *neural* threshold. The psychophysical threshold refers to the minimal level of stimulus energy that will just elicit an overt (behavioral) response from a subject: "I see the light" or "I hear the sound" or "I feel the vibration." The neural threshold, by contrast, is the minimal stimulus energy that will just elicit a specified change in the neural activity being measured.

Comparing the response of a monkey neuron to the stimulus intensity which corresponds to the human threshold may give insight into what it is about the neural response which ultimately permits us to detect the stimulus. For example, it may be that at this stimulus intensity the neural unit first begins to fire, sending action potentials to a second-order neuron. Or it may be that the unit, or perhaps a whole population of receptors which has been firing spontaneously all the time, suddenly changes response patterns in some way.

Intensity: Strength of Sensation vs. "Strength" of Neural Response

Whenever there is a sensation, it has an intensity. There are several ways that intensity can be coded within the nervous system, and it is up to sensory physiologists to discover how such coding is effected in each modality. The frequency of action potentials is one way that a neuron can code the intensity of the stimulus. Individual nerve fibers can produce action potentials up to a rate of about 1000 per second. Also, the stimulus can be effective over a wider area of the sensory surface as the intensity increases, bringing additional receptors into excitation. This *recruitment* of additional receptors is augmented by the excitation of receptors with higher thresholds. Thus, intensity is also coded by the number of active neural units in a given area. One might predict then, based upon the recruitment of fibers, that the geometrical size of the stimulus would appear larger as the intensity of stimulation increases. For example, the apparent sizes of stars vary this way, with the more intense stars appearing larger, even though the physical sizes of the images on the eye are constant. This sensory phenomenon occurs only over a small range of intensity because of *lateral inhibition,* a topic to be discussed more fully later.

In order to discover what the intensity code is, stimuli of different intensities are applied to human subjects and to animals. Humans may estimate the intensity by assigning numbers they feel are proportional to the stimulus magnitude. Then a monkey is subjected to the same range of stimuli and the activity of several receptors is recorded and studied to determine the relation between the strength of the stimulus and the amount and kind of neural activity. Thus, for example, a human might report the strength of the sensation to be directly proportional to the magnitude of the stimulus. In turn the animal's neurons might fire more and more often per unit of time as the stimulus intensity increases. Therefore, it may be that in humans as well as monkeys, it is the frequency with which a receptor fires that provides the information which the brain interprets as stimulus intensity.

Space: Location and Movement

The neural processes that encode location and movement on a sensory surface are generally more complex than those encoding quality and intensity, so we will devote proportionately more space to these concepts. First a detailed consideration of the way receptors provide information about the size of an area being stimulated and the distance between two separate stimuli on a sensory surface is necessary.

The Receptive Field

A receptor neuron on a sensory surface will encode information only from a certain area of the sensory surface which is called the neuron's *receptive field.* Let us consider some examples. Sometimes an afferent fiber might be excited by only one receptor or receptor ending. In the cutaneous system Pacinian corpuscles have single myelinated fibers attached to them. One would imagine that such a fiber could monitor stimuli only in a small area. Yet a Pacinian corpuscle, buried deep below the cutaneous surface, is sensitive to tissue vibration and thus has a receptive field of fairly large area on the skin. The receptive field is large due to the diffuse spread of vibratory mechanical energy from the skin surface to the corpuscle's location. A bone in a limb is a good conductor of vibration; consequently, tapping your elbow is likely to excite every Pacinian corpuscle in your arm, hand, and shoulder.

However, in general, more than one receptor ending is connected to an afferent nerve fiber. In the cutaneous system, a single afferent fiber may be excited when any one of several hairs is touched. In this case, the receptive field for that fiber is the area on the skin with a set of sensory receptor endings distributed around the base of hair follicles under the skin.

For units at higher levels in the nervous system sometimes receptive fields become larger than at the periphery. For instance, there are many more hair follicle receptors (or in general, sensory endings) present on the sensory surface than there are nerve fibers present in the nerves leading to the central nervous system. The connection of many receptors to a single fiber is called *convergence.* In a human eye there are at least 130 million light receptors, but only one million nerve fibers carry information from all these receptors to the brain. In the extreme, tens of thousands of receptors in the olfactory epithelium converge upon one cell in the olfactory bulb through an interposing neuron. Thus one cell in the olfactory bulb has a large receptive field due to its anatomical connection with the many cells which converge upon it.

The Size of Receptive Fields Varies. Within a given sensory surface, there is a functionally ordered arrangement of receptive field sizes. Small receptive fields are found in that part of the sensory surface which is used for highly discriminative functions. For example, in the fingertips the receptive fields for encoding mechanical movement of the skin are small. The receptive fields become increasingly larger for neurons encoding the mechanical movement of the skin of the arms and back. Similarly, the central region of receptive fields of neurons receiving projection from the fovea of the eye are small. Visual information from the fovea (a small area in the retina composed of many receptor cells tightly packed together) which is transmitted by afferent fibers with small receptive fields provides the large amount of visual data necessary to make fine visual discriminations. In contrast, the receptive fields of neurons in the periphery of the retina are relatively large. This results in a decrease in the sharpness of the images constructed on the basis of information sent from the periphery of the retina.

If a receptive field of a given neuron is small (less area on the sensory surface is effective in stimulating that neuron) the neuron will excel in signaling information about the spatial attributes of the external world. For example, since the spatial discrimination in the cutaneous system is very well developed at the tip of the fingers, a Braille reader can use his fingertips to discriminate embossed points separated by only one-tenth of an inch.

Receptive Fields Overlap. Another principle relating spatial information to the neurons of the sensory system was recognized by Adrian and Matthews (1927a; 1927b). While recording from a nerve with many fibers, they recognized that a stimulus to any single point on the sensory surface would excite many of the fibers. Recall that any single sensory unit has a receptive field; i.e., the neuron is excited from a specific spatial area. Hence if a stimulus at one point on the sensory surface excites many fibers, we must conclude that the receptive fields of adjacent neurons overlap. Anatomical examination also shows that the terminals of adjacent axons projecting from receptors in the sensory surface intertwine with each other extensively (Figure 5-4). This principle of *partially shifted overlap* of receptive fields is of great importance in spatial discrimination.

As a result of the overlap between receptive fields, even a localized stimulus will often excite more than one nerve fiber. Therefore the processing mechanism of the brain can compare the pattern of input from many axons and, on the basis of more information than is available from a single nerve, create a sharper mode of the stimulus than would otherwise be possible. In particular this system allows the brain to detect spatial details smaller than the diameter of a single receptive field (by locating the stimulus in the relatively small area of overlap between two or more receptive fields). It should be quite obvious that spatial discrimination will be a function of the number of receptors in an area, the number of fibers running to the brain, and the number of analyzer

Figure 5-4 Schematic diagram of two areas with different innervation density. In the left-hand drawings two receptive fields for two afferent fibers are shown at the sensory surface. Afferent fibers distribute to the receptive field area on the sensory surface. The two fibers have partially overlapping distributions. The circles represent both the receptive field and the area of anatomical distribution. The right-hand drawings illustrate a similar area on the sensory surface, but one which has a higher innervation ratio, i.e., more neurons per unit area. Again the receptive fields overlap extensively. This area would have greater spatial discriminatory power than the sensory surface illustrated above. The increased innervation ratio increases the mass of the peripheral nervous system; moreover, more of the central nervous system is given over to areas with high innervation ratios.

neurons in the brain devoted to decoding the information it receives from that area. The surface of our hands is richly innervated compared with an equivalent area on our back; in addition, there is a great deal of neural tissue in our brains whose responsibility it is to decode input from tactile receptors in our hands, and a relatively tiny amount which analyzes the small amount of information sent from our back. As a consequence we can discriminate between similar tactile stimuli much better by touching them with our hands than by having these stimuli brought in contact with our backs.

Similarly, the amount of human brain tissue assigned to vision at or near the fovea is greater than the quantity of brain subserving the rest of the visual field combined. However, such a visual field representation is not characteristic of all animals. In frogs all parts of the retina have equal receptor density; no part has become densely packed with receptors for more acute vision. Birds, however, sometimes have two regions specialized for acute vision: one where receptors are closely packed allowing the bird to have fine visual acuity in the direction of flight and another directed to the ground where food may be found.

Receptive Fields Have Areas of Different Sensitivity. An examination of the sensitivities of the sensory surface is important for understanding spatial perception, since the changing sensitivity within receptive fields helps account for movement detection. The sensitivity of the sensory surface to stimulation can be expressed by the quantity of stimulation necessary to elicit a threshold response. Here we will discuss the sensitivity of a receptive field in terms of the amount of stimulation needed to increase the rate of firing in the associated neuron.

In Figure 5-5, the schematic diagram illustrates a neuron in the sensory surface with its connection to another neuron in the nervous system. Neurons communicate, exciting or inhibiting one another by release of a chemical transmitter. In the diagram the site of chemical release is shown by the forked symbol, while the site of sensitivity to the released chemical transmitter is shown by the solid dot. This region is the synapse, a word that means *connection*. The first synapse is always excitatory, shown by the E in Figure 5-5.

Let us apply a specific stimulus to point 1 on the receptive field of the neuron in Figure 5-5. If sufficient, this stimulus will generate a certain increase in the frequency of action potentials sent by the second order neuron (say 14 impulses per 100 milliseconds instead of 10). The exact same stimulus at point 2 generally will induce a smaller change or no change at all in the rate of firing. The receptive field grows still less sensitive to stimulation at more peripheral sites. Looked at another way, the further from the receptor itself the greater the strength of the simulus required to excite the receptor and its related neurons to any given level of message sending. The receptor does not code events occurring on exactly one spot on the sensory surface because "spillover" of energy from nearby areas is present. For example, a cone in the retina in practice

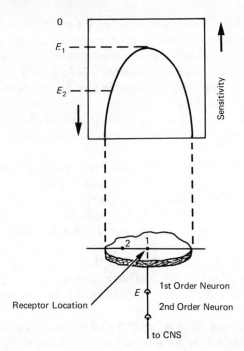

Figure 5-5 A schematic diagram of the sensitivity of a receptive field. This illustration shows a receptor at location 1 in the sensory surface. The receptor membrane of the first order neuron excites the neuron, leading to a release of an excitatory transmitter at the first synapse, marked E. The second neuron in the chain of neurons is called the second order neuron. The area of the sensory surface which is effective in excitation of the neurons is dependent upon the energy of the stimulus. The sensory surface is maximally sensitive at the position of the receptor. The energy (E) necessary to excite the first order neuron is much larger at position 2 than at position 1. The extent of the area which will influence the neuron (the receptive field) is dependent upon the amount of energy in the stimulus. Notice that the energy scale is plotted downwards, i.e., sensitivity is the inverse of required energy.

does not have a receptive field on the retina equal to the anatomical size of the cone. Light, which is directed at neighboring cones, is scattered in the eye and retina, and this scattered light falls over a relatively larger region. The upper part of Figure 5-5 illustrates a sensitivity profile across a receptive field which is representative of many modalities where the center of the receptive field is maximally sensitive.

Figure 5-6 illustrates the receptive field of a third-order cell called a *ganglion* cell actually studied in the visual system of a cat. The receptive field of the cat ganglion cell is used to demonstrate the roughly circular shape of these receptive fields.

The receptive field was determined in three different ways to obtain Figures 5-6a, 5-6c, and 5-6d. The eye was focused on a small spot of light flashing against a constant intensity background. Only if a spot of

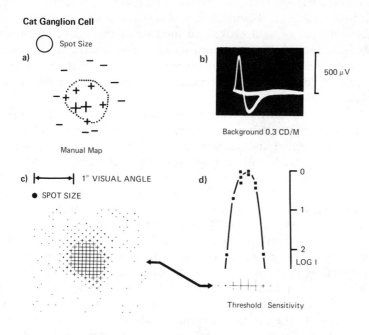

Cat Ganglion Cell

a) Spot Size — Manual Map

b) 500 μV — Background 0.3 CD/M

c) |———| 1° VISUAL ANGLE — ● SPOT SIZE

d) 0 1 2 LOG I — Threshold Sensitivity

'Random' Spatial Sequence
200 Ms. On, 200 Off.
Computer Presentation
Spot I LOG Above Minimum Threshold

Figure 5-6 The receptive field of a cat retinal ganglion cell. Figure 6b illustrates an action potential of a cat retinal ganglion cell. It is a photograph of several superimposed sweeps of an oscilloscope showing both the action potential and the noise level of the recording apparatus. The recording was made with a tungsten microelectrode placed just outside the cell.

The receptive field was determined in three different ways to obtain Figures 6a, 6c, and 6d. The eye was focused on a small spot of light flashing against a constant intensity background. Only if a spot of light is on a particular part of the visual field will the rate of action potentials increase. In Figure 6a a plus sign (+) was drawn on a coordinate grid at the location where the light evoked an increase in the rate of action potentials; the minus signs represent locations where the light decreased the rate of action potentials. Following Kuffler (1953), the cell is classified as an "on center" receptive field with an antagonistic surround.

The receptive field of the same cell was determined in more detail (Figure 6c) with an automated technique which permitted the collection of data from 481 locations (Scobey, 1968). The dense area in Figure 6c was generated from overlapping crosses whose size is proportional to the number of action potentials counted when the light shone at corresponding points on the retina. The sensitivity of the receptive field across a diameter of Figure 6c is more obvious in 6d. The receptive field was maximally sensitive to flashing light in the center.

light is on a particular part of the visual field will the frequency of action potentials change. In Figure 5-6a, a plus sign was drawn at the location where the spot of light evoked an increase in the frequency of action potentials; the minus sign represents a location where the spot of light decreased the frequency of action potentials. This cell is an "on center" receptive field with an antagonistic surround (originally described by Kuffler, 1953).

Figure 5-6b illustrates the action potential of this cell. It is a photograph of several superimposed sweeps of an oscilloscope showing both the action potential and the noise level of the recording apparatus. The recording was made with a tungsten microelectrode positioned outside of the ganglion cell.

The receptive field of the same cell was determined in more detail (Figure 5-6c) with an automated technique which permitted the collection of data from 481 locations (Scobey, 1968). The dense area in 5-6c is generated from overlapping crosses whose size is proportional to the number of action potentials counted when the spot of light was at a corresponding point in space. The sensitivity of the receptive field determined across a diameter of Figure 5-6c is more obvious in 5-6d which shows how the receptive field was maximally sensitive to flashing light in its center. In the cat and the primate, all sensory neurons leading to the visual and cutaneous cortical areas have receptive fields which are roughly circular. Abstraction of other stimulus patterns occur at the cortical level and will be discussed later.

Spatial Summation and Field Sensitivity. From the threshold sensitivity profile in Figure 5-6d, it can be seen that a receptive field profile is dome shaped. In other words, the receptive field is most sensitive to stimulation in a small region about the center; responsiveness falls off as the stimulus moves out of the central area. Within the central zone spatial integration of stimulation can occur. For example, if a tiny spot of light centered on the receptive field is made somewhat larger, the rate of firing in the associated neuron *increases.* This relation, called *spatial summation,* holds over a limited area within the central region of a receptive field and is characteristic of all sensory systems.

Having discussed some of the properties of sensory systems in general, we will now examine in detail the relation between the activity of tactile receptors and the sensations of touch and pressure.

The Origins of Sensation from Our Largest Sensory Surface

Touch the back of your hand with a pencil point. What do you feel? There is a sensation of initial contact, then one of steady pressure, and each carries a sense of *intensity.* Also, there is a clear sensation of the location of the pencil upon the skin, the *spatial aspect* of the stimulus. Move the pencil slightly and touch again. The detection of small changes in location is possible, and also the small size of the pencil point stands apart from the sensation of total width of the indentation. The sensory inhibition and movement-sensitivity mechanisms accounting for these sensations have been discussed. Now let us turn our attention to the peripheral correlates of tactile *quality* and *intensity*.

Mapping Qualities in the Sense of Touch

By meticulously applying a sharp pencil point lightly to the back of the hand you will notice some spots where you can feel cold sensations very clearly, others where warmth is noticeable, places where touch is especially distinct (above the hair follicles), and places where nothing in particular is felt. All of these sensations seem to originate in the skin. Years ago psychologists mapped and described these sensory spots very carefully. Now it is possible to explain many of those early findings neurophysiologically.

The hairy skin has many receptors sensitive to different forms of physical energy as opposed to the retina which encodes primarily the presence or absence of light. The hairy skin has receptors sensitive to mechanical deformation of the skin, touch receptors, temperature sensors—one set of receptors underlies sensations of cold, another warm sensations—and pain receptors. Although pain is perhaps the least understood, it appears that some receptor types can signal pain which is the result of tissue damage whether from thermal, chemical, or mechanical energy.

Light mechanical deformation of the skin excites one set of nerve cells while steady deformation of the skin stimulates other receptors. Touch your hand repeatedly and notice the onset and offset of the touch, a transient feeling that does not endure along with the sensation of pressure.

Lightly move the point of your pencil across a textured surface. You feel the desk and also the vibration deep in your hand. Put your hand on your throat and hum a note. You can feel vibration inside your hand again, yet very little distortion of the *skin* is occurring; but another mechanoreceptor, the Pacinian corpuscle deeper in the hand and wrist, is responding. Let us now consider a more formal set of experiments which will help to explain the relation of neural elements to touch.

An Experiment on Sensations from the "Touch" Receptors

The object of this experiment is to record electrical events from single monkey nerve fibers and to compare the data with human sensations. Surgical dissection of an anesthetized monkey's arm exposes the nerve, a fairly small white strand somewhat larger than a pencil lead. Using a microscope, the tough sheath encasing the nerve fibers is removed and a small bundle of fibers is carefully pulled off with microscopically sharpened and polished forceps. The bundle is placed over an electrode connected to an amplifier which amplifies the microvolt-sized action potentials from the fibers so that there is sufficient power to drive visual displays, computation equipment, and a loudspeaker. The loudspeaker permits us to hear the response from the nerve bundle. Before employing highly precise mechanical stimulation, the skin is stimulated by patting, rubbing, and tapping the monkey's forearm and hand to find out which types of mechanoreceptors are in the bundle. At this time we also find which areas of skin produce neural responses, the receptive fields of the neurons. The bundle of fibers on the electrode produces a burst of sounds from our loud speaker as the action potentials pass the electrode in their all-or-none fashion when we stimulate the skin. Trying to locate one receptor sensitive to a particular aspect of touch is a demanding and difficult job because each bundle of nerve fibers carries action potentials from many different receptors located in many different sites on and below the skin. However, finally a single fiber is isolated which fires when one spot on the skin of the monkey is stimulated in a particular way.

Repeated experiments with single fibers show that some respond throughout steady indentations of the skin, others only to the onset and offset of indentation; these two well-defined classes are labeled "slowly adapting" and "rapidly adapting." As our experiment proceeds, we continue to select small bundles of fibers and try to isolate the responses of single fibers electrically, mechanically, and by careful dissection. By investigating which aspect of touch a fiber responds to, we can eventually formulate a complete picture about the different populations of fibers that convey tactile information from the skin. Then by studying the human sensations elicited by the same stimuli, the sensory function that each class of fibers serves can be inferred. This procedure has been used with substantial success by Mountcastle and his colleagues at the Johns Hopkins School of Medicine.

Pressure Receptors

Pressure receptors are sensitive to *steady* pressure. Sustained indentation in the receptive field produces an enduring response which only slowly fades or "adapts." The frequency of discharges depends on how hard the skin is pressed. Pressing hard also causes the fiber to respond rapidly whereas pressing with less emphasis causes it to respond more slowly. Stretching the skin also excites this receptor.

Photographs of the neural response of a typical slowly adapting fiber, taken during an experiment by Harrington and Merzenich (1970) which involved stimulating the hairy skin of a monkey, are shown in Figure 5-7. Beneath each burst of impulses from the nerve fiber, shown by oscilloscope traces on the left, the position of the stimulating probe appears. Notice that the amplitude of the impulses remains fairly constant in all-or-none fashion. When the probe indents the skin 32 microns, the fiber responds with only one action potential. A 64-micron indentation (about the thickness of a heavy piece of paper) produces a large number of action potentials. The frequency of action potentials increases as the amount of skin indentation increases (Figure 5-7). This information provides you eventually with the sensation of how much your skin is being indented. A graph showing the number of action potentials for each indentation is shown to the right in Figure 5-7, and its form is very similar to a graph showing the *intensity* of human sensation in response to these same stimuli. In other words, how strong a tactile sensation feels is a function of how rapidly a tactile receptor fires. Once more indent your skin by placing a pencil on your arm. Try to appreciate that the sensation of pressure you feel is merely a result of trains of action potentials cascading up nerve fibers to your brain.

The "Contact" or "Rapidly Adapting" Mechanoreceptors

Some fibers respond just to the initial indentation of the skin or perhaps only to contact of the stimulus probe with the skin. These respond very briefly, and may also fire a burst of impulses when we remove the stimulus. The "rapidly adapting" receptor ceases to respond after the movement ceases, even if steady pressure is still being applied; therefore, these receptors signal the central nervous system only that a mechanical *change* occurred.

We will find several varieties of these rapidly adapting units as we sample fibers with our electrode. One fiber isolated in dissection might discharge only when we move any of three or four hairs or contact the skin over a follicle. The absolute amount of movement of the hair or the skin over the follicle would not be important to this unit—only change. Generally, more than one hair follicle is connected to each of these hair-associated afferent fibers. Anatomical study of the hair follicle demonstrates that there are sensory fibers innervating the bases of the hairs that are responsible for encoding this information.

Since the hair follicle receptor is only effective in signaling a change in the position of the hair or skin, an effective way to stimulate rapidly adapting skin receptors repeatedly is to indent the skin and stimulate with vibration. This produces a continuous and easy-to-measure response. Smooth sinusoidal vibration is usually used, and the stimulus then conveniently has only two parameters, frequency and amplitude.

Figure 5-8 illustrates the response of a fiber subserving a rapidly adapting hair-associated receptor in the skin as the skin was vibrated with a small rounded probe (from Merzenich and Harrington, 1969). The top record in Figure 5-8 is a photograph of an oscilloscope showing the action potentials travelling up the fiber past an electrode. The second record illustrates the varying position of the probe as it vibrates the skin. At the beginning of the record the probe moves about a millimeter to indent the skin. At this time there is a burst of impulses on the fiber.

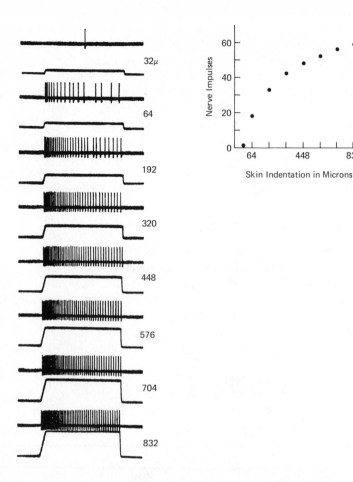

From "Neural coding in the sense of touch: Human sensations of skin indentation compared with the responses of slowly adapting mechanoreceptive afferents innervating the hairy skin of monkeys" by T. Harrington and M. M. Merzenich, in *Experimental Brain Research* **10**, 1970.

Figure 5-7 The neural response in an afferent fiber to skin indentation. On the left is a series of photographs from an oscilloscope which illustrates the action potentials on a fiber serving the hairy skin of a monkey. The action potentials passing the electrode are shown immediately above a record of the depth of probe indentation. The response to 8 different values of indentation lasting $\frac{1}{2}$ second are shown. When the skin was indented 32 microns (the first pair of records) only one action potential passed by the recording electrode. With greater indentations there are more action potentials. That is, indentation intensity is encoded in the nervous system as the frequency of action potentials. On the right of the figure the number of action potentials that were evoked by the different skin indentations is presented as a graph. The curve could be described mathematically as a power function and can be shown to fit the sensation function of man. This record also illustrates a common property of many first order afferent fibers: without stimulation, the spontaneous activity is low.

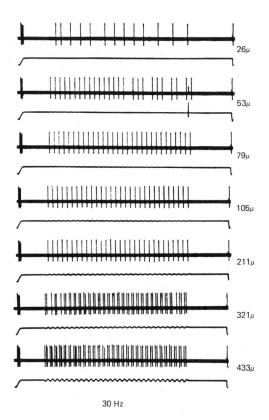

30 Hz

From "The sense of flutter vibration evoked by stimulation of the hairy skin of primates: Comparison of human sensory capacity with the responses of mechanoreceptive afferents innervating the hairy skin of monkeys" by M. M. Merzenich and T. Harrington, in *Experimental Brain Research* **9**, 1969.

Figure 5-8 The response of a hair receptor's afferent fiber to a vibratory stimulus on the skin. The upper line of each pair of lines shows action potentials passing a recording electrode along a single axon. The lower line is a record of the position of a probe which indents the skin. First, the probe moves about a millimeter to indent the skin—just above, on the upper line, a burst of action potentials is seen; then the fiber is quiet. This is called a "rapidly adapting" response. The length of the record covers one second of time. After $\frac{1}{5}$ second, the probe begins to vibrate at 30 Hz. The amplitude of vibration is given at the right of each pair of records. For the first pair of lines, the peak-to-peak amplitude of vibration was 26 microns. This was above the vibratory threshold and the fiber propagated a train of 13 action potentials. When the amplitude of the vibration was between 53 and 211 microns, the response of the fiber was entrained to the frequency of the stimulus. That is, the fiber responded with one action potential per stimulus cycle. At higher amplitudes the response was a function of the probe frequency but two action potentials resulted from each cycle of the probe. Rapidly adapting axons can give a maintained discharge to vibratory stimuli. These axons signal the contact, vibration, or movement of stimuli.

The records represent one second. After one-fifth of a second, the probe is made to vibrate at 30 Hz. Evidently 30 Hz was a very effective stimulus for the receptor. When the amplitude of the vibration was only 26 microns, about one-thousandth of an inch, a few action potentials were generated; the threshold for vibratory stimulation is therefore even less than this. As the amplitude of the vibration increases (see successive pairs of records), the intermittent response per cycle of vibration changes to perfect entrainment at 79 microns; that is, one action potential for each push of the probe results. This entrainment remains well defined until the amplitude of vibration reaches 321 microns. At this amplitude the receptor doubles its firing rate. "Tuning plateau" is the phrase used to describe the range over which the cell's response is well entrained with stimulus amplitude. The amplitude of vibration where this "plateau" begins will be the basis of our comparison between different receptor types in the next figure and is called the "tuning point" (see Mountcastle, 1968).

The Mechanoreceptors Subserving Touch and Vibration

In Figure 5-9 (from Merzenich and Harrington, 1969) groups of tuning points collected from some 500 mechanoreceptors, of the three prominent types in monkey hairy skin, illustrate how each type's "threshold" changes as vibratory frequency is changed. The hair-associated, quickly adapting fibers just mentioned are shown in the central section of Figure 5-9. Each dot in the figure represents the tuning point of one single fiber at a particular frequency of vibration. Tuning points could be determined over a range of about 5 Hz to 200 Hz for some hair-receptor axons, but the range of 20 to 50 Hz is optimal for entrainment. In the figure, tuning points from two other classes of fibers are shown, the delta QA fibers and the Pacinian fibers.

The solid line in each graph is the human vibratory threshold curve obtained by asking people whether they could or could not feel these same vibration stimuli. The hair receptors of the monkey have a threshold which is about the absolute threshold of the human in the low frequency range, but not in the higher frequency range. However, the Pacinian fibers respond to a threshold vibration close to the human threshold in the high frequency of vibration but not at the low frequencies (compare Figures 5-9b and 5-9c).

From Figures 5-9b and 5-9c, it is obvious that no single fiber population detects vibration in a way that could account for human threshold; thus there is not a single population of receptors accounting for touch. Mountcastle *et al.* (1967) demonstrated that in the glabrous finger-printed skin, high frequency vibration was subserved by the exquisitely sensitive Pacinian corpuscle deep underneath. The tuning points in Figures 5-9b and 5-9c, matched with the human threshold curves, show that this is also the case in hairy skin.

Oddly, the fibers termed "Delta QA" in Figure 5-9a are *too* sensitive at low frequencies to account for human sensory ability because, as the

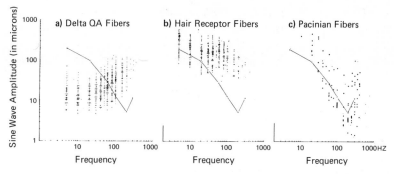

From "The sense of flutter vibration evoked by stimulation of the hairy skin of primates: Comparison of human sensory capacity with the responses of mechanoreceptive afferents innervating the hairy skin of monkeys" by M. M. Merzenich and T. Harrington, in *Experimental Brain Research* **9**, 1969.

Figure 5-9 Comparison of electrophysiological data from monkey with the human threshold. Figures 19a, b, and c illustrate the thresholds of three classes of fibers found in a cutaneous nerve serving the hairy skin of a monkey. Each dot in three scatter diagrams represents one measurement. The amplitude of the vibrating probe was plotted as a function of frequency for equivalent responses on the axon. That is, the dot in the scatter diagram is the minimum amplitude that will give one action potential per stimulus cycle. This amplitude of vibration was called the "tuning point." The clusters of "turning points" of axons subserving the hair follicle receptors (Figure 9b) and the Pacinian corpuscle (9c) are clustered about the solid line, which is the absolute sensitivity of the human skin to vibration. The hair receptors and Pacinian corpuscles account for the sensitivity of the human hairy skin to vibratory sensibility.

The hair receptors have sufficient sensitivity to account for the low frequency end of the vibratory scale while the Pacinian corpuscles have sufficient sensitivity to account for the high frequency end. The receptor system encoding vibratory sensibility in the human hairy skin has two receptor types. Figure 9a illustrates data from another class of receptors found in the hairy skin of monkey. These receptors responded to a lower amplitude of vibration than the human's subjective report of threshold. The function of these receptors is unknown and discussed in the text.

figure shows, they exceed human performance. Their function may be other than sensory.

Merzenich and Harrington (1969) performed an additional set of experiments which further demonstrates the dual neural nature of touch on hairy skin. They selectively stimulated either the skin receptors or the Pacinian corpuscles in humans and measured vibratory threshold in each condition. To achieve this, first they stimulated skin over large muscles. The muscles minimized the amount of vibration reaching the Pacinian corpuscles which were largely deep underneath, and thus skin receptors were the major contributors to sensation. Low frequency vibratory sensitivity was still good, but high-frequency sensitivity suffered. Then they stimulated only the Pacinian corpuscles but not the skin receptors by anesthetizing the skin receptors with cocaine. Now low frequency vibration sensitivity virtually disappeared but sensitivity to high frequencies was unchanged.

How Does the Brain Know the Intensity of Vibration?

In Figure 5-8, increasing vibratory amplitude smoothly does not cause a correspondingly smooth increase in the firing frequency of quickly adapting receptors as increasing pressure did in slowly adapting fibers. The receptors seem to signal only the frequency of the vibrating stimulus or some multiple of it with their own frequency. It is apparent that a single fiber probably doesn't tell the brain about intensities of either high frequency or low frequency vibration. Most likely a whole population of such receptors is required, signaling perhaps that more fibers at a distance from the stimulus or fibers of higher threshold are being brought into play as intensity is increased.

Neural Space

Having examined in some detail the behavior of tactile receptors, we will now follow receptor signals (primarily from photoreceptors) through the nervous system and study the effect that the neural structures and processes have in shaping our perceptions. Physical energy, once transduced into a neural signal, travels through *neural space*. The concept of neural space permits a distinction between the physical world and an organism's encoded representation of that world. Neural space embraces the neurons, the anatomical interconnections between neurons, and the signals passing over those interconnections in a nervous system. The connections between neurons at synapses and the type of interaction, excitatory or inhibitory, are very significant components of the concept because they enormously affect what happens to an initial receptor message and thus ultimately what an animal perceives.

Higher-Order Receptive Fields

Let us develop the concept of neural space by explaining the nature of receptive fields for higher-order neurons. If a second-order neuron had only a single connection from a single receptor neuron, it would have the same spatial relation to the sensory surface as the receptor cell (see Figure 5-5). In this artificial example, the second-order neuron carries all the available information from the sensory surface. The spatial relation of the two receptive fields of the first-order and second-order neurons would precisely overlap one another. Actually, such is not the case. Except in limited instances, every first-order neuron sends signals to many second-order neurons and each second-order neuron receives signals from many receptors. This would indicate that fields for second-order neurons might be much less well defined than receptive fields of first-order units. Neural recording, however, shows that this is not true. A certain quality of neural space prevents this kind of blurring of sensory signals.

Sensory Inhibition in Neural Space

All sensory systems are affected by the phenomenon of *sensory inhibition* which works to contain or funnel the information along certain neural pathways. Inhibition is caused by the release of a transmitter substance by one neuron at the synapse which affects the neuron to which it is connected and makes it less probable that the neighboring neuron will fire. In the process it may actually sharpen the perception of differences in stimulation between adjacent areas. Sensory inhibition can be understood by following the pathways that the neural signals traverse. Figure 5-10 shows tentative schematic interconnections of the primate retina in the left-hand section. The nerve from the eye connects to the lateral geniculate nucleus (LGN) of the thalamus, deep in the brain. The connections at the cortex are less well understood and not illustrated. The receptors sensitive to light are labeled as R in the figure. These receptors send messages to two cells, a horizontal (H) and a bipolar (B). Likewise, the bipolar cell connects to two cells, the amacrine (A) and the ganglion cell (G). The neural signals encoding light are slow nonpropagated permeability changes in the majority of the cells in the retina; i.e., R, H, B, and A. The ganglion cell G propagates an action potential along the optic nerve to the LGN. The primary cell of the LGN relays action potentials to the cortex. The secondary cell receives feedback from the axons projecting to the cortex. The LGN can be modulated by central influences, the cortex, and the reticular activating system, RAS. Consider a spot of light centered on retinal receptors 2 and 3 in Figure 5-10. The information that light is present is transferred through bipolar cells B_2 and B_3 to the ganglion cell G. The axon from the ganglion cell is functionally connected to all the receptors but will signal the presence of light only at receptors 2 and 3. The ganglion cell does not signal the presence of light on location 1 or 4 because these receptors, through their associated cells, do not excite B_2 or B_1 or the ganglion cell. The horizontal cell H, has an inhibitory (antagonistic) effect on B cells. Receptors 1–4 all stimulate the H cell, thus inhibiting the B cells. But receptors 2–3 also excite the B_2 and B_3 cells and therefore a restricted spot of light on R_2 and R_3 is more effective, possibly causing the ganglion cell to fire. Similarly, at all synaptic levels, elements side by side inhibit each other's influence and thereby limit neural transmission to certain restricted stimuli. Sensory inhibition serves to limit the kind and extent of stimulation of the sensory surface which will be encoded within a single afferent channel. It is the central spatial sharpening mechanism of all sensory systems which helps reject "energy spillover."

Temporal Inhibition

A special synaptic relationship, called *reciprocal synapsing*, is illustrated in Figure 5-10 where the bipolar cell (B) synapses with the amacrine cell (A), and then the amacrine impinges back upon the bipolar cell. This

Retina

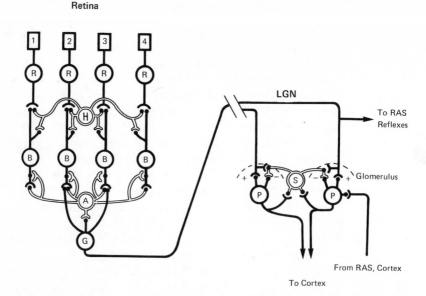

Figure 5-10 Schematic interconnections in the visual system. Schematic interconnections of the retina are shown on the left. The nerve from the eye connects to the lateral geniculate nucleus (LGN) of the thalamus, deep in the brain. Tentative schematic connections for the primate are illustrated. The connections at the cortex are less well understood and not illustrated. The receptors sensitive to light are labeled *R* in the figure. These receptors connect to two cells, the horizontal (H) and the bipolar (B). Likewise, the bipolar cell connects to two cells, the amacrine (A) and the ganglion cell (G). The neural signals encoding light are slow nonpropagated permeability changes in the majority of the cells in the retina, i.e., R, H, B, and A. The ganglion cell G propagates an action potential along the optic nerve to the LGN. The primary cell of the LGN (P) relays action potentials to the cortex. The secondary cell (S) receives feedback from the axons projecting to the cortex. The LGN can be modulated by neural influences from the cortex and the reticular activating system (RAS).

feedback pathway back to the bipolar is very short and probably inhibitory, blocking the bipolar cell immediately after it has excited the amacrine and ganglion cell. In addition, the amacrine cell spreads its inhibitory effects and may quickly decrease the effectiveness of the other converging bipolar cells. Similar reciprocal synapses are seen in the olfactory pathways and have been shown to be inhibitory. Reciprocal synapses could be the anatomical basis for the short duration of the transient type ganglion cell response when an image is placed or displaced on the retina.

Spatio-temporal inhibition may perhaps explain certain phenomena of mechanical sensation. In Figure 5-11 the intensity of sensation in response to an indentation of the skin is schematically represented. The location of maximum sensation is found directly over the location of the

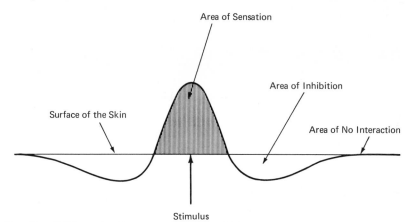

Area of Sensation

Area of Inhibition

Surface of the Skin

Area of No Interaction

Stimulus

From "Neural inhibitory units of the eye and skin" by G. von Békésy, in *Journal of the Optical Society of America* 50, No. 11, 1960.

Figure 5-11 Sensory inhibition. The application of the stimulus on the sensory surface produces an area of sensation and a ring of inhibition, called an *antagonistic surround.* A stimulus applied in the area of inhibition produces less of a sensation than if it were presented alone. The area of inhibition occurs because of lateral neural interconnections that inhibit the neurons from nearby areas on the sensory surface. At increasing distances from the point of stimulus application, the inhibition diminishes and the sensory surface resumes its normal sensitivity.

stimulus. A considerable amount of "energy spillover" occurs at the skin surface due to the stretching of the skin and waves or ripples of skin motion that spread out from the stimulus location. However, the lateral connections in the nervous system, similar to those illustrated in Figure 5-10, produce an area of inhibition called the *antagonistic surround* about the point of stimulus application where sensitivity is decreased, effectively "funneling" the signal. The perceptual mechanism is thereby automatically given more precise information about the location of the stimulus.

The subject of sensory inhibition is treated in detail by the Nobel laureate Georg von Békésy, in a 1967 monograph. Von Békésy explored the psychological effects of time delays between two stimuli presented in different modalities, and found that sensory inhibition is common to all sensory systems. The reader can easily convince himself that inhibition is relevant to his own experiences: touch the middle finger to the index finger and find that two contact sensations are felt and localized on the contacting surfaces. The neural impulses from these areas arrive at the cortex at the same time because of equal conduction distances. Inhibition is produced by an interneuron and takes a millisecond to develop fully. Now, if the neural signal of one area precedes the other, the prediction is that the first signal would inhibit the following signal, reducing its magnitude en route to the cortex. Touch your index finger to your cheek or lips. The conduction distance from the head to the central nervous system is shorter, and the signal from the lips precedes the neural signal from the fingers. Repeated light touches to the lip or cheek are bright and sharp, and localized on the lip or cheek, not on the index finger. The situation can be reversed by repeating the procedure on body points more distant from the central nervous system than the fingers, such as the leg. In this case the sensation is clearest on the finger. This example emphasizes that one role of sensory inhibition is to accent the spatial location of the stimulus.

Another common experience illustrating sensory inhibition occurs when an elbow is placed on the arm rest or the window while driving a car. The vibration of the car is clearly felt only at the elbow even though your passenger can verify that your entire body vibrates. The vibration passes up the arm relatively slowly as a traveling wave, sequentially stimulating the entire tissue of the body. Sensory inhibition from the first surface stimulated inhibits the other locations and probably accounts for this sensory phenomenon.

The Thalamus: A Neural Space System Using Inhibition

In this discussion of neural space we have followed the signal into the central nervous system. For the sense of vision, the retina itself must be considered part of the central nervous system because in the development of the body the neurons of the retina originated as an outpocketing of the brain. Other sensory systems have similar relationships with the sensory surface but their messages travel to the spinal cord and brainstem rather than to a relatively peripheral part of the central nervous system, like the retina. From the brainstem, most afferent neurons project to the *thalamus,* which is a large structure with specific divisions for visual, auditory, and cutaneous systems. Since it is a major relay station for sensory information in the brain, we will consider it in terms of the properties of neural space discussed thus far.

Thalamic Sensory Inhibition

One function of the thalamus is sensory inhibition, to further sharpen the temporal and spatial properties of signals from the sensory surface as they ascend to the cortex. Although the actual mechanism is still controversial, the anatomical wiring diagram shown in Figure 5-10 is perhaps most likely. As the figure shows, there are two cell-types in the thalamus, the primary cell (P) and the secondary cell (S). Let's follow the neural events which occur when a ganglion cell in the retina fires relaying action potentials to the thalamus. These messages reach the P cells located in highly organized regions of the lateral geniculate nucleus (LGN), itself a structure within the thalamus. Inputs from the G cells excite the *P* cells. The *G* cell also synapses on the *S* cell which normally inhibits the *P* cell. The *G-S* synapse is from one axon to another (instead of from one axon to the cell body of the neighboring neuron) and is therefore called an *axo-axonal synapse*. Such synapses are almost certainly inhibitory in nature. In other words, the *G* cell input initially excites the *P* cell and blocks the *S* cell. This makes it probable that the *P* cell will fire sending its message to areas in the cerebral cortex. However, the *P* cell's axon forks such that at the same time that it is relaying information to the cortex it is stimulating the *S* cell. This cell as we have noted inhibits the *P* cell which sent the message in the first place (as well as other *P* cells with which the *S* cell synapses—see Figure 5-10). At first glance this appears to be a ridiculously complicated way to conduct neural business. Why has such a complicated design system featuring so many feedback and inhibitory channels evolved?

Because certain kinds of information are not likely to be biologically valuable. This system is designed to be maximally sensitive to *spatial changes* in light intensity. Changes in our visual field often signal movement. Movement may mean a predator is present or a prey moving or a mouth moving as someone talks or an eye brow lifting—all highly significant events to attend to. The thalamic system is generally designed to relay information about biologically relevant events—to alert the animal to changes in the environment because these changes may demand changes in behavior.

In still more complicated ways the intricate relationships among *G* and *P* and *S* cells promote the preferential transmission of information about edges and changes in light intensity in bands of light to the brain. The system is designed to heighten contrasts between objects and the background, making it possible to detect their presence in the environment. Thus the thalamus does not give equal weight to all information it receives from the ganglion cells but instead selects some and blocks others in ways which make biological sense.

The Thalamic Volume Control

The second function of the thalamus is to regulate the signals bombarding the cortex. During sleep, input from the *reticular activating system* on the brainstem effectively decreases the excitability of the thalamus and thus reduces the efficiency of neural transmission in the LGN between the ganglion cell and the cortex. For example, during sleep the mean level of spontaneous discharges of geniculate cells decreases and the sensitivity of LGN cells to flashing light falls by a factor of ten. Likewise, the cortical response to vibration of the hand decreases during a change in the attention of a monkey. Therefore, the brain itself has a method of regulating the information from the sensory surface. This efferent control of input, which is a major feature of sensory systems, probably allows selective attention to different sensations.

In addition to regulating sensory input at the thalamic level, efferent control is also important at more peripheral levels. For example, in the auditory system, the tensor tympani muscle can be contracted to reduce the energy to the inner ear by modification of the sound transmission through the middle ear bones. In addition, the hair cell receptors themselves receive synaptic contacts from neurons on the CNS. Figure 5-1 illustrates two synapses at the hair cells of the organ of Corti. One synapse transmits from the receptor cells to the afferent neuron. The other synapse is the axon from the CNS terminating on the receptor, presumably to modify the receptor function. All these controls on sensory input are important because the brain can only cope with so much information at any one time. Moreover, making a decision on a particular issue may occur more efficiently if the brain can simply block extraneous inputs for some time. Cortical control of this sort helps insure that the brain will be capable of responding most effectively to the most important stimuli present in our environment.

Representation of Movement in Neural Space

An effective stimulus, in any sensory system, evokes a response with two characteristic components. At the onset of the stimulus there is an initial transient burst of action potentials which may then either quickly stop or decrease to a steady rate that is maintained for the duration of the stimulus. As a result, physiologists have classified neurons into rapidly and slowly adapting types. The distinction is determined mainly by the absence or presence of the detectable steady-state rate of firing. The rapidly adapting neural units subserving touch, the hairs and Pacinian corpuscles for example, have no maintained discharge, and respond only at the onset and offset of stimulation. Consequently these receptors cannot signal steady pressure on the skin. Even at this peripheral level, the nervous system has abstracted onset and offset of the stimulus as having primary importance. An example often used to illustrate this fact is the adaptation to clothing. There is a clear sensation when the clothes are put on but this disappears and normally we are unaware of the garments after they are in place. The presence of the clothes is no longer detected and this can be correlated with the absence of action potentials on the nerve fibers.

Pain is a sensation which is without adaptation. For the nervous system to achieve this, action potentials must be present for the duration of the stimulus. The pain threshold, as determined by slowly raising the temperature of the skin, occurs at 45 degrees Celsius for man, and remains constant for hours. An initial transient phase for pain can be demonstrated by sudden immersion of the arm into water that has a temperature slightly below the threshold for steady pain. Entering a hot bath we have all experienced an initial flash of bright pain which rapidly disappears, demonstrating the early transient component.

From the previous discussion we can appreciate that a stimulus which moves on the sensory surface sequentially stimulates receptors, resulting in transient bursts of signals in each, and is easily felt, heard, or seen, depending upon the sensory surface in question. The initial burst of action potentials begins the process of lateral inhibition. The first transient burst of action potentials inhibits the parallel channels representing neighboring parts of the sensory surface. Consequently, the transient part of the neural signal is most useful in spatial localization. To illustrate the effectiveness of the early transient, let us consider some neurophysiological data from the visual system of the monkey.

Figure 5-12 illustrates two actual visual receptive fields and the responses of lateral geniculate neurons to flashes of light over time (Scobey, 1968). In Figure 5-12a, the receptive field maps (A and E) were obtained by exploring 481 locations on the retina and plotting a cross whose size is proportional to the number of action potentials evoked by a 200 millisecond flash. This neuron of the thalamus responded with a short burst of action potentials when the light was turned on. The duration of the light is shown by the dark bar below Figures 5-12 b–d and f–h. The ordinates of these figures are the rates of neuronal firing with each large tic mark representing 100 action potentials per second. Notice that the response to a 25 millisecond flash (Figure 5-12d) is almost identical to the response to longer flashes of 200 (Figure 5-12c) and 400 milliseconds (Figure 5-12b). Therefore, this response is rapidly adapting and is characteristic of the responses of axons subserving the peripheral visual fields of the monkey visual system. Neurons subserving the fovea are slowly adapting, firing for the duration of the flash.

Figures 5-12e–f illustrate the response of another class of neurons in the visual system of all vertebrates. About half the neurons of the visual system respond when a spot of light is turned on in the center of their receptive fields. These neurons are said to have "on-center" receptive fields. The other half of the neurons respond when the spot of light turns off, and have "off-center" receptive fields. In general, the organization of the receptive fields is similar, roughly circular in shape, with an antagonistic surround. Like the "on-center" neurons, the "off-center" neurons respond only briefly to their effective stimulus—a decrease in illumination. Both of these neuron types project to the cortex, signaling that a change has occurred at the retinal surface.

The physiological events just described have their perceptual counterparts in humans. We have seen that LGN neurons may not fire in response to prolonged stimulation from the retina. If the reader stares fixedly at a part of a period on this page he may experience "tunnel

Duration of Response

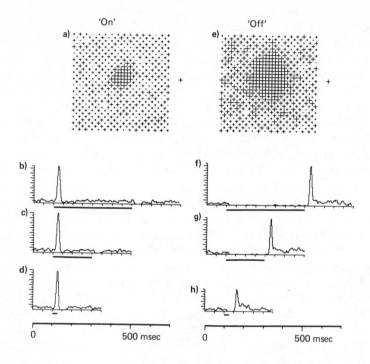

From a dissertation submitted to the Johns Hopkins University by R. P. Scobey, 1968.

Figure 5-12 Responses of the primate visual system. Two neurons, representative of the two rapidly adapting types found in the lateral geniculate nucleus of the monkey thalamus, were used to obtain the data illustrated in this figure. Figures 12 a–d illustrate the receptive field and responses of an "on center" neuron. Figures 12 e–h illustrate the receptive field and responses of an "off center" neuron. The receptive field maps (a and e) were obtained by exploring 481 locations on the retina and plotting a cross whose size is proportional to the number of action potentials evoked by a 200 millisecond flash. A calibration cross representing 10 action potentials is to the right of each figure. The "on center" receptive field was derived from a count of the number of action potentials during the presence of light while the "off center" receptive field was derived from a count of action potentials for 200 milliseconds after the flash.

Below each receptive field map is a histogram of action potentials giving the number of action potentials as a function of time. Figures 12 b–d illustrate a rapidly adapting response to the onset of the small spot of light that is independent of the duration of the stimulus used. The black bar underneath each response illustrates the duration (25, 200, 400 milliseconds) of the light flash. The "on center" neuron responded only to the onset of the stimulus while the "off center" neuron responded only to the stimulus turning off. Each large tic on the ordinate represents 100 action potentials per second.

vision" with objects at the periphery of the visual field fading or disappearing. By reducing eye movements to a minimum, images on the peripheral part of the retina move very little. This results in more or less constant stimulation of ganglion cells and translates perceptually into loss of vision.

If an image is held *absolutely* motionless on the retina, perception of it quickly disappears as we have discussed. This is fortunate. For example, blood vessels on the retina move with the eye and thus are not normally seen. However, if a light is shined into the eye from one side and rapidly moved back and forth to keep the shadows of blood vessels in motion on the retina, then they can be seen. Another way of demonstrating the loss of perception of images held constant on the retina is to place on an eye a contact lens on which an object has been drawn. Since everywhere the eye moves the object moves as well, the image of the object will disappear within a few seconds. If the human eye failed to move, we would be restricted to seeing things only briefly after head movements and to being blind to all but moving objects the rest of the time. Elsewhere in the animal kingdom this happens.

For example, a frog must have a totally different appreciation of the physical world than man. The frog's eye is fixed in the head, rendering the world about it immobile and featureless. Only when something in the immediate surroundings moves or changes intensity will its visual sensory system respond with neural signals. A caged frog surrounded by dead flies dies of starvation. Other animals, including cats and primates, have eyes which are always in motion. Even when humans try to focus absolutely steadily on one point, the eye drifts with rapid involuntary flicks in position and also has a high-frequency tremor. The movement of the eye itself prevents our visual system from adapting.

The human perception of movement is a basic visual function, appearing a few weeks after birth. Movement detection is also the last visual function to disappear and the first to reappear in the cure of pathological visual disturbances. Movement in the peripheral vision of man, as in other animals, attracts attention and results in movements of the body, head, and eye to place the image on the fovea to obtain best resolution. Many prey animals freeze and remain immobile when threatened by a predator, an adaptive response which reduces the possibility of detection by predators' visual systems.

Because receptive fields of sensory neurons are generally quite large, it is not obvious how animals can detect relatively small displacements of images. However, humans can align the two segments of a broken line to an accuracy of less than the equivalent width of a *single cone* in the retina and can detect movements just as small. Although peripheral visual fields do not resolve spatial details, this part of the sensory surface with large receptive fields is especially sensitive to small movements. In the following paragraph this apparent paradox is explained.

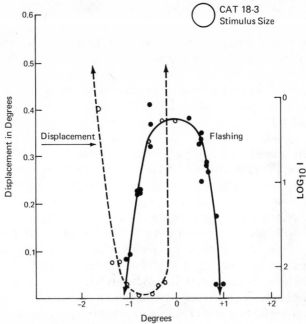

From unpublished data by R. P. Scobey and J. M. Horowitz, 1970.

Figure 5-13 The sensitivity of retinal ganglion cell of the cat to displacement of a constant intensity image. The filled circles and curve illustrate the sensitivity of a retinal ganglion cell of the cat to a small stationary flashing light. The ordinate of this curve is shown at the right. The receptive field was about two degrees of arc in diameter. The dotted line with the open circles, superimposed on the threshold sensitivity curve, represents the displacement threshold curve. A constant intensity spot of light was displaced quickly from left to right and adjusted for the minimum distance which would just excite the cell. This minimum distance was plotted as the ordinate on the left-hand side of the figure. The cell was most sensitive to movement along the border of the receptive field center. The cell could encode a movement as small as 0.01 degrees of arc even though the receptive field was two degrees of arc in diameter at the intensity of the moving spot of light. Thus, a retinal ganglion cell with a large receptive field can encode a small displacement of the image as the image moves through a region of changing sensitivity. The background intensity was 0.3 cd./m².

Figure 5-13 illustrates the sensitivity of a cell's receptive field to stationary flashing light and the threshold of that same cell to movements of a constant-intensity image (Scobey and Horowitz, 1972). The solid line in Figure 5-13 illustrates the threshold sensitivity along a diameter across the receptive field to a flashing spot of light. (See also Figures 5-5 and 5-6.) The dotted line in Figure 5-13 represents the distance that a constant-intensity image must quickly move in order to barely excite the retinal ganglion cell. This minimum distance is called the *displacement threshold* and is plotted as the ordinate with the scale on the left-hand side of the figure. Notice that the cell is not maximally sensitive to the displacement of the image in the center of the receptive field but is most sensitive in a relatively small region along the edge of the receptive field. This particular retinal ganglion cell responds to movements as small as 0.01 degrees of angular displacement on the retina even though the receptive field is two hundred times larger. The reason it can do so is a result of the difference in sensitivity to light intensity in the receptive field. For example, as a dot of light moves from the edge toward the center of the field, the cell's response may change enormously—from just barely firing to firing a great deal. Likewise, a bright dot moving from the center to the edge will cause the cell to fire quite differently. The brain can use this change in rate of firing within *one* receptive field to present us with the perception of movement. This movement sensitivity is not related to the size of the receptive field. A small movement in a large receptive field can be quite easily detected if the receptor cell and related neuron change their response.

The Cortex: An Integrating Center of the Neural Space

Different sensory systems originate from separate sensory surfaces which respond more or less exclusively to only one form of physical energy. These sensory systems have independent pathways to the cortex. For example, the visual system channels information from the retina to the lateral geniculate nucleus to the rear of the cortex where the primary visual sensory area is located.

About as many fibers leave the thalamus as enter it. However, an enormous number of cortical neurons process the information provided by each fiber coming from the thalamus. Thus each point on the sensory surface projects through the thalamus to many cells in the cortex. These cells abstract the different properties of the stimulus from afferent signals.

Each incoming fiber enters perpendicular to the surface of the brain and distributes fibers to a group of cells. From these cells activity spreads up and down—lateral excitation of the cortex is limited again by inhibition and so the cortex may be functionally arranged in columns of cells. This organization is not seen on gross inspection of the anatomy of the brain, but has been found by microelectrode recordings. These have demonstrated that a cell column is not precisely circular but has an irregular cross section of approximately half a millimeter and contains hundreds of cells. The column, measured from the surface of the cortex, varies in length depending on the thickness of the cortical mantle, and is roughly a few millimeters deep.

One reason for the explosive proliferation of cells at the cortical mantle is that the cells at cortical levels become relatively specific, responding to specific spatial attributes of stimuli at the sensory surface. Movement on the sensory surface is one such attribute. Some cells in the cortex will only respond to movement in a given direction on the sensory surface. Similar neural units have been found in the visual and somato-sensory systems. In cat, monkey, and presumably man, the process of complex signal analysis occurs in the cortical mantle.

Ontogenetic Development of Sensory Systems

Perception is not strictly determined by the anatomical and physiological aspects presented thus far. To a degree, a useful sensory system must be flexible. As an organism develops, its sensory equipment must allow it to sustain a coherent real world in the face of disruptive influences. Growth is a problem because bones change length, and important distances such as those between the eyes or ears are altered; and, as we grow, our eyes and ears become much farther above the ground. Correlations between visual and auditory direction must be consistently reorganized during growth.

Sensory isolation during animal development will modify certain perceptual aspects to the point where the sensory system appears to be inadequate or malfunctioning. Even the sense of pain does not develop normally without continuing stimulation throughout development. Thompson and Melzack (1956) raised dogs in isolation, depriving them of as much sensory stimulation as possible. Thereafter the dogs did not have the normal appreciation of pain. On presentation of a match flame, the dogs would sniff it without the normal reflexive withdrawal of their noses at the pain of burning. Riesen (1950) and others have kept animals in various states of sensory deprivation during different phases of their development and have found that normal interaction with the environment is essential if perceptual facility is to develop normally. If animals are deprived of light or even of contrast patterns of light anatomical degeneration of the neural structure can result. Normal anatomical development and function of the brain are dependent upon its use.

Even if enough stimulation is present to ensure anatomical growth, perceptual mechanisms may not develop normally. If an animal lives with a completely homogeneous, nonpatterned visual field, then normal development still is impeded and such animals fail in many tests of visual perception. This is also noticeable in humans who have had cataracts, clouded lenses in their eyes, from birth. Light reaches the retinas, but it is homogeneous, nonpatterned light. When these patients finally obtain surgical attention their vision can be corrected optically so that adequate images are formed on the retinas, but visual perception is far from adequate without weeks and months of training. Such patients fail on surprisingly simple tasks of form perception. They know when some kind of form is present in the visual field, but can't tell squares from triangles or recognize such things as knives and forks even though they have always been able to recognize these things tactually without difficulty. However, the complete story of perceptual development remains complex and is somewhat puzzling. Ingenious experiments by Richard Held and Alan Hein (1963) and others have delineated another important factor which must be present for the normal facilities of visual perception to develop. Held and his colleagues have allowed young animals to spend periods of time growing up with patterned light stimulation but have deprived them of the opportunity to correlate this stimulation with any of their own movements. In one experiment kittens were paired such that one received periods of normal vision in which he walked around and around the bottom of a circular tank. This kitten's movements activated a gondola in which another kitten rode, receiving exactly the same stimulation without producing any vision-correlated movements. The walking kittens developed normal visual ability, while the passively stimulated kittens failed to develop even such basic visual functions as blinking in response to an approaching object or reaching the paws out to contact a surface onto which they were lowered. In a subsequent experiment, Held allowed kittens to change places between walking and riding periodically. When each was changed from the active to the passive condition, an eye patch was changed from one eye to the other. Held wished to explore the possibility of developing kittens with both an "active" eye, with which they might see normally, and a passive eye through which normal visual functioning might be impaired. These results were in fact accomplished; however, it is not yet evident how they should be interpreted. It appears that part of this effect might be due to supression of visual information. That is, kittens might have ceased to attend to and utilize information from passive stimulation because it wasn't correlated with anything of import. Yet even this would be an interesting discovery about the perceptual mechanism.

That there is something about self-produced movement, per se, that is indispensible in the formation of normal visual function is also suggested by studies of Stratton (1897), Held and Hein (1958), and others in which humans adapt to visual rearrangements such as inversion of the visual world. Here the movement in and experience with the visual rearrangements are not adequate precursors of adaptation unless the movement is self-produced. Thus Held and Hein (1958) found that human subjects wearing displacement prisms (lenses that displaced the visual image horizontally) made smaller errors in an eye-hand coordination task if they moved their own arms instead of having them moved by the experimenter. Exactly how self-produced movement operates to calibrate and coordinate the sensory systems for accurate perception and sensation is still a puzzle, and much work remains to be done in this interesting area.

In closing this chapter now, we recognize that only a small segment of the knowledge on sensory systems has been presented. The omission of names of the many investigators is in no way intended to slight their contributions, but to attempt to improve the readability of material which is often difficult to comprehend. The references given at the end of the book will guide the reader to additional discussions and to some of the original papers.

6   Reflexive Behavior

Benjamin L. Hart

There is a great deal of interest in how parts of the central nervous system mediate and integrate various elements of a behavior into coordinated patterns. One has only to read recent accounts in the popular press of investigations into the biochemistry of learning, the neural control of behavior, or the control of behavior by drugs to realize that the science of the biological basis of behavior is currently a very active field of research. In fact the possibility of understanding the biological basis of behavior has had the serious attention of scientists and thinkers at least since the time of the famous philospher René Descartes. In 1649 Descartes formulated what is probably the first comprehensive mechanistic explanation of animal behavior by introducing the concept of reflexive control of behavior. Descartes reasoned that animals had no mind or soul and were simple machines or automatons with hydraulic neuromuscular systems geared into a changing environment. Descartes drew a close analogy between statues in the royal gardens which made motions and sounds in response to underground streams of water and the control of animal behavior. Of animal behavior Descartes said:

And truly one can well compare the nerves of the machine that I am describing to the tubes of the mechanisms of these fountains, its muscles and tendons to divers other engines and springs which serve to move these mechanisms, its animal spirits to the water which drives them, of which the heart is the source and the brain's cavities the water main. (1972, p. 22)

Descartes imagined that nerves leading to and from the brain were filled with fluid which was confluent with fluid in the ventricles of the brain. Sensory stimuli caused movement of the fluid in nerves which in turn agitated fluid within the ventricles. This caused fluid to flow out nerves to muscles, filling the bellies of muscles making them fatter, resulting in contraction and hence movement of the trunk or limbs. To Descartes all animal behavior was reflexive in the sense that the effects of sensory stimulation were "reflected" back out to muscles from the brain. Considering that electricity as well as the nerve impulse were unknown at this time, one cannot help but marvel at Descartes' ingenuity.

Research supported in part by Grant MH-12003 from the National Institute of Mental Health. Much of the work benefited from the technical assistance of Charles M. Haugen and David M. Peterson.

The difference between man and animals, according to Descartes, was that in man a soul existed within the pineal body—the pineal body being strategically located on the roof of the third ventricle so as to both perceive fluid movements in sensory nerves arising from sensory stimulation and to initiate fluid movements in motor nerves resulting in muscle contraction. Activity involving the soul in the pineal was of course conscious; all else was purely reflex.

Although Descartes was not a scientist and his theoretical formulations were based solely on philosophical conjecture, it would appear as though he had a profound influence on later scientific work. Since the time of Descartes, various scientists and thinkers have attempted to deal with biological bases of behavior by separating behavioral acts into either unconscious, reflex-like, automatic responses mediated by wired-in neural connections—or learned and conscious responses mediated by acquired neural connections.

The famous neurophysiologist Charles Sherrington, whose work in the first part of this century opened the door to much of the current research on the physiology of behavior, greatly developed the concept of the dichotomy between conscious behavior and reflex activity. One of the experimental procedures used by Sherrington (1906) was the decerebrate preparation—usually a cat or a dog in which the cerebral hemispheres were surgically isolated from the rest of the brain. Such an animal, which is capable only of reflexive behavior, exhibits responses which are more than just meaningless immutable movements in response to specific stimuli. Rather, each movement seems to carry an obvious meaning. For example, the decerebrate dog can walk or run according to the speed of a treadmill on which it is placed; it will scratch parts of the body that have been tickled; if a thorn or a needle is stuck into the foot, the foot that might be injured by such a stimulus is withdrawn while the other legs "push" the animal away from the stimulus through extensor movements; milk placed in the mouth is swallowed and unpalatable solutions such as acid are rejected; water placed on the leg results in a shaking of the leg as if to remove the water. Furthermore, all of these reflexes or behaviors are not identical each time they occur. A reflex response to a weak stimulus will often be weak and to a strong stimulus more intense and longer lasting. Sherrington emphasized that this type of reflex action accounts only for a small part of the behavioral abilities of the intact animal.

In contrast to the dichotomy between conscious and reflex behavior emphasized in Sherrington's approach, Ivan Pavlov popularized a completely different approach. Using the salivary reflex and the "conditioning" (which will be discussed later) of this reflex as a model, Pavlov considered that all animal behavior was no more than a chain of reflexes—either innate reflexes or conditioned reflexes.

Since the work of Sherrington and Pavlov there has been a great deal of disagreement and confusion about the nature of the reflex and the role of these reflexes in behavior. Even for Sherrington (1906) the study of reflexes became so complex that he concluded that the simple reflex was an improbable but useful fiction. As Efron (1966) has pointed out, the vast majority of present-day writers in the field of neurophysiology and neuropsychology no longer attempt to define the reflex. As a matter of fact, there is no readily acceptable anatomical or physiological definition of the reflex, let alone a conceptual or behavioral definition.

Before going further it may be useful to make comparisons of behavior at three types or levels of organization. At one level of organization is learned behavior which can be clearly differentiated from reflex activity. This is behavior that requires the conscious activity of higher brain structures and which is almost completely dependent upon environmental influences.

At another level of organization are species-typical behavioral patterns or response tendencies which are largely inherited and undoubtedly involve conscious activity at high levels of neural organization. These behavioral patterns create some difficulty in defining or delineating reflexive behavior. Ethologists have described complex behavioral sequences in a number of mammalian and avian species. Often, in dealing with maternal, aggressive, or mating functions, the responses are so stable or stereotyped as to be frequently referred to as *fixed action patterns.* Can we call species-typical behavioral patterns, such as the retrieval of young pups by a mother rat or fighting among male deer during the rut, reflexes in the same sense that we refer to the pain withdrawal response or swallowing as a reflex? We might be tempted to classify behavioral processes which involve the cerebral cortex as nonreflexive and processes which do involve the cerebral cortex as reflexive: but, since birds have no cerebral cortex, we would then have to take the unacceptable position that all behavior in birds is reflexive.

At a third level of organization we have behavior which we will call reflexive. Reflex behavior does not necessitate conscious activity of higher portions of the brain, but rather involves sensory and motor neurons as well as interneurons of the lower brain stem and spinal cord. Reflexes in general are mediated by neural elements that are relatively well circumscribed within the central nervous system. In contrast to more complex behavioral activities, with reflexive behavior we are able to define with a fair degree of precision the stimulus which evokes the response. The above characterization of reflexive behavior is admittedly quite restrictive in that it excludes a number of apparent reflexes such as the salivary reflex and the eye-blink reflex in which the forebrain is involved. But at least we can be certain we are examining responses which are clearly no more complex than reflexes.

The Study of Reflexes

A number of researchers believe that the neural mechanisms involved in animal behavior are so complex that a fruitful area of neuropsychological investigation involves trying to understand the neurology and adaptive significance of the simplest reflexes. One particular advantage of this approach is that the concept of the reflex provides a convenient level of analysis to which more complicated behavioral responses may be reduced.

For the purpose of studying reflexes, Sherrington's decerebrate preparation has even turned out to be too complex, so investigators have turned to studying just one or a few reflexes mediated in a small, isolated part of the nervous system. Such approaches usually involve isolating and studying a portion of the central nervous system that mediates the reflex; and, often, comparing the response mediated by the isolated part with some complete behavioral sequence. For example, the lower part of the spinal cord is an easy area to isolate surgically, and in working with this structure one can be sure that conscious activity is eliminated from the responses studied.

Surgically isolating a section of the central nervous system (CNS) for study means that all the neural connections coming into the section from the rest of the CNS are severed. This sudden deprivation of normal neural input, much of which is tonic (constant), is believed to throw the isolated neurons into a state of physiological shock. This shock results in depression of activity and thus depression of reflexes mediated by the affected neurons. When the isolated part of the CNS is the spinal cord, the physiological shock is called *spinal shock*. Most spinal neurons seem to recover from this shock and continue to function. Simple reflexes usually gain normal strength within a few hours or days, but complex reflexes do not appear normal for several weeks. The duration of spinal shock is apparently related to the degree of encephalization (the localization of behavioral functions in higher brain areas) of the species involved. For example, in rats simple reflexes such as the leg withdrawal reflex may recover within a few hours following transection of the spinal cord, and recovery of complex sexual reflexes from spinal shock may take two weeks. In dogs and cats, the simplest reflexes may not obtain complete strength for several days, and complex sexual reflexes may not appear normal for six weeks. In primates, reflexes take even longer to recover from spinal shock.

Animals in which the spinal cord has been transected in order to isolate part of the nervous system for a special study are referred to as *spinal animals*. Studies on spinal animals of various species and on human patients in which the spinal cord has been traumatically severed by some injury, have provided us with a good deal of information about reflexes in general. Experimental animals in which the spinal cord has been transected for study are generally given a great deal of nursing care to make certain that they remain healthy, active, and alert. These animals are not in pain; they seem to be relatively happy and eagerly respond to the experimenter's attention and care.

Some Examples of Reflexive Behavior

Before considering the role that some reflexes have in various behavioral patterns, let us briefly examine some possible reasons why animals with very advanced nervous systems still have reflexive behavior which is characteristic of more primitive animals. Perhaps the chief reason why reflexes have survived in mammalian evolution through numerous changes, advances, and growth in the central nervous system, is that this form of behavioral organization achieves a sort of neural economy. Some reflexes are associated with protection of the organism from harmful objects while others, such as sexual reflexes, are basic to the reproductive success of individual animals. Having essential aspects of behavior mediated by reflex function assures that a critical response will be made to a specific stimulus without the animal having to learn a biologically appropriate action or even being consciously aware of its behavior. Other reflexes appear to have adaptive value to higher organisms because they mediate routine functions, thus leaving higher areas of the brain free to deal with other activities.

Let us first consider a number of reflexes that have evolved presumably because they are important in protecting animals from serious injury. You are surely familiar with some. For example, the eye-blink reflex occurs in response to an object lightly touching the cornea of the eye. An irritating particle that remains on the surface of the cornea evokes the *lacrimation reflex* causing a flow of tears which tend to wash away the irritating particle. Sneezing and coughing reflexes resulting in an explosive outburst of air are caused by irritating substances or particles in the respiratory tract. Most people have also experienced the pain withdrawal or *flexion reflex*. This reflex, which has received a good deal of attention from experimental neuropsychologists, is evoked by a painful stimulus to the digits or limb. The response is immediate withdrawal or flexion of the limb from the painful stimulus. Spinal frogs exhibit this reflex when the back leg is pinched. If an irritating substance such as a drop of acid is placed on one of the legs of the spinal frog, the opposite leg will scratch or rub the site of the irritation.

The back legs of spinal dogs and cats exhibit a strong flexion reflex when the toe is pinched. If the stimulus is particularly intense the leg remains flexed for several seconds. This is called after-discharge, because the stimulated limb is held in a flexed position long after the stimulus is terminated. In a spinal dog or cat, pinching a toe not only triggers the flexion reflex, it also evokes simultaneously an extension of the opposite leg, the *crossed extension reflex*. The value of both reflexes for dogs and cats is quite obvious. If a sharp object such as a thorn or sharp stone is stepped upon, the limb which is likely to be injured is immediately withdrawn and the limb of the opposite side is more fully extended so as to support the animal in a standing position and to help push the animal away from the harmful stimulus (Figure 6-1).

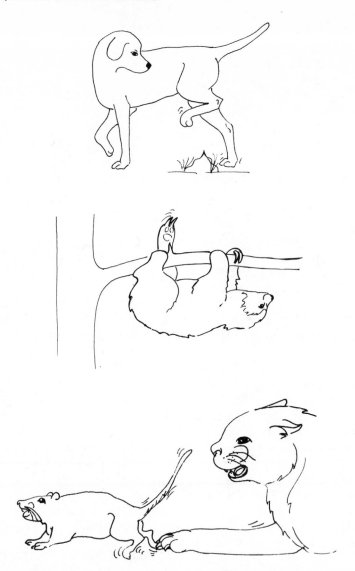

Figure 6-1 Protective reflexes. The same stimulus—a sharp object stuck into the foot—evokes a different form of the reflex in three different animals. In a) the stimulus of a sharp thorn in the foot of a dog evokes flexion of the stimulated leg and extension of the opposite leg allowing the animal to protect its stimulated foot and avoid falling. In b) a protruding branch poking into the back foot of a two-toed sloth evokes extension of the stimulated leg and contraction of the opposite leg allowing the animal to protect the stimulated leg without falling from the tree. In c) a sharp stimulus to the back foot of a rat from the claws of a potential predator evokes rapid kicking movements of both back legs simultaneously which evidently contributes to the animal's chances of escaping from the predator. These are all spinal reflexes and in spinal subjects movements of the back legs resemble those seen above in normal animals.

The specificity of this reflex in protecting an animal from injury becomes apparent when we look for species differences in the way a protective reflex may be useful. Consider the locomotor pattern of the two-toed sloth. This animal spends almost all of its time upside down moving about or clinging to tree branches. The flexion and crossed extension reflexes which serve the dog and cat quite well would be of little value to the sloth. Esplin and Woodbury (1961) had an opportunity to study spinal sloths: they found that when the toe of the animal was pinched, instead of flexion of the stimulated limb there was extension, and instead of extension of the opposite limb there was flexion. This makes good sense for the sloth because if there is likely to be an injury to its foot it would probably be due to sharp sticks projecting from the branch along which it is moving. The most useful reflex would be extension of the foot which might be injured away from the tree branch plus the stronger flexion of the limbs on the opposite side to help the animal push the stimulated paw away from the sharp object and to support the weight of the animal (Figure 6-1).

Consider next a small animal such as a mouse or rat. A painful stimulus to the back legs of these animals may represent the teeth or claws of a predator rather than a thorn or other sharp object encountered accidentally. Thus a well-organized flexion and crossed extension reflex pattern is not likely to be very helpful in dealing with a predator. When one actually pinches the back foot (or tail) of a spinal rat, one sees vigorous shaking or rapid alternate extension and flexion movements of both legs. The value of this particular reflex may be to enable the small mammal to shake itself free of the claws or teeth of a predator (Figure 6-1).

These three animals provide good examples of species specificity in reflexive behavior. The same stimulus—a painful jab at the toe or skin of the foot—causes in each of these three animals a different protective response related to the dangers the animal is most likely to encounter in its own environment.

Quite a different group of reflexes are those related to the routine functions of feeding and elimination. These involve secretion from salivary glands, swallowing reflexes, jaw-closing and jaw-opening reflexes, and a variety of others associated with elimination. When the urinary bladder of spinal female dogs and cats is manipulated, or when the area of the external genitalia which comes in contact with urine as it is voided is stimulated, the spinal subject assumes a squatting posture which is the same as that of normal female dogs and cats during urination. The apparent value of this reflex and the squatting posture is to hold the body so that there is a minimum of soiling of the skin by the urine being voided. This reflex cannot be evoked in male dogs, which is not surprising since adult male dogs do not urinate in the same way females do.

In spinal male and female dogs stretching the wall of the rectum or anus (which normally occurs during defecation) evokes a squatting

Figure 6-2 Clasp reflex of male frogs and toads (amplexus). The reflex as seen in this illustration is usually evoked by the large rounded thorax of the female. In spinal animals this reflex of the front legs is evoked by any object which approximates the size and shape of the female.

posture which is the same as that which customarily occurs when normal dogs defecate (Kellogg, et al., 1946). The neural complexity of both the urination and the defecation postural reflexes is evidenced by the involvement of the striated muscles of the trunk and legs as well as the smooth muscle of the urinary or digestive tract; this means that the reflexes contain neural elements from both the somatic motor and autonomic nervous systems.

The type of behavior in which reflexes are particularly crucial is sexual behavior. An interesting example is the clasp reflex (or amplexus) of frogs and toads (Figure 6-2). During amplexus the external fertilization of ova takes place. This response, which lasts frequently as long as two or three hours, occurs in male frogs during the mating season when the skin on the inside of the front legs and belly is stimulated. The clasp reflex has been studied in frogs in which the hind brain, which mediates the reflex, has been surgically separated from higher structures of the brain (Hutchison, 1964; 1967; Hutchison and Poynton, 1963). Transected males were able to distinguish between normal and abnormal clasp objects, maintaining a clasp only when the object provided the correct sensory input. Males of one species were even able to make minor adjustments from an abnormal position on the female.

Some of the best examples of the role of reflexes in sexual behavior in mammals are the sexually receptive postures exhibited by females of some species during estrus. These postures, which are mediated by the

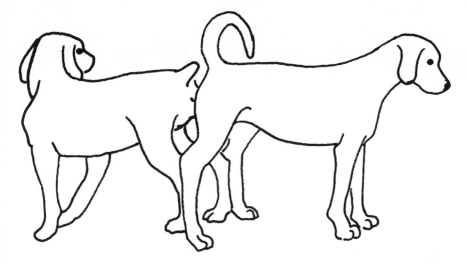

Figure 6-4 A copulatory reflex of male dogs. The genital lock which occurs in dogs during mating is mediated as a spinal reflex in the male. Part of the penis becomes so engorged within the female genital tract that the dogs remain locked together for approximately 10-30 minutes because of the reflex which maintains the engorgement of the penis.

spinal cord, occur in response to tactile stimulation by the male. In the female rat the sexually receptive posture takes the form of the *lordosis* seen in Figure 6-3 (Hart, 1969a). In the female cat the receptive posture consists of elevation of the pelvic and perineal area, treading of the back legs, and holding the tail to one side (Figure 6-3). The treading of the back legs and tail position (but not pelvic elevation) appear to be spinal reflex responses to tactile stimulation of the anogenital and perineal regions. In the dog the receptive posture of the female is characterized by a lateral curvature of the hind quarters together with arching of the tail to the side (Hart, 1970). This reflex is usually evoked by the tactile stimulation of the female by the male's sexual investigation of her genitalia, as seen in Figure 6-3.[1]

The sexual reflexes of female dogs and cats are apparently facilitated by administration of estrogen (Hart, 1970; 1971b), suggesting that when the female comes into estrus the increased secretion of estrogen not only enhances the female's total responsiveness to the presence of the male but also facilitates the occurrence of the sexual receptive posture at the spinal level.

Male mammals, which have a more active role in mating than females, do not exhibit static sexual receptive postures. Several types of

---

[1] The receptive posture may also occur without contact from the male. Although the response is the same we would not consider this a reflex since the response is then activated from the brain rather than from a local stimulus.

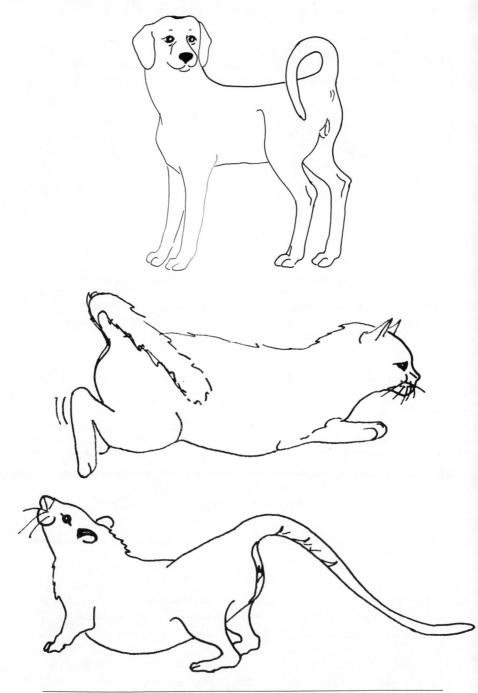

Figure 6-3 Sexual reflexes of some female mammals. The special postures of the sexually receptive female, dog, cat, and rat are evoked both
in the normal and in the spinal animal by tactile stimuli around the genital
region.

masculine reflexes have been studied. Reflexes studied in the male represent, for the most part, the consummatory responses of erection and ejaculation. In spinal male rats, erections and rapid penile movements have been observed (Hart, 1968a). An interesting aspect of sexual reflexes in male rats is that there appears to be a spinal timing mechanism which limits the occurrence of the reflexes to periods of two or three minutes, the time necessary to inseminate a female.

In male dogs two ejaculatory reflexes have been described which correspond to the two behavioral patterns that dogs exhibit during copulation (Hart, 1967a). One of these reflexes is the intense ejaculatory reaction which occurs immediately after the male penetrates the female and lasts about 15 to 30 seconds. The other reflex is that which is responsible for locking of the two dogs together for approximately 10–30 minutes in what is called the genital lock or tie (Figure 6-4). The duration of genital lock reflex is probably the longest of any reported reflex so far studied.

In both male rats and male dogs there appears to be marked facilitation of the duration or intensity of sexual reflexes by the male gonadal hormone *testosterone* (Hart, 1967b; Hart, 1968b). The significance of this will be discussed later, but it may be pointed out now that when male mammals are castrated and gonadal hormone support is removed, changes in sexual behavior are gradual. During this gradual decline in sexual activity those aspects of the sexual behavior which seem to be impaired most readily and most severely are the consummatory responses of erection and ejaculation (Young, 1961). Thus in males, not only is testosterone necessary for maintenance of general sexual interest and motivation, but it is also critical for maintaining the spinal reflex functions essential for copulation.

Animals spend a good deal of time just standing still or walking. Even these simple activities involve complex interactions of opposing sets of flexor and extensor muscles and related neural circuitry, and thus it is of interest to determine at which levels in the nervous system of mammals this behavior is normally mediated. As noted previously, Sherrington (1906) observed that a decerebrate dog or cat is capable of standing or walking when placed on a treadmill, and that the speed of walking is related to the speed of the treadmill. But can standing and walking reflexes be mediated at the level of the spinal cord? Again, Sherrington found that dogs and cats in which the spinal cord had been transected in the lumbar region could stand when their back legs were placed beneath them, but his observations suggested that the spinal preparation was not capable of reflex walking. Since the time of Sherrington's work, a number of investigators who have studied spinal carnivores have arrived at conflicting opinions about the existence of an integrated walking reflex. Cate (1940), in reviewing this problem, suggested that the apparent walking activity shown by some spinal dogs was simply a reflection of the ability of the spinal animal to hold the posterior

From Hart, B. L. (1971a), "Facilitation by strychnine of reflex walking in spinal dogs," *Physiology and Behavior* **6**, 932–934.

Figure 6-5 Walking spinal dogs. Photographs on the left show one of the few spinal dogs that was capable of walking in the normal spinal condition. Most spinal dogs like the one shown on the upper right did not walk and normally dragged their back legs behind them. When given a subconvulsive dose of strychnine, this dog was capable of walking (lower right).

region hoizontal (using the front legs as a fulcrum and lowering the head as a counter-weight) while walking on the front legs. However, Cate describes his experience with one spinal dog in which the subject, after raising his hind quarters off the ground by lowering its head and contracting its spinal musculature, was then able to support the weight of its hind quarters through extension of the hind limbs and to perform walking movements. Shurrager and Dykman (1950) who studied the development of a walking reflex in ten kittens and one puppy, transected at a very early age, found that the animals showed a surprising walking ability including jumping and running for short distances. However, these workers felt that adult spinal dogs and cats were incapable of reflex walking. A recent study (Hart, 1971a) has added an interesting note to this line of experiments. In this study, a walking reflex was observed in only a small proportion of adult spinal dogs. However, when a subconvulsive dose of strychnine was given, nonwalking spinal dogs became capable of reflex walking (Figure 6-5). Since it is known that strychnine is a suppressor of neural inhibition, its facilitation of the walking reflex apparently resulted from its action in suppressing intraspinal tonic inhibition of the walking reflex within the cord itself.

The study of the walking reflex is interesting because it suggests a kind of *disinhibitory process* that integrates the walking reflex with other locomotor activities.[2] Under this interpretation, the neural elements mediating the walking reflex are tonically inhibited, suppressing the walking reflex, so that it will not be easily evoked and interfere with the animal's locomotor coordination. However, when the animal is walking steadily, control centers in the brain send messages to the lower part of the cord blocking neurons in the cord that normally inhibit the walking reflex.

The Neural Bases of Reflex Systems

Let us now consider some aspects of the mechanisms involved in the neural mediation of reflexes and the integration of these reflexes into complex behavioral sequences. The very simplest of reflex organizations is the two-neuron reflex arc of the knee-jerk reflex. When the *patellar tendon* (the tendon just below the kneecap) is gently tapped, sensory receptors in the muscle attached to the patellar tendon are suddenly stretched. The axons of these receptors convey nerve impulses evoked by the stimulation to the spinal cord where a synapse is made directly upon the motor neurons that innervate fibers in the muscle from which

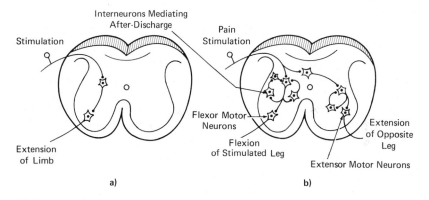

Figure 6-6 Neural organization of two reflex systems. In a) the two-neuron reflex arc mediating the knee-jerk reflex is represented. Sensory neurons directly activate motor neurons causing contraction of extensor muscles in the leg. In b) the more complex neural organization of the flexion (leg withdrawal) reflex of carnivores is represented. Sensory neurons activate interneurons which in turn excite flexor motor neurons causing flexion of the leg on the same side. Interneurons are also activated which excite extensor motor neurons on the opposite side causing extension of the opposite leg. The small interneurons off to the side play a role in maintaining activation of the reflex system after activity of the sensory neurons has subsided; thus the response continues after sensory stimulation has subsided resulting in what is referred to as *after-discharge*.

[2] This will be discussed more fully later.

the sensory impulses arose. This causes a synchronous contraction of muscle fibers in the extensor muscles of the knee (Figure 6-6).

The knee-jerk reflex illustrates two components of the neural organization of reflexes. The *sensory component* includes the sensory receptors, long axons carrying nerve impulses to the CNS, and the terminations of the sensory neurons within the CNS. The *effector component* includes the motor neurons, axons carrying nerve impulses to the muscles, and the neuromuscular junction through which muscle contraction is activated.

The organization of all reflexes except the knee-jerk involves more than two neurons. In the flexion reflex, for example, a painful stimulus excites sensory endings of many neurons innervating the skin. A burst of nerve impulses travel to the spinal cord over the sensory components of the reflex. These axons synapse upon *interneurons* which in turn make synaptic connections to the effector component of the reflex, resulting in flexion of the stimulated limb (Figure 6-6). You will remember that in carnivores painful stimulation of a foot not only triggers withdrawal of the stimulated limb but also extension of the leg on the opposite side. This means that input from the sensory component of the flexion reflex must stimulate nerves in the CNS that relay messages to the motor neurons controlling the extensor muscles on the opposite limb (Figure 6-6). This *central component* that controls and integrates complex aspects of reflexes consists of many highly organized sets of interneurons. The activity of these central interneurons is also apparent in the after-discharge that is responsible for prolonged withdrawal of the dog's leg after termination of the painful stimulation that provoked withdrawal in the first place.

Integration of the activity in a reflex system with that of other parts of the CNS involves, on the sensory side, neural connections conveying information from the sensory component of the reflex to higher neural structures such as the thalamus and cerebral cortex. These connections are not properly part of the reflex but certainly play an important role in coordinating reflexive behavior with other behavioral patterns.

Another important factor in understanding the rule of reflex function in behavior is to realize that in normal animals reflexes are not invariably activated simply by presentation of an appropriate stimulus. Several studies have shown that many reflexes, especially sexual reflexes (Beach, 1967), are tonically or constantly suppressed through inhibition from higher areas of the central nervous system. One of the reasons why such inhibitory control may have evolved is to prevent the unintentional activation of a reflex when it might be disadvantageous to the animal. For example, activation of a sexual reflex through accidental stimulation of the penis in a male hunting dog or cat that was moving through heavy underbrush would be quite disruptive. During mating the sexual reflexes are of course released from this inhibition and they occur at the appro-

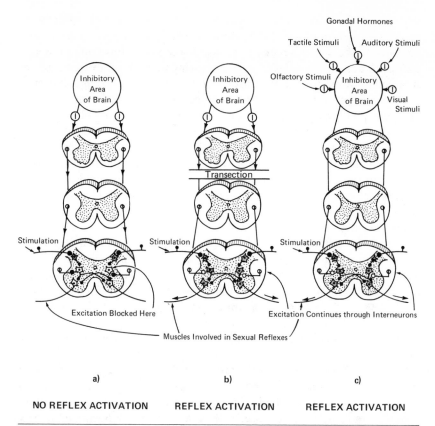

Figure 6-7 Model of a sexual reflex system which is controlled by an inhibitory system of the brain. In a), the normal state, the reflex system cannot be activated because of supraspinal tonic inhibition of the inter- neurons. Activation of sensory neurons is able to excite interneurons (shaded) to the point where inhibitory influences come in. In b) transection of the spinal cord cutting the descending inhibitory axons removes supra- spinal inhibition of interneurons. Activation of sensory neurons now excites all neurons in the reflex system. In c) supraspinal inhibition is also removed because of sensory stimuli which are believed to indirectly or directly suppress the brain inhibitory area. This is the process (disinhibition) which is believed to occur during sexual excitement and allows the sexual reflexes of the male (erection and ejaculation) to be very easily evoked by stimula- tion of the penis.

priate moment. The presence of a sexually receptive female near adult male rats or dogs may lead to the activation of an inhibitory mechanism in the male that blocks the neurons normally inhibiting the male's sexual reflexes (Beach, 1967; Hart, 1967a; 1968a), (Figure 6-7).

Another type of inhibitory control of a sexual reflex has been ana- lyzed by Beach (1966). It was known from the work of Boling et al. (1939)

that neonatal female and male guinea pigs, for a short time after birth, react to tactile stimulation of the rump by exhibiting a posture essentially identical to the lordosis reflex of sexually receptive adult females. Beach cites evidence from a series of studies showing that this reflex comes under supraspinal inhibitory control after two to three weeks of age and later can be evoked only from adult females in estrus. Beach argues convincingly that in the neonate (before a tonic inhibitory system has formed) the lordosis reflex is vital for survival. In very young mammals of altricial species, including dogs, cats, rats, and others, it is necessary for the mother to lick the infant's perineal region if reflex elimination of urine and feces is to occur since the young do not defecate or urinate spontaneously. The lordosis reflex comes into play because when the mother licks the dorsal rump of the infant (in her routine grooming of the young) the evoked reflex elevates the perineum, thus maximally exposing the excretory orifices. In response, the mother then licks the area immediately around the orifices, and this induces defecation and urination by the infants. At about the same time that young guinea pigs began to eliminate without maternal stimulation, the lordosis reflex seems to come under supraspinal inhibitory control. It never again normally appears in the males. In females, Beach hypothesizes, the same lordosis reflex again plays a crucial role in mating behavior as the reflex is then released from inhibition by estrogen.

The control of reflexes by the processes of inhibition and disinhibition (blocking of inhibition) may be related to evolutionary changes in brain size and complexity. As the brains of mammals became markedly more complex and larger than those of their ancestors, it appears as if new brain structures were added on top of existing, more primitive components. The control of some behavioral functions, particularly reflexes, remained within older parts of the CNS. Behavioral functions of the older parts of the CNS were apparently brought under control of the newer additions through the process of tonic inhibition (Diamond, Balvin and Diamond, 1963).

How Reflexes are Altered

The Effects of Hormones

First let us recognize that there are a number of factors which can change the intensity or duration of reflexes by acting on the sensory, central, or effector components of the reflex. These include general fatigue of muscles, adaptation of the peripheral sensory receptors resulting from excessive stimulation (this is different from habituation of reflexes which we will discuss later), and general anesthesia of the animal. We will not consider these further but will rather discuss physiological factors that may alter the reflexes in some basic way and that may also be related to important changes in behavior patterns to which the reflex contributes.

We have already mentioned that sexual reflexes of males and females may be altered by changes in circulating levels of gonadal hormones. Recall that when a sexually active adult male animal is castrated there is a decline in sexual activity, but that all aspects of sexual behavior do not change uniformly. Usually the consummatory responses of erection and ejaculation, representing spinal reflex function, are impaired sooner than other changes in sexual behavior. It is common to see such animals mounting a female just as persistently after castration as before, but without the ability to penetrate, apparently because of a loss in the strength of reflexes controlling erection. Such changes in the consummatory responses of intact animals have been paralleled in experimental studies of spinal male animals. These experiments have shown that sexual reflexes can be strengthened or weakened by withdrawal or administration of testosterone. A particularly striking example is the reflex which is related to the genital lock of dogs, which in intact males is approximately fifteen minutes in duration (with a range of ten to thirty minutes). Following castration there is a decline in the duration of the genital lock over the next few weeks to about five minutes with a range of about one to eight minutes (Hart, 1968b). In spinal male dogs with normal testosterone levels, the genital lock can be simulated with appropriate technique (Hart, 1967a). In these animals there is a decline in the simulated lock following withdrawal of testosterone that is very similar to that seen in nonspinal subjects following castration (Hart, 1968b).

Changes in quantitative aspects of sexual reflexes of spinal male rats as a result of withdrawal or administration of testosterone have also been demonstrated (Hart, 1967b). These studies on both dogs and rats do not actually differentiate between the possible influences of the hormone on neurons of the central component of the reflex system or on genital receptors of the sensory component. This question is an important one because it is known that in some animals, particularly rats and cats, there is a marked atrophy of genital papillae which project from the surface of the penis following castration. Although these papillae are probably not related to receptors that evoke sexual reflexes, they do demonstrate that peripheral changes occur following castration. In order to distinguish between the possible action of the hormone on sensory and central components, crystalline testosterone propionate was implanted into the lower spinal canal of spinal male rats that had been castrated and were receiving no other hormone treatment (Hart and Haugen, 1968). Sexual reflexes in these animals, which were previously depressed because of the lack of testosterone, were activated by this procedure. In this experiment there were controls demonstrating that there was very little leakage of testosterone into the peripheral vascular system which could act on the sensory component of the reflex. This experiment suggested that testosterone in facilitating sexual reflexes (at least of rats) acts on the spinal cord rather than on the peripheral receptor or effector portions of the reflex. Additional evidence points to the same conclusion. In one study (Cooper, 1970), electrophysiological recordings were taken from the dorsal nerve of the penis of male cats while applying tactile stimulation to the epithelium over the penis. Male cats that had been castrated and in which penile papillae were markedly degenerate were compared with animals that had not been castrated. No differences could be found in the intensity or pattern of sensory nerve impulses traveling over the dorsal nerve of the penis to the spinal cord.

Just as gonadal androgen seems to play an important role in maintaining optimal functioning of sexual reflexes in the adult male, the ovarian hormone estrogen appears to have an important role in facilitating sexual reflexes in some females (Hart, 1970; 1971b). Experimental evidence indicates that the local effect of estrogen is a contributing factor in directly activating the reflex aspects of the receptive posture in dogs and cats. However, experiments with female rats (Hart, 1969a) have not demonstrated any effect of hormones on the lordosis reflex. Thus there may be some species differences in this regard.

In addition to the influence that hormones may have on sexual reflexes of the adult, there is some evidence that hormones may also play a role in the neural organization of sexual reflexes in the developing neonatal nervous system. For example, most male rats castrated on or before the fifth day after birth, and then not injected with testosterone until adulthood, show a high degree of sexual motivation when tested with receptive females as indicated by the frequent occurrence of mounts and copulatory intromissions. But apparently these animals do not complete the mating sequence; they do not exhibit the ejaculatory response. On the other hand, most male rats castrated after ten days of age do exhibit the complete mating pattern culminating in an ejaculation. In an extension of these findings, a study by Hart (1968d) showed that male rats castrated at four days of age and given testosterone in adulthood, as expected were not capable of ejaculating. When tested for sexual reflexes after spinal transection, these animals displayed an impairment of sexual reflexes. Males castrated on day 12, who in adulthood (after testosterone replacement) exhibited the complete mating sequence, also showed normal sexual reflexes after transection. Thus, in addition to the influence that gonadal androgen has on maintaining adult sexual reflexes, it also appears as though it plays a role in the neural development of sexual reflexes in newborn animals.

The Effects of Alcohol

Having discussed how certain hormones may influence sexual reflexes, the question arises as to the possibility of other substances influencing sexual reflexes. There is a popular conception that alcohol has two effects on sexual functioning: it increases sexual desire by lowering social inhibitions, but it diminishes sexual capability or potency, i.e., the ease with which the consummatory sexual reflexes are activated. Although there is some clinical evidence that alcohol diminishes sexual potency (Abraham, 1926), Shakespeare has described the same phenomenon more colorfully in *Macbeth:*

Macduff: *What three things does drink especially provoke?*
Porter: *Marry, Sir, nose painting, sleep, and urine. Lechery, Sir, it provokes, and unprovokes; it provokes the desire, but it takes away the performance.*

Animals apparently show the same phenomenon. When rats (Hart, 1969b; Dewsbury, 1967b) and dogs (Hart, 1968e) are given a strong dose of alcohol before mating, they show an impairment in mating ability while still demonstrating a strong sexual interest in the female. In spinal dogs (Hart, 1968e) and rats (Hart, 1969b) a similar dose of alcohol markedly impaired sexual reflexes. These results thus support the view that while alcohol intoxication may not abolish sexual motivation, it does markedly diminish sexual potency or capability. Thus the drug has a negative effect directly on sexual reflexes.[3]

In spinal dogs a strong dose of alcohol abolished the standing reflex, but with the same dose of alcohol, intact male dogs remembered the location of a behavioral testing chamber, showed considerable sexual interest in receptive females, and were capable of walking although they exhibited marked uncoordination. These observations suggested that perhaps some of the impairment of motor function associated with alcohol intoxication may be a reflection of depressed or abolished postural reflexes (Hart, 1969b). This conjecture was initially proposed by Dodge and Benedict (1915) as a result of their findings that in human subjects moderate doses of alcohol depressed reflexes much more than psychological functions such as free association and memory.

Learning and Reflexes
Can the neural substrate which mediates reflexes be altered as a result of experience? To put the question a different way, is it possible for some elementary types of learning to occur in reflex systems? Let us adopt a broad definition of learning such as a change in behavior as a result of practice or experience; there are then two possible types of learning which could occur in reflex systems. One type is *habituation* of the response and the other type is *reflex conditioning* (sometimes referred to as *classical conditioning* or *respondent learning*). Both types affect an underlying response system which is innate or unlearned. In habituation, the response to a stimulus is decreased as a result of repeated stimulation that causes a change in the *central* component of the reflex. The change in response is not due to a decrease in receptor activity as a result of receptor adaptation. Nor is it due to a reduced effector response resulting from fatigue of muscles or interference with activity at the junction between

---

[3] However, because sexual reflexes are tonically inhibited by higher nervous activity, mild doses of alcohol may enhance performance through disinhibition without affecting reflex activity directly.

nerve and muscle. In reflex conditioning, a neutral or conditioned stimulus which does not initially evoke a reflex response is paired in time with a stimulus which does (unconditioned stimulus), until the conditioned stimulus, presented alone, evokes the response. There is considerable evidence that both of these types of learning can occur in spinal reflex systems. Let us first consider the habituation process.

In intact animals habituation occurs with stimuli that, for example, initially cause a startle response or orientation response to the stimulus. Upon repeated presentation of the stimulus the organism shows a decreasing startle reaction or a weaker orientation response until eventually there is practically no response to the stimulus. Thus birds by a highway may no longer startle and fly off when a car appears.

Experimentation with habituation of spinal reflexes is not just a recent research interest. Sherrington (1906) observed that repeated mechanical or electrical stimulation resulted in a decrement of both the scratch reflex and the flexion reflex of spinal dogs. Although Sherrington referred to the change as a type of fatigue, he was clearly working with the habituation phenomenon because he was able to rule out both receptor adaptation and muscle fatigue. Prosser and Hunter (1936), working with spinal rats, observed a decrease in reflex leg and tail movements as electrical or mechanical stimulation was repeated.

Since these initial studies there have been a number of investigations of various aspects of the habituation process using sophisticated electrophysiological techniques and different spinal preparations. Habituation of the flexion reflex has been demonstrated in spinal rats (Griffin and Pearson, 1967), cats (Kozak et al., 1962; Spencer et al., 1966a; 1966b), and dogs (Nesmeinova, 1957). Quantitative relationships between various stimulus parameters and the degree of habituation of spinal reflexes has been studied in spinal cats (Groves, Lee, and Thompson, 1969). These workers have shown that habituation occurs more rapidly and to a greater degree for higher frequencies of stimulation and for lower intensities of stimulation. Thompson and Spencer (1966) have shown that the habituation process which occurs in spinal reflexes shares all the characteristics of habituation that have been thus far studied in intact animals. Some of these characteristics are: 1) the decrease in a response to a particular stimulus is related in a characteristic way to the number of stimulus presentations; 2) if the stimulus is withheld the response tends to recover over time—this is referred to as spontaneous recovery; 3) habituation becomes progressively more rapid with repeated series of habituation training and spontaneous recovery; 4) the more rapid the frequency of stimulation the quicker the habituation; 5) the weaker the stimulus the more rapid the habituation.

Since it has been rather well established that habituation can occur within a reflex system, another question is where, within the system, habituation is taking place. Conceivably there could be habituation of some part of the sensory or the effector components, or alternately habituation could be taking place within the central components. Neurophysiological studies (Kugelberg, 1952; 1962; Griffin and Pearson, 1968; and Wickelgren, 1967) indicate that the actual changes which occur during the establishment of habituation are in the interneurons of the central component of the reflex system rather than in the sensory or motor components.

What adaptive role does habituation play in altering the behavior of an animal? In the case of behavior involving the cerebral cortex, habituation allows the organism to adapt to its environment in the sense that it comes to ignore repetitive stimuli associated with routine environmental stimuli, thus allowing the organism to focus more carefully on new stimuli. Although the issue is less clear in the case of reflex habituation, it seems probable that it complements cortical habituation. However, it would be useful to know more about the process in different reflex systems.

Let us consider next the research on reflex conditioning. The experimental paradigm which has been used is the same as that employed with the intact animal. Let us review the process briefly. A known reflex system such as the salivary reflex or the eye-blink reflex is used. The stimulus that normally evokes the reflex is referred to as the *unconditioned stimulus.* In the conditioning process a neutral stimulus which does not evoke the reflex is presented to the organism followed immediately by the unconditioned stimulus which then triggers the response. Over a number of such pairings, the neutral stimulus eventually takes on some of the response-producing properties of the unconditioned stimulus and is capable of producing the response alone. This learned connection—the connection between the previously neutral stimulus and the reflex—is temporary or reversible in the sense that it can be extinguished if the neutral stimulus is never again paired with the unconditioned stimulus. The same paradigm has been used by workers studying the flexion reflex of the back legs of spinal animals.

Shurrager and his associates (Shurrager and Culler, 1938; 1940; 1941; Shurrager and Shurrager, 1941; Dykman and Shurrager, 1956) were probably the first to apply this approach. They used very young spinal dogs and cats and reported that the flexion reflex could be conditioned by pairing a light touch or mild electric shock to the tail with a strong electric shock to the leg (which was the unconditioned stimulus). After a sufficient number of pairings, the touch or shock to the tail alone came to elicit the flexion reflex. They achieved the same results in their spinal preparations, including extinction of the conditioned response, as other investigators have observed for conditioned reflexes in intact organisms.

However, other investigators attempting to replicate their results were unable to demonstrate conditioning of spinal reflexes (Forbes and Mahan, 1963; Kellogg et al., 1946; 1947). With the exception of a study by Franzisket (1951) on the conditioning of the flexion reflex in spinal frogs, there was little further interest in this work until quite recently when new research techniques became available. Using electrophysiological techniques and a cautious experimental design, Buerger and Dawson (1968) have presented further convincing evidence for the existence of conditioning of the flexion reflex in spinal kittens. These investigators have also presented evidence pointing to interneurons as the locus of the reflex conditioning process (Buerger and Dawson, 1969).

It should be emphasized that although investigators studying conditioning of spinal reflexes have used the experimental paradigm developed by Pavlov for the conditioning of the salivary reflex, the phenomena are fundamentally different. Conditioning of the salivary and some other cranial reflexes involves the cerebral cortex; in addition, the subject must be conscious. Conditioning of spinal reflexes presumably involves only elements within the reflex system and obviously does not require the conscious participation of the subject.

Questions regarding the roles that conditioning of spinal reflexes may play in normal behavior must remain open. At the present time we do not know what reflexes can be conditioned and we know very little about the situations under which conditioning of such reflexes may occur.

Conclusion: The Concept of the Reflex and Animal Behavior

In the beginning of this chapter we noted that for centuries man has pursued his interest in explaining animal and human behavior in mechanistic or physiological terms. Fearing (1930), in observing several decades ago that the concept of the reflex played an important role in psychological theorizing, mentioned that the discrete stimulus-response attributes of the reflex arc system made it easy to visualize and diagram this physiological model of behavior. It was fashionable for awhile to extend the reflex concept as an explanatory mechanism to all kinds of animal behavior. However, it was not long before many scientific leaders agreed with Judson Herrick (1924) in his observation that the concept of the reflex was not a master key, competent to unlock all of the secrets of the brain and mind.

As our perspective of the role of reflex action in the total neurological control of behavior has taken on a more balanced form, our concept of reflexive behavior itself has changed. Some reflexes, such as those involved in sexual behavior, are much more complex than earlier investigators believed. Also, we can no longer view reflexes in quite the invariant and predictable form as before. As we have seen, some reflex systems are readily influenced by hormones and drugs. From other experiments we have seen that a primitive type of learning, as shown in the habituation experiments, occurs in reflex systems.

The history of the concept of the reflex together with recent research brings up a timely concern in regard to reflexive behavior. Even though reflexes have been studied longer than any other physiological system related to behavior, we are a long way from a satisfactory understanding of the role of reflex action in behavior. Probably because of experimental procedures used by neurobiologists to artificially isolate and study reflexes, there has been a tendency to conceive of reflexes as separate or isolated units of behavior. Although it is necessary to isolate the neural elements mediating a reflex in order to study them, the isolated reflex is an atypical situation and only indirectly related to the understanding of integrated responses of the intact organism. Fearing (1930) made this point several decades ago and went on to emphasize that one of the main problems for future investigations was analysis of the reflex in relation to all of the concomitant events in the integrated nervous system. Fearing found it particularly necessary to warn against the arbitrary distinction between voluntary and involuntary action which, as he noted, came from the Greek physiologists and was translated into an artificial neuropsychology by Descartes.

In the future, isolated reflexes will be studied to more fully explore the types of learning that may occur at the reflex level and to understand how neural elements of some reflex systems may be influenced by hormones or drugs. Some reflex systems will serve as useful models for research aimed at elucidating basic neuropsychological processes that may be applied to the central nervous system in general. From the behavioral standpoint the most important work will be related to understanding how reflexes are integrated into the total behavioral systems of normal, functioning organisms.

7 Courtship and Mating

Gordon Bermant and Benjamin D. Sachs

Every animal begins life as part of another animal. Every animal cell contains some information about building another. Reproduction is the process by which this information is put to work and its potential realized. Some species are quite capable of reproducing nonsexually: a single individual produces cells that give rise by themselves to other complete individuals. Thus the hydra creates numerous offspring by budding them off from specialized cells in one region of its body, and the single-celled paramecium may reproduce by splitting in two. The physiological mechanisms involved in these two reproductive sequences are quite dissimilar, as are the environmental circumstances in which they occur. But they are identical in the very important sense that the genetic information acquired by the offspring comes from only one parent.

But most of the animals we are familiar with do not reproduce in this way. Instead *two* individuals, male and female, each contribute some genetic information in the creation of offspring which are therefore genetically different from both parents. *Sexual reproduction* is extremely widespread, present in every animal phylum, and among our own sub-phylum of vertebrates it is a rule normally broken only by a few groups of fishes and reptiles. It may also be utilized by the same species that normally reproduce nonsexually; for example, both paramecia and hydra shift to reproduction by dual parentage under certain environmental circumstances. Thus, sexual reproduction achieved through dual parentage is one of the most ancient and widespread adaptations of animal species (Bermant and Davidson, 1973).

Why should sexual reproduction be ubiquitous? If you think about it, it seems a peculiar way to reproduce compared with the much more direct way of simply making a copy of oneself. As noted in Chapter 1, evolutionary explanations of biological phenomena must be made in terms of relative reproductive success. If we want to know why animals reproduce sexually, we must ask why it is to the species' reproductive advantage to have offspring in this manner. The answer seems to be that sexual reproduction leads to the controlled generation of variation in one's progeny (Williams, 1966). Producing young that are similar, but with a

Benjamin D. Sachs' participation in this chapter was in part supported by the National Institute of Child Health and Human Development Research Grant HD-04048.
194

range of differences, is likely to be advantageous because the environments in which they will grow up are likely to be variable from place
to place and season to season and year to year. The fact that sexual
reproduction basically throws together all sorts of combinations of genes
in the union of eggs and sperm means that parents will produce offspring
each with an unique genotype, sharing some features in common with
each parent and each sibling, but different nonetheless. The creatures
that develop will therefore exhibit a range of phenotypic differences.
These different phenotypes may exploit a range of different environments
and, on the average, reproduce more successfully than if each were a
carbon copy of one parent.

   If this argument is correct, one would expect nonsexual reproduction
to occur in highly stable conditions in which one genotype is likely to
be very successful. As Williams (1966) points out, plants that reproduce
both asexually and sexually do so in such a way that their asexually
reproduced offspring grow up right around them in conditions where the
parent's genotype has already proven itself. For example, one form of
asexual reproduction in plants involves the dropping of leaves which later
develop into new individuals right in the vicinity of the parent tree. The
same plant reproduces sexually through the production of pollen or seeds
that are likely to be dispersed a very great distance from both parents.
By producing a variety of genotypes the plants take advantage of the
fact that their "offspring" will surely encounter a variety of conditions
after dispersal. Similarly with animals, one would predict that asexually
reproducing creatures would live in conditions which were relatively
constant and produce young that were unlikely to scatter widely into a
variety of habitats. In fact, there are relatively few animal species that
reproduce *only* asexually; many have the option of sexual reproduction,
which they exercise in times of environmental change (Bermant and
Davidson, 1973).

   Although sexual reproduction is extremely common (a reflection of
the fact that there is so much temporal and spatial variation in environments), it does pose some problems for the individuals that employ it.
The chief and most obvious problem is how to insure that the male and
female gametes will get together to unite and form a new complete
genotype. When two organisms are involved in the reproductive process,
there is more room for error than if each is reproductively independent.
There are several fundamental *behavioral requirements* made on animals
operating in this system that are absent under conditions of single
parentage.

   First, the animals must be able to get physically close to one another. If they are not normally physically close enough to allow sperm
to be transferred, then they must be able to move to the proper position.
Of course some animals are sessile, i.e., attached permanently to a single
spot. These animals must depend on water-borne currents to carry their
gametes together.

   Second, it is not generally sufficient that an animal move about at random; its movements must be guided or directed so as to maximize the likelihood that it will encounter a proper partner. An animal moving about in its environment may encounter numerous other animals, only some of which are suitable reproductive mates. *The proper mate is an adult animal of the same species but opposite sex who is in breeding condition.* From this description we can see some of the errors that must be avoided if the system is to function properly. Consider for example an animal such as the goldfish, which sheds its gametes into the water and hence relies on fertilization taking place externally. If the goldfish releases gametes when it is alone, or in the presence of a juvenile or a same-sexed goldfish, or a fish of another species, then from the biological point of view the fish has wasted its gametes and the reproductive system has failed. It is thus a general behavioral requirement that an animal be able to locate a proper mate.

Having met the first two requirements, the animal must satisfy the demands of a third, more subtle task. We can understand this requirement by putting it first in human terms. During the early stages of a potentially reproductive relationship, a man and a woman work to discover each other's intentions. These discoveries are important because further participation in the reproductive sequence will involve exposure and vulnerability for both of them. In our own case the exposure and vulnerability are primarily to social forces. In American society, for example, an unmarried girl who engages in sexual intercourse with the understanding that her partner's intentions include the continuation of the relationship on a nonsexual basis as well, and later finds that he has no such intentions, may come away from her experience feeling both used and threatened. And if she continues to make the same mistake in assessing her partners' intentions she may eventually find herself in a very unfortunate social position, one from which she will find it difficult to move into marriage, the socially sanctioned foundation for completion of reproduction.

This example is sufficient to convey what we mean by the importance of discovering intentions. How does this idea relate to the behavior of other animal species? To begin with, the dangers to which other animals expose themselves by participating in the reproductive process are even more direct and obvious than they are in our own case. One danger is from physical attack. A clear example is seen in the case of the hamster. A female hamster will attack and bite a male that approaches her when she is not sexually receptive (Payne and Swanson, 1970). If the male is to remain unscathed, he must be able to discern the female's condition. Numerous other examples, particularly among birds and fishes, show that the behavior of an animal during the early stages of the reproductive process often serves to allay hostile or fearful responses to the potential mate (Hinde, 1970). This very important point will arise again later, but for now it is enough to emphasize that the common elements of exposure and vulnerability produce a similarity between the human need to discover the intentions, and the nonhuman need to resolve the conflicting behavioral tendencies, that exist during the early stages of reproductive activity. This is the third behavioral requirement of reproduction through dual parentage.

For a number of animal species there is still a fourth behavioral requirement for the successful completion of reproduction. Recall that the definition of a "proper mate" included a reference to the animal's *breeding condition.* This condition has two aspects. First, each animal in the pair must be fully *behaviorally receptive* to the reproductive responses of the other. The development of receptivity is partially under automatic physiological control, and partially under the control of the physical and social environment. The relative importance of internal and external factors in producing receptivity varies from species to species. In many species the social interactions between members of a potentially reproductive pair are very important for the development of full receptivity. The second criterion of breeding condition is that the mates be *physiologically capable* of releasing gametes at the right time. This capacity is partially independent of behavioral receptivity and in some cases is dependent upon stimulation from the partner. For example, female cats and rabbits become behaviorally receptive at regular intervals during the year. But ovulation depends upon copulatory stimulation from the male. In order for reproduction to be successful, therefore, the male is required not only to deliver the sperm but to deliver it along with stimulation adequate to trigger ovulation. The point of these examples is to show that in several species there is a behavioral demand on one or both members of a potentially reproductive pair to produce in the other member the physiological capacity to complete the reproductive process.

The fifth requirement we will mention is as obvious and basic as the first: both members of the pair must know how to get their gametes together. This would be a trivial point were it not for the general fact that no two animal species join their gametes in exactly the same way. For example, the copulatory sequences of mice, deer mice, hamsters, guinea pigs, and rats are all quite distinct. Among mammals, the species differences are greater for the males than the females, but the point remains in these cases as in all others that the deposit of sperm into the vaginal passage demands a substantial amount of whole-body coordination from each of the individuals and substantial cooperation between them. Successful insemination is unlikely to occur accidentally. It is of great interest, then, to discover how animals acquire the information that allows them to meet this final behavioral requirement.

The thoughtful reader will have noted that these five behavioral requirements are not completely independent of each other and that they are not the only ones that could be listed. But they do give a clear idea of the biological significance of courtship and mating behaviors, *for these behaviors are just the ways that every species has evolved to cope with the requirements of dual parentage.* The distinction between courtship and mating is one between the earlier and later stages of a social relationship that leads to the fertilization of ova. Taken together, courtship and mating constitute the first major segment of the reproductive process. The second segment of the process, the development of the fertilized egg and the care given to it by its parents, is discussed in Chapter 8 of this book.

The remainder of this chapter contains two sections, one on courtship and one on mating behavior. Each section is divided into two parts, the first of which deals with some physiological aspects of the behaviors discussed and the second with the function of the behaviors in their natural environments.

Courtship

In many animal species there is a substantial period of time between the first encounter of a male and female and the first copulation or spawning. During this time both individuals are likely to engage in a sequence of activities which apparently contributes to eventual mating and, in some species, to the formation of an enduring pair bond. There are two fundamental questions to ask about these courtship activities. One is what causes them, and the other is what functions they serve.

The Causal Basis of Courtship: Ethological Theory

One approach to questions about the motivation and physiology of courtship is provided by ethological theory, which holds that three response tendencies—fighting, fleeing, and mating—may be activated in male and female prior to copulation. From the competition among these tendencies within the organism emerge those responses we recognize as courtship. (We cannot here review more than the basics of this theory. For a more detailed summary, see Hinde, 1970; Tinbergen, 1953b; Morris, 1956, reprinted in Morris, 1970.)

Several types of behavioral evidence, not all independent of each other, have been marshalled to support the ideas that in the sexual activities of animals more than just sexual motivation is present, and that the plurality of motivations may not only coexist, but may be in conflict.

Primacy of Attack in Courtship
The initial responses of males toward females are often indistinguishable from the agonistic responses of males toward other males. For example, the bowing coo of male ring doves is the common initial behavior toward females as well as males (Barfield, 1971a). So too the fluffed posture of male chaffinches is initially given to females entering upon a male's territory as well as to intruding males (Hinde, 1953; Marler, 1956). These attack postures are particularly likely to occur when the male and female of a species look very much alike; the agonistic behavior tends to be

modified or eliminated as the female, by her reactions to the male, makes known both her sex and her state of receptivity.

Oscillation or Alternation Behavior
Part of the courtship behavior of the male three-spined stickleback is a "zigzag" display consisting of alternate approaches (zigs) toward and withdrawals (zags) from the female. A "zigzag walk" is also typical of a stage of courtship in the chaffinch. Among many gulls the courting male alternates postures of pointing toward and facing away from the female. These and other examples are taken as evidence for a conflict between competing tendencies. A similar explanation has been given for the courtship display of many galliform birds (e.g., turkeys, quail). The male circles the female in a manner called waltzing or strutting, and the circling is viewed as the result of or compromise between the competing tendencies of approach and withdrawal.

Presence of Attack Components in Copulation
In many animals, responses characteristic of attack on conspecifics or prey are also a part of the copulatory act. Among many primates, females or subordinate males express their submission by presenting their hindquarters toward the dominant animal. Dominant males may then mount and thrust in a manner very similar to the initial behavior of normal copulation. Among the cats and some other carnivores such as minks and ferrets, the male bites the female's neck during copulation. Since this type of behavior occurs primarily in predators, it is probably not a coincidence that the killing bite is directed at the same region of the neck.

Correlations between Acts in a Sequence
If one act is always followed by attack and another act is always followed by copulation, then the behavioral tendencies or motivations underlying those acts are clear: aggression and sex, respectively. However, if a display is variably followed by attack, escape, or copulation, then competing tendencies are assumed to be involved in the display. Through a statistical technique called *factor analysis,* Wiepkema (1961) analyzed the correlations among the sequence of courtship acts displayed by the male Bitterling, a fish. Many of these behaviors fell into one of three categories (factors), labeled as aggressive, sexual, and nonreproductive. Other behaviors had high "factor loadings" in more than one of the categories and were assumed to have multiple, possibly competing, tendencies underlying them. A fuller description of the bitterling's behavior is given later in the chapter.

Intention Movements, Displacement, and Redirection Activities
These responses play an important role, not only in the evidence for conflict in courtship, but also in the discussion of the evolution of courtship and other social behaviors.

Intention movements are the initial components of normal behaviors. For example, the leg flexion and head elevation shown by birds prior to flight are called intention movements. Occasionally intention movements occur without the rest of the sequence. The failure to "follow through" on intention movements is believed to be the result either of a conflict between tendencies, or an incomplete activation of one tendency.

Displacement activities are actions that are inappropriate to the context in which they are observed. Preening or eating in the context of a sexual or aggressive encounter might be viewed as displacement activities especially if they do not occur in their normally complete form. Though there is doubt about the origin of displacement activities, one frequent view has been that they emerge when other tendencies are in conflict or thwarted.

Redirection activities refer to appropriate acts directed at inappropriate objects. This concept is roughly equivalent to the Freudian concept of displacement, e.g., the man who kicks his dog instead of the person who angered him.

These terms may be clarified by an excerpt from an ethological analysis of the Chilean Great Skua, a bird in the gull family (*Laridae*). This passage also contains an illustration of the ethological view of threat and courtship as an interaction of drives.

Sometimes a Chilean Great Skua will peck violently at the ground instead of attacking an opponent. This appears to be a form of redirected attack. . . . Redirected attack pecking at the ground sometimes develops into a more complex and spectacular performance. A bird may not only peck downward, but also seize bits of vegetation . . . in its bill and pull them very vigorously. . . . The grass-pulling of the Chilean Great Skua is much rarer than ordinary redirected attack pecking at the ground. I only saw it performed by a very few birds during the most lengthy and violent boundary disputes. The circumstances in which grass-pulling occurs would suggest that it is produced when the attack drive is approximately as predominant over the escape drive as during ordinary redirected attack pecking at the ground, but both the attack and escape are actually stronger than they are during all or most ordinary pecking at the ground. . . . The Chilean Great Skua sometimes assumes a "bill-down" posture which may be related to grass-pulling and ordinary redirected attack pecking at the ground. . . . It is probable that [these] bill-down postures are simply low intensity indications of a desire to peck at the ground, i.e., "intention movements" of pecking at the ground. . . . Skuas will sometimes do brief lateral head-shaking movements during hostile encounters. These movements are very similar to, or identical with, head-shaking movements which occur during ordinary cleaning or "comfort" activities. At first sight, they appear to be "out of context" in hostile situations; and they are the sort of patterns which are sometimes considered to be "displacement activities". This is by no means the only possible interpretation of such movements, however. (Moynihan, 1962, p. 4–6)[1]

Analysis of the pairing behavior of the Chilean Great Skua revealed that

it includes many hostile and even more ambivalent patterns, produced by activation of the attack and escape drives, with or without simultaneous activation of the sex drive or drives (and possibly some other types of motivation such as nest-building), in addition to those patterns which seem to be overt expressions of the sex drive(s) alone. (ibid, p. 28)[1]

The Physiology of Courtship

Although there is some disagreement and ambiguity among the ethological theories of courtship concerning the exact nature of the interactions among fighting, fleeing, and copulating tendencies, most agree that courtship results from an interaction among these tendencies. However, the evidence for these theories is with few exceptions derived from strictly behavioral studies in field and laboratory. Despite a large body of literature on the neural and hormonal control of sexual and aggressive behaviors, few studies bear directly on the question of courtship as an interaction among competing tendencies. This lack of evidence results partly from the fact that so few physiological studies have been designed specifically to test the conflict hypothesis. Experiments that are limited to observing only sexual or only aggressive behavior can be of only marginal relevance. An additional limiting factor is that few neuroendocrine studies have been done on the kinds of animals from which the conflict hypothesis was derived, namely, birds and fishes. Most of the physiological work has been done on mammals, notably laboratory rats and mice, in which courtship is negligible or absent. We are also limited by the fact that most of the physiological and behavioral work has been done on male animals, although in many courting species the female engages in soliciting behaviors that, in principle, should be accountable for by the conflict theory of courtship.

Before considering the physiological implications of the conflict theory of courtship, it will be useful to digress briefly to consider a closely related ethological theory and an experiment that was designed to test that theory. Briefly stated, it is believed that threat behaviors result from a conflict between tendencies to fight and to flee, two of the same tendencies believed to be present in producing courtship responses. Brown and Hunsperger (1963) electrically stimulated subcortical areas in the brains of cats and observed the agonistic (fight, fear, threat, escape) responses that resulted from stimulation in different areas. Although they were able

---

[1]Reprinted with the permission of E. J. Brill and Martin Moynihan.

to elicit "pure" escape in a large "outer zone" of the brain stem and forebrain, stimulation within this zone always yielded either threat behaviors alone or attack preceded by threat. They did not find any area that, upon stimulation, yielded attack alone. Simultaneous stimulation of escape and threat areas tended to increase threat responses or delay escape responses. However, because they were unable to find a "pure" attack site, Brown and Hunsperger were unable to do the experiment of simultaneous stimulation of "pure" attack and "pure" escape. Their failure to find "pure" attack sites, their finding that attack was always preceded by threat, and the nature of the observed interaction between simultaneous threat and escape stimulation, led Brown and Hunsperger to doubt the accuracy of the ethological theory that threat represents an interaction between attack and escape. In particular, they doubted that threat is properly viewed as resulting from simultaneous arousal in, and mutual inhibition between, pure attack and pure escape areas. While it has been argued that Brown and Hunsperger's data are not so incompatible with the conflict hypothesis as they suggest, there is general agreement that their reasoning from ethological theory to necessary physiological consequence was sound, and serves as a model for other experiments in the area.

We will now attempt to specify certain relations between neuroendocrine variables and the ethological theory of courtship. To simplify the logic somewhat, we will deal only with the conflict between fighting and copulating, and will omit the hypothesized tendency to flee.

First, one may expect that the tendency to fight and to copulate would be under similar tonic hormonal control. This conclusion follows because unless both aggression and mating are potentiated at the same time, i.e., during the mating season, they could not be activated simultaneously in the way ethological theory asserts would bring about courtship behavior. On this point there is ample evidence that, at least for males in a broad range of species, both fighting and copulatory behaviors require the presence of male gonadal hormones (androgens). Castration reduces the tendency of males to fight or to copulate, and treatment with androgens following castration restores the sexual and aggressive tendencies. There is, however, considerable variation among species in the time that sexual and aggressive behaviors are maintained after castration. In general, carnivores and primates maintain these behaviors after castration for a longer time than do other mammals, especially if they have had precastration experience in these behaviors; we return to this point later in the chapter.

Second, one may expect to find well-defined neural systems in which electrical or chemical stimulation would elicit fighting alone or copulation alone. Data collected from several mammalian species show that electrical stimulation in restricted areas of the forebrain and the forebrain-midbrain border will lead males to attempt copulation (Caggiula, 1970). Humans have reported sexual fantasies and sensations during electrical stimulation of these same brain areas (Heath, 1964; Moan and Heath, 1972). However, these data have only marginal relevance to the conflict hypothesis because the experiments were generally structured so that sexual behavior was the most probable response. That is, the animal experiments were structured so that receptive females were available, but other males, food, and water, were generally absent. Thus we cannot yet be certain that the areas stimulated could properly be called "pure" copulation areas. With respect to "pure" aggression areas we are even less informed, since a good many of the studies available, especially those using cats and rats as subjects, have actually not studied intraspecific aggression, but rather what may be called predatory aggression—attacks upon mice, for example, even if the mice are not then eaten.

The search for "pure" fighting or copulation sites raises the question of how specific a function is served by restricted brain areas. This problem has received a great deal of attention through the years, most recently through some experiments with rats reported by Valenstein and his colleagues (1970). What made these studies unusual was that the subjects had available to them a "cafeteria selection" of stimuli: food, water, and wooden dowels. When the rats were stimulated electrically in the lateral hypothalamus, Valenstein *et al.* observed the commonly reported stimulation-bound activities, e.g. eating, drinking, and gnawing on the wooden dowels. Most of the animals fixed on one or another of these responses whenever the stimulation was turned on; from this one might have inferred a functional specificity for each animal's electrode site. However, after each rat had fixed upon a particular stimulation-bound activity, that rat's "preferred" stimulus was removed. Under these conditions many of the animals eventually switched to another stimulation-bound activity, and they maintained that new activity even when the originally preferred stimulus was returned to the testing cage. From this result Valenstein *et al.* concluded that there was probably a great deal less specificity of function than had previously been thought. Although this interpretation has been questioned (Wise, 1968; Roberts, 1969), there seems little doubt of the importance of making several response possibilities available to the subject if conclusions about specificity of function are to be made with confidence. Thus, studies seeking fighting or copulating sites in the brain ought to test the subjects with males as well as with receptive and unreceptive females as goal objects.

Third, if courtship results from an interaction of neural activity in "sexual" and "aggressive" systems in the brain, then courtship should result when these systems are stimulated simultaneously, especially in the presence of a suitable female to be courted. On this point again the available evidence is minimal. There simply is no appropriate study using electrical stimulation. One relevant experiment used small implants of hormone into the brains of capons and obtained promising but not conclusive results. Untreated capons demonstrated no aggression, no copulation, and no courtship. Capons implanted with testosterone in the preoptic area of the brain copulated, but did not aggress or court. Other capons with testosterone implanted in the lateral forebrain, showed aggression but not copulation or courtship. Finally, a few capons with testosterone implants in both the lateral and preoptic areas exhibited courtship prior to copulation; this is the result predicted by ethological theory. However, this finding needs to be replicated with a larger number of animals (Barfield, 1969).

Similar experiments using ring doves (*Streptopelia risoria*) have yielded different results (Barfield, 1971a; Hutchison, 1971). In these animals the full pattern of courtship behavior, as well as aggression and copulation, could be restored in castrated males by implantation of androgen into any of several brain sites. Whether the difference between the effects in domestic fowl and in doves stems from differences in their social structure, in degree of sexual dimorphism, or some other factors is not yet clear, but the difference again points up the value of comparative studies and the problem of generalizing from research on only one species (Barfield, 1969).

Fourth, whenever stimulation of a particular neural site is found to yield full courtship behavior, it should be possible to demonstrate that this site is neurally downstream, that is, receives neural input from other sites that, when stimulated, elicit only copulation or aggression. This conclusion follows from the ethological theory of courtship, but no clear supporting data are yet available, though the experiments on capons and doves mentioned above are clearly relevant. One would expect to find an integrating site in the capon where a single implant of androgen would lead to full courtship displays. Similarly, one would expect to find sites in the dove where pure copulation or pure aggression could be stimulated.

Neither the lack of available data, nor the few experiments with negative results, are adequate grounds for dismissing the ethological hypotheses about courtship. Negative results are especially difficult to interpret, for the failure to find what one was looking for does not necessarily mean that it does not exist. It may mean only that one was looking in the wrong place or with the wrong tools. Only when a considerable body of data is available will we be able to judge adequately whether the theory that courtship arises out of a conflict among competing response tendencies will bear fruit in understanding the physiological basis of courtship. The notion of conflict, Hinde (1970, p. 361) reminds us, ". . . is not used to refer to a hypothetical or physiological state, but to incompatible response probabilities." Presumably the physiological basis of the conflict thus defined may vary among species, as even the available data on capons and doves suggest.

Having discussed some of the approaches to problems surrounding the motivation and physiology of courtship behavior, let us now consider some ecological aspects of these activities.

The Functions of Courtship

One of the most striking aspects of reproductive behavior is the enormous variability and complexity of courtship responses displayed by existing species. In fact, courtship behavior may be more variable than any other class of responses. Why should this be the case? We can approach an answer to this question of diversity in courtship behavior by considering some of the obstacles to successful reproduction faced by each species, and some of the factors that may influence the way in which these obstacles are overcome.

Location and Attraction of Mate

As we have already noted, the most obvious problem for a species to solve is the physical separation of male and female organisms. If males and females live together in closely knit groups at all times, then the problem of locating a mate during the breeding season is relatively trivial. In such diverse groups as schools of fish and troops of baboons, there is little courtship. Among some species of seals, the females do not really locate or choose a male, but rather they locate each other and places to bear their young. They congregate in groups in certain areas, and these areas are defended by bull seals against each other. When the females become sexually receptive they are usually mated by the male that occupies the territory on which the females give birth. Courtship may also be absent among some solitary, noncolonial species, due to reliance on secondary factors to bring potential mates together. The males and females of some flies, for example, meet because they are attracted to the same feeding sites, where they come into contact with each other.

Paradoxically, there are other species, including many birds, that have a group social organization through most of the year, yet, as the breeding season approaches, disperse over a considerable area. As a result of this dispersion the males may engage in elaborate courtship displays in order to attract females, perhaps the very same females that they grouped with during the nonbreeding season. This paradox alone would suggest that courtship serves other functions than mate location, for if that were the only function, then it would be more economical of the energy of animals to mate directly in the colonial situation.

For most animals in which the two sexes are independent organisms, mating requires the broadcast of signals that have the primary effect of informing potential partners of the broadcaster's location and readiness to mate. These signals may be given by the male (e.g., the songs of many birds) or the female (e.g., the airborne chemicals of the monarch butterfly) or both (e.g., facial expressions in *Homo sapiens*).

The range over which the signals need to be effective will vary with the social structure and ecology of the species. For example, a solitary species inhabiting woods or the floor of grasslands cannot rely on visual cues and may use olfactory or auditory cues. When potential mates have been brought into proximity by the long-range broadcast, there may be a shift in stimulus modality appropriate to the lesser distance between mates, for example, to visual and tactual cues.

Location of Mating and Nesting Sites

Many courtship displays advertise not only a potential mate, but also a place where copulation, nest formation, and rearing of young can be carried out. This is particularly true in egg-laying animals in which the eggs may be fertilized and incubated at the same location. In some species, demonstration of the nest site, or of the nest itself, or of nest-building activities *per se,* form an integral part of the courtship sequence. This type of behavior has been thoroughly studied in the three-spined stickleback (Tinbergen, 1951). In this fish, the male establishes a territory and builds a nest on it. Females that enter the territory are courted by a "zigzag dance," the zigs being swimming movements toward the female, the zags being movements away from the female toward the nest. The nest is approached more closely by the male and female with successive zigzags until the male places his nose into the nest opening; a fuller description of this behavior is given below.

A similar behavior has reached astonishing proportions in many species of the bower birds of Australia and New Zealand. The males build elaborate structures (bowers) of vegetation and mud, often enhancing the visibility of the bowers by incorporating bright shells or flowers into the construction. The males then exhibit these nests by posturing at the entrance. Copulation occurs within the bowers, but the female then leaves and builds a maternal nest nearby. It is interesting that there is an inverse relation between the elaboration of a bower and the brightness of the male's coloration.

Synchronization of Reproductive Condition and Mating Behavior

Most vertebrate species do not breed the year round. It has often been noted that vertebrate breeding seasons and hatching or gestation periods are coordinated in ways that lead to the occurrence of birth or hatching under optimal environmental conditions, e.g., in the spring for many species living in temperate zones (Bermant and Davidson, 1973). This relationship is one example of an adaptive synchronization of the reproductive process with the environment that surrounds it. Two other synchronizations are required for successful reproduction. First, as we have already mentioned, males and females must become physiologically and behaviorally ready to mate at approximately the same time. Second, the motor responses of the two sexes also must be synchronized during mating if fertilization is to result. An important function of the courtship behaviors of many species is the development and maintenance of these social synchronizations.

A careful analysis of the role of courtship in promoting reproductive readiness and behavioral receptivity has been conducted by Daniel Lehrman and his associates using the ring dove (*Streptopelia risoria*) (e.g., Lehrman, 1965). A description of ring dove courtship and parental behaviors appears in Chapter 8 of this book; here we want to emphasize the interaction of behavioral and physiological variables that characterize the dove's reproductive cycle. For example, a female dove can be brought to a moderate level of reproductive readiness (as measured by the development of her oviduct) by exposing her to optimal conditions of lighting, temperature, and so on; however under these circumstances she will ovulate only rarely. If, however, under these circumstances she is placed with a male dove who shows appropriate courtship behavior, then the female undergoes a series of physiological changes which promote ovulation and the tendencies to engage in copulation, nest-building, and incubation. The facilitation of these processes in the female is directly related to the duration of her exposure to the male's courtship display (Barfield, 1971b). Males that have been castrated do not exhibit courtship behavior; females placed with castrated males do not ovulate. If male sex hormone (androgen) is injected into castrated males, they will show some courtship behavior; within limits, the amount of courtship exhibited increases with increased doses of androgen. Males with relatively little androgen circulating in them have less effect on the female's reproductive physiology than males whose androgen levels have been completely restored (Erickson, 1970). As we shall see in Chapter 8, this socially-produced effect on physiology is one of several that occur during the dove's reproductive cycle. Continuing reciprocity of stimulation is characteristic of the reproductive behavior of many species.

In many species the postural and temporal adjustments required for copulation would be simply impossible without the cooperation of both members of the mating pair. Courtship is often used to solicit cooperation from the female. In the guppy, a fish in which true copulation exists, the courtship is directed not at forming a pair-bond, but rather at soliciting the female's participation in copulation. Even in species where no true courtship seems to exist, short term stimulus-response interactions may be used to assure synchrony at the moment of copulation. In order for a male rat to achieve insertion the female must stand still and elevate her hindquarters in the position called lordosis. If the female resists or even ignores the male's attempts to mount, then he cannot achieve insertion. Occasionally one sees a female resist a male's mount attempts with great vigor, and yet if he succeeds in clasping her flanks between his arms she may exhibit lordosis, and he may succeed in copulating (see below for discussion of the importance of sensory integrity). In other species the mutual adjustments of posture required for copulation may be much more elaborate (e.g., birds that copulate in flight), and in some it is the female that plays the active role.

Intrespecific Reproductive Isolation

Our modern biological concept of the animal species is based upon a criterion of reproductive behavior: a species is "the largest and most inclusive . . . reproductive community of sexual and cross-fertilizing individuals which share a common gene pool" (Dobzhansky, 1950). Inclusion within a species is dependent upon the potential, under natural conditions, for interbreeding. Similarly, the biological distinction between two species is based upon the lack or failure of interbreeding, again under natural conditions:

Species are more unequivocally defined by their relation to nonconspecific populations ('isolation') than by the relation of conspecific individuals to each other. The decisive criterion is not the fertility of individuals but the reproductive isolation of populations. (Mayr, 1963)

Separate, closely related species that occupy the same geographical area (i.e., are *sympatкic*) do not interbreed; the mechanisms which operate to prevent successful interbreeding are known as *reproductive isolating mechanisms*. There are several reasonably complete discussions of reproductive isolating mechanisms in the literature (Bermant and Davidson, 1973; Mayr, 1963; Spieth, 1958); here we want to emphasize only that in many cases it is the species-specificity of courtship behavior that serves to keep two closely related species from interbreeding. One of the most interesting examples of this function of courtship occurs among the sympatric species of fiddler crab (genus *Uca*) that live on the beaches of tropical America. The male crabs use their outsized claw (cheliped) in a more or less elaborated "semaphore signal" courtship

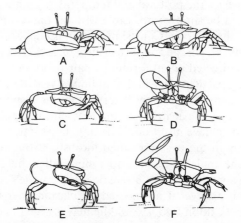

From "Basic patterns of display in fiddler crabs (*Ocypodidac*, genus *Uca*)," by J. Crane (1967), *Zoologica*, **42**, 69–82. Photo from New York Zoological Society.

Figure 7-1 Species signaling patterns. Three species of fiddler crab genus *Uca* exhibit different, species-specific cheliped motions during courtship. Illustrated in a) and b) are *Uca rhizophorae;* in c) and d), *Uca signata;* in e) and f), *Uca samboangana*.

display. As is shown in Figure 7-1, different species have distinctively different signaling patterns. Females from one species are generally unresponsive to the signaling of males from another species (Crane, 1941; 1957).

Rarely is there only one mechanism at work in the separation between two species. In one case, however, which involves two species of acridid grasshoppers, the only barrier to hybridization rests in the females' discriminations of auditory differences in the "songs" (stridulations) performed by males in breeding condition: the songs of the two species have different average frequencies and tempos (Perdeck, 1958). As might be expected, this barrier is not completely effective, with the result that viable fertile hybrids are found in zones of overlap. However, the rate of hybridization is apparently not large enough to produce a significant fusion of the species' gene pools.

The Formation of Pair Bonds

At least theoretically, any male of a species should be able to breed successfully with any female; indeed the genetic advantages of sexual reproduction would be maximized, other things being equal, if there were random breeding among the members of a species. In fact there are always factors operating in natural populations that restrict mating opportunities. Some of these factors are themselves part of the basic "machinery" of social organization within the species. In fact one of the major functions of courtship is the production of a more-or-less long lasting bond between a pair of individuals that reduces their exposure to others. There is a substantial diversity in the variety of "mating systems" that result from this activity; it is difficult to give a systematic account of the relations between mating systems and other variables in a species' existence, although attempts at systematization have been made, as we shall see. First we shall present some examples of different mating systems.

Most mallard ducks form monogamous pairs in the fall and winter after communal displays of groups of males to single females. However, the pairs that form in the fall do not copulate until the spring, and copulation is then preceded by displays between male and female that are different from the displays shown during initial pair formation (Bastock, 1967). Copulation is followed by a period of egg-laying and incubation. Pairs break up during the incubation period or just afterwards, but the same ones may be reformed in the following year. In this case the male will not engage in the displays that normally precede pairing (Weidmann, 1956; cited in McKinney, 1961). The mallard, then, is a seasonally monogamous species in which copulation occurs only after a considerable period of nonsexual social interaction.

Ring doves, as studied in the laboratory, also form monogamous pairs (Lehrman, 1965). The bond is formed during a characteristic courtship in which male and female cooperate in locating a nest site and building a nest. Copulation, however, occurs without a preceding display, sometime during the nest building period. Both sexes participate in the incubation of the eggs and in brooding and feeding the young. The bond lasts at least until the young are fledged and may continue through repeated cycles.

Among wild Norway rats and many other rodents, males and females are promiscuous within the limits set by the social structure of the colony (Calhoun, 1963). There is no pair formation and no proper courtship, though there may be extensive precopulatory responding. Unless there is a dominant male in the burrow system, many males may copulate with each receptive female. The female may also solicit copulation by orienting her hindquarters toward the male and darting about in front of him. Generally, copulation occurs within a minute or so after the initial contact between a male and a female at the peak of her receptivity.

The mating system of some species involves a changing pattern of male-female relations, so that terms like monogamous and promiscuous are inappropriate for describing the whole pattern. For example, among savanna baboons (DeVore, 1965), each female may copulate with several males as she approaches maximal receptivity around the time of ovulation. However, in the days immediately before and after ovulation the dominant, or *alpha,* male may reserve the female for himself. Thus during the course of an estrous (menstrual) cycle, the female is promiscuous. Over the course of several females' cycles the males are also promiscuous. Yet, at the time of each female's maximum fertility she is in an enforced monogamous relation with the alpha male.

The cataloging of the enormous variability in mating systems, by itself, offers no scientific appreciation or understanding. For these we need to know the sources of the variability. Since closely related species may have rather different mating types, and distantly related species may have very similar mating types, the correlation between the taxonomic position of a species and its mating type will be imperfect. We must then look to other origins of the variability in mating types. One way to phrase the question is: what selective pressures during evolution have conveyed adaptive value to the different mating systems? One approach to answering this question is to look for correlations between mating systems and other features of the ethology and ecology of the species.

In one such recent analysis, Orians (1969) started with the fact that females of all species produce relatively few eggs, while males produce very large numbers of sperm. A consequence of this discrepancy is that males can produce more offspring by mating with more than one female, but females will rarely produce additional offspring by mating with more than one male. A further corollary of the sex difference in gamete production is that errors in mate selection are more serious when the female makes them. For example, a single infertile mating may eliminate a season's contribution to the population by a female, but an infertile mating by a male may cost only a small fraction of his potential contribution. Because mating errors are much more costly for females, it is likely that females will exercise stronger mating preferences than males. Due also to the sex difference in gamete production (as well as other factors), polygamy (more than one mate) is more likely to take the form of polygyny (more than one "wife") than of polyandry (more than one "husband").

Factors other than the relative number of gametes include the role of the male in rearing young, the amount and variety of potential food, the availability of nesting sites, and the number of predators. From examination of all these factors Orians generated a number of broad principles. For example:

1) Polyandry should be rare among all animal groups. We have already seen the reason for this assumption.

2) Monogamy should be rare among mammals but predominant among birds. This principle follows from the lesser parental role of the male in most mammalian species. The exceptions, that is, the monogamous mammals, are primarily those groups (felines, canines) in which the male does play an important parental role, usually in bringing food to the mother and young and defending them against predators.

3) Polygyny should be more common among precocial (mobile-at-hatching) birds than among altricial (immature-at-hatching) species. This principle follows again from the parental role of the male, which is reduced if the precocious young are able to feed independently soon after hatching.

The knowledgeable reader may note exceptions to these and the other principles suggested by Orians. However, such exceptions do not necessarily detract from this attempt to bring order from the chaos of mating systems. On the contrary, the discovery of lawful exceptions can generate additional principles. For example, Orians noted that most species of ducks constitute exceptions to the third principle, since they tend to be monogamous and precocial, with the male offering little or no parental care. However, "the prevalence of monogamy . . . may be the result of the advantage of pair formation on the wintering ground and rapid initiation of breeding which give a stronger advantage to monogamy for the males than would otherwise be the case." (Orians, 1969, p. 597). Significantly, tropical ducks tend not to be monogamous.

Fuller analyses of the correlation between social organization and ecological variables in birds and in primates have been made by Crook (1965; 1970) and Rowell (1969). One important point that emerges from these analyses is that we may err if we characterize a particular social organization or mating system as typical for a species. Within a single species, groups living in different ecosystems may have rather different social organizations.

This sketch of the various functions of courtship has had two major points. First, courtship activities are diverse both in form and function. There is no single reason why animals engage in these behaviors and most importantly, the ultimate "goal" of courtship may be substantially more than just copulation. Courtship is the foundation for pair bonds, which may be vital for the survival of the offspring (see Chapter 8) and the reproductive success of individuals. Second, the differences between species in courtship often make ecological sense. That is to say animals differ in courtship behavior because the environmental pressures operating on their ancestors were different. This leads to the evolution of responses suited to meet particular ecological demands confronting different species.

Mating Behavior

Courtship concludes with mating, during which the members of the courting pair join their gametes. We have seen that there is marked diversity among species in the details of their courting behaviors, and that the reasons for this diversity are both physiological and ecological. Differences between species also extend into the behavioral details of mating behavior; here again we may look both to the physiology of the animals and to their environments for an understanding of these differences.

The Causal Basis of Copulation

The focus of this section will be on the data and theory of copulatory behavior that have been developed in laboratory studies. Most of these studies have been pursued from the viewpoint of the physiological psychologist. The emphasis has been on the relation between the precise measurement of copulatory behavior under carefully controlled conditions, with an eye toward determining behavioral capacities and discovering the physiological, particularly neural and hormonal, correlates of the behavior. Rodents bred for generations in the laboratory (such as rats, mice, and guinea pigs) have been the preferred subjects of investigation, though data have also been collected with cats, dogs, and several species of monkeys. Although the practice of relying on rats for behavioral research is occasionally criticized (e.g., Beach, 1950; Lockard, 1971), the fact is that rats have been, for several reasons, almost ideal subjects for physiologically-oriented studies of copulation. We will make this point clear by example later. For now we will emphasize only that, by using the rat, students of animal sexual behavior have been able to relate their results on the one hand to the substantial amount of purely physiological information already collected on this species, and on the other hand to the psychological theories of drives and rewards that have been based on studies of conditioning and learning in rats. Just as classical genetics relied heavily on the fruit fly and biochemical genetics relies heavily on one intestinal bacterium (*Escherischia coli*), so physiological psychology has relied on the rat.

The copulatory behavior of large domesticated mammals is of obvious economic significance, and in recent years it has come under increasingly systematic study (Cole and Cupps, 1969; Fraser, 1968; Hafez, 1968, 1969). These animals, like the laboratory rodents, cats, and dogs have been bred for generations in environments designed for human convenience. Moreover, they have been selectively bred to meet particular human requirements: high milk output, tender meat, soft wool, and so on. As a result the ethological significance of their behavior is relatively small, and most studies of their copulatory behavior have been designed to provide information relevant to maximizing reproductive efficiency (e.g., Hulet *et al.*, 1962, a, b, and c). There have also been a few experiments in which the methods or theoretical issues have been relevant to those of the psychological laboratory (e.g. Beamer *et al.*, 1969; Bermant *et al.*, 1969; Schein and Hale, 1965), and others that have investigated the neural and hormonal correlates of copulation (e.g., Clegg *et al.*, 1958; Clegg *et al.*, 1969).

Factors Influencing the Copulatory Behavior of the Male Rat

Assume that an adult male rat has been placed in a cage with a female rat that is sexually receptive; what will the animals do? To begin with, the male may do nothing at all with the female, no matter how long he is left with her or how many opportunities with receptive females he has. Some male rats, like some males of other species, simply do not copulate when given the opportunity. The percentage of males in a given population that fail to copulate depends on many factors.

First, the genetic stock from which the animals come can affect the proportion of noncopulators. In the authors' experience, for example, hooded (black and white) males of the Long-Evans strain are more likely to copulate than are albino males of the Sprague-Dawley strain. Of course the phrases "genetic stock" or "genetic background" do not explain why there are more noncopulatory males in one strain than in the other; rather they point out that part of the eventual explanation will deal with inherited physiological characteristics that influence copulatory behavior.

It might be, for example, that all noncopulating males have insufficient male sex hormone (androgen) circulating in their bloodstream; as we will see below, removal of this hormone by castration leads eventually to the cessation of copulatory behavior. If this hypothesis is true, then noncopulating males ought to copulate if they are given, by injection, additional amounts of male hormone. Whalen, Beach, and Kuehn (1961) performed this experiment on noncopulating male rats and reported that only one out of six males penetrated the female after treatment, and none ejaculated. The hypothesis was false for these animals. However, this result did not rule out any possible faulty relationship between hormones and the nervous system as a cause of noncopulation. As Whalen *et al.,* pointed out, it could be that the nervous systems of these males were insufficiently sensitive to the effects of androgen circulating in the blood. We will discuss this point further below.

Another area in which to search for the causes of noncopulation is the early history of the noncopulators. We will see below that raising animals in isolation from others can have profound effects on sexual behavior. However, this variable cannot account for all failures to copulate, because some animals raised in completely "normal" environments are noncopulators.

Further, we might look to the conditions of the environment in which the male is expected to copulate. For example, the light may be too bright (rats prefer to mate in darkness), or the room may be too noisy, distracting the male. Even under conditions of relative darkness and quiet some males require several separate opportunities before they will show their first copulatory responses. Thus Whalen *et al.,* (1961) found that only 54% of 129 males achieved penetration of the female during their initial exposure; after a total of five tests an additional 32% had achieved penetration, for a total of 86%.

Finally, we can see that it is extremely difficult, if not impossible, to infer the causes of noncopulation after the fact. An individual non-copulator may have reached his state by any one of a series of pathways, which can involve physiological, historical, and environmental factors to varying degrees. The lack of simple explanations for this behavioral disposition in the rat should probably be borne in mind when considering explanations of abnormal sexual behaviors in humans.

Let us return to the pair of rats in the cage, and assume further that the male has had prior copulatory experience. In this case the male will approach the female almost immediately. He may sniff around her face, push into her flank with his nose, or move directly behind her and put a paw on her hindquarters. At this point the fully receptive female will either dart away from the male with a characteristic quick scurry or else elevate her hindquarters and deflect her tail; (this arched position of the female is called *lordosis*). In either case, the next move of the male is to approach the female from behind and mount her. Having mounted, the male begins to thrust his pelvis against the female. The first few thrusts are typically rapid and shallow; then the male executes one much more vigorous thrust and immediately pushes himself up and away from his partner. The deep thrust and dismounting occur as one smooth behavioral sequence. The deep thrust occurs so rapidly that the untrained human observer can easily miss it.

A better understanding of what the male has done can be gained by taking slow motion photographs of the behavior while the animals are connected to an electrical circuit that operates only when the penis contacts the vaginal opening. This electrical circuit can then be used to operate various measuring instruments. In one experiment, for example (Bermant, 1965), when the circuit was closed a light went on in the field of view of the camera. From the films it was possible to determine that the brief, shallow thrusts of the male were not associated with penetration of the female. Rather, these thrusts served to orient the penis to the location of the vagina. When that orientation had been achieved, the male immediately executed the single deep thrust and withdrawal response. Normally the penis is inserted in the vagina for approximately one-fourth second (250 msec) (Bermant, Anderson, and Parkinson, 1969). This brief penetration is called an *intromission*. The male does not emit sperm or semen during an intromission.

After completing an intromission, the male will not approach the female again for approximately 40–60 seconds. During this interval (called the *intercopulatory* or *interintromission* interval) the male engages in a number of different activities, of which grooming is the most obvious. In particular, the male manipulates his erect penis and scrotum with his mouth and forepaws. Then he may walk around the cage sniffing at the walls, or lie quietly; in any case he directs very little, if any, behavior toward the female during this period of time (Dewsbury, 1967a). The interval terminates when the male again approaches and mounts the female.

The male does not succeed in achieving an intromission every time he mounts the female. In some cases he pushes away from her before making his single deep thrust; this occurrence is called a *mount*. Failure to achieve intromission is probably due to the failure of the penis to contact the vaginal opening during the period of shallow thrusting. It has been shown experimentally that if the male's penis is anesthetized, he will continue to mount and thrust against the female but will not be able to achieve penetration (Adler and Bermant, 1966; Carlsson and Larsson, 1964; Sachs and Barfield, 1970).

After the male has achieved a number of intromissions (usually between 8 and 15) and mounted the female once again, his thrusting behavior shows a marked change. To begin with, his shallow thrusting appears to be performed more vigorously; and when he achieves his first deep thrust he does not immediately jump back from the female. He may continue to thrust deeply three or four more times. His final deep thrust is accompanied by a convulsive tightening of the muscles of his hind-quarters and limbs. He continues to grip the female tightly for about five seconds, then slowly dismounts. These actions of the male are the behavioral signs of ejaculation.

After his first ejaculation the male will not approach the female again for approximately five minutes. This period of time is called the *post-ejaculatory interval*. The interval is concluded when the male achieves his next intromission. The number of intromissions preceding the second ejaculation is almost always less than the number preceding the first. The copulatory behavior leading to an ejaculation is called an *ejaculatory*

First Series

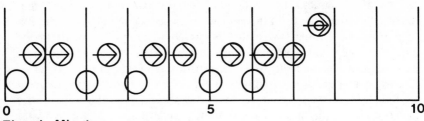

0   5   10
Time in Minutes

Second Series

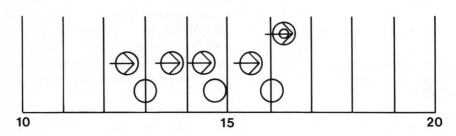

10   15   20

Third Series

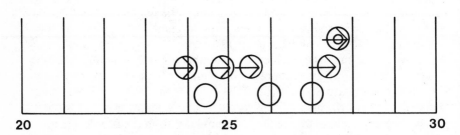

20   25   30

⭘ Mount   ⊖ Intromission   ⊖ Ejaculation

From "Copulation in Rats" by Gordon Bermant, in *Pyschology Today,* July 1967. Copyright © Communi-cations/Research/Machines, Incorporated.

Figure 7-2 The time course of normal copulatory behavior in the rat may involve five or more ejaculatory series (three are shown here). The post-ejaculatory intervals become progressively longer. The time from the intro-duction of the female into the mating arena until the first mount without penetration is called *mount latency.* The time from her introduction until the first penetration is called *intromission latency.* The time from the first intro-mission to ejaculation is called *ejaculation latency.*

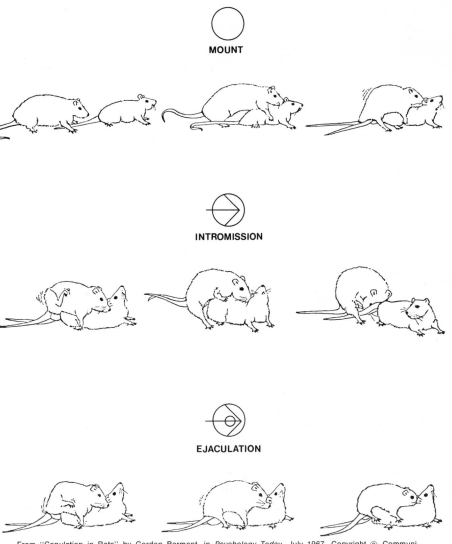

MOUNT

INTROMISSION

EJACULATION

From "Copulation in Rats" by Gordon Bermant, in *Psychology Today,* July 1967. Copyright © Communications/Research/Machines, Incorporated.

Figure 7-3 Mount without intromission, intromission, and ejaculation. The symbols correspond to those used in Figure 7-2.

series. Given access to just one female, a typical male rat (of the Long-Evans strain) will achieve between three and five ejaculations in an hour. If additional females are made available toward the end of the period, the male may achieve one or two additional ejaculations.

A summary of the rat's copulatory pattern is given in Figure 7-2; drawings of the postures are shown in Figure 7-3, and Table 7-1 provides

Table 7-1 Means and Standard Deviations of Different Sexual Behaviors During the First and Second Ejaculatory Series, Based on 74 Male Rats.

Behavior	Mean (*in seconds*)	Standard Deviation (*in seconds*)
Mount latency[1]	30.7	90.7
Intromission latency	66.5	104.7
Mount frequency		
First series	5.6	4.8
Second series	2.8	2.7
Intromission frequency		
First series	11.2	4.1
Second series	5.3	1.4
Ejaculation latency		
First series	514.1	314.3
Second series	219.3	105.5
Intercopulatory interval		
First series	51.5	37.5
Second series	44.8	25.0
Postejaculatory interval	335.8	68.5

[1] The definitions of mount latency, intromission latency, and ejaculation latency may be seen in the caption to Figure 7.2.

quantitative information collected from a large number of males.

We may use this description of copulation in the rat as a basis for a brief comparison with the copulatory behavior of other laboratory rodents: the laboratory mouse, guinea pig, and hamster. These animals share one major feature of copulatory behavior: all the males typically achieve more than one intromission before they ejaculate. However, the rates at which they achieve these intromissions, the durations of each intromission, and the numbers of intromissions achieved before ejaculation, all vary from species to species. In the hamster, for example, each complete intromission lasts for two to three seconds; the average delay between intromissions is approximately 10 seconds; approximately 10 intromissions are required for the initial ejaculation; the male pauses only about 35 seconds after his first ejaculation; and he may achieve nine ejaculations or more before his postejaculatory interval extends for 15 minutes (Beach and Rabedeau, 1959). Thus, per unit time, the hamster is a more active copulator than the rat. A major distinction between the copulatory behavior of the laboratory mouse and the other rodent species is in the number of deep thrusts accompanying each penetration. Depending upon the genetic strain of the male, he will achieve between 7 and 36 penetrations of the female before he ejaculates. During each of these penetrations he may execute 15-20 deep thrusts; each thrust lasts approximately 0.7 second. Thus each deep thrust of the mouse lasts more than twice as long as the single deep thrust of the rat's intromission response (McGill, 1962). Male guinea pigs also exhibit more than one deep thrust per intromission. In one experiment, for example, males achieved ejaculation after an average of 6.4 penetrations, which include an average of 29.1 thrusts, or 4-5 thrusts per penetration (Gerall, 1958). Finally, the guinea pig and mouse, unlike the rat and hamster, typically achieve only one ejaculation during a testing session with a single female.

The multiple intromission pattern of these rodents contrasts strongly with the virtually instantaneous ejaculation achieved by some other species, for example the wild ungulates and domesticated species such as the sheep. Given their susceptibility to predation, it is not surprising that in their natural environments rodents copulate under cover of darkness or within a burrow. Moreover, the penile stimulation required before ejaculation is acquired in relatively short bursts of activity, between which the animals are separated and capable of escape from a predator if necessary.

From the viewpoint of the physiological psychologist, the pattern of the male rat's copulatory process highlights some questions about the control of copulatory behavior which can be applied, with only minor modifications, to the behavior of males of any mammalian species. These are the questions: 1) What are the determinants of the initial intromission? In other words, why does the male begin to copulate when he does? 2) Given the initial intromission, what determines the rate at which the male will achieve subsequent intromissions? 3) What is the mechanism by which the occurrence of the several intromissions facilitates the eventual occurrence of ejaculation? 4) What determines the durations of post-ejaculatory intervals? 5) What is the process by which the occurrence of ejaculation alters the probability, number, and spacing of subsequent intromissions? In this chapter we will concentrate on the first of these questions; readers desiring more information may consult Bermant and Davidson (1973).

Determinants of the Initial Copulatory Response

In the earlier discussion of noncopulating males we mentioned that no single explanation could account for the copulatory failures of all animals. Even under apparently completely adequate conditions of rearing and testing, some males will fail to mate. Nevertheless, the results of numerous experiments have given us information about the variables that are important in determining the likelihood that a male will mate on a given day. In considering these variables we must distinguish between continued failures by virgin animals and failures by animals that have already demonstrated copulatory competence. Here we will consider experiential

Table 7-2 Effects of Social Isolation on Copulatory Behavior in Male Guinea Pigs[1]

| | Percentage of Males Displaying: | | | |
Rearing Condition	sniffing, nuzzling, etc.	mounts	intromission	ejaculations
Isolated condition	100	30	30	30
Social condition	100	100	100	100

[1]Modified from Valenstein *et al*, 1955

and hormonal determinants and briefly discuss the importance of sensory integrity. Because our space is limited, we will focus our attention on factors affecting the male.

Experiential Determinants: Social Conditions of Rearing

The occurrence of adequate copulatory behavior in adulthood is partially determined by the social conditions in which the male grows up. Effects of post-weaning social isolation on sexual effectiveness in adulthood have been shown clearly in the male guinea pig (Valenstein, Riss, and Young, 1955). Table 7-2 shows data collected on two groups of 10 males each. Members of each group were weaned at 10 days of age, then assigned either to an isolation condition or a social condition; the social condition consisted of living with a female guinea pig of the same age. Tests for copulatory behavior were begun when the males were 77 days old. Seven weekly tests were administered.

The data in the table show that all 10 of the males raised with a female showed completed copulatory sequences during the mating tests, while only 3 out of the 10 isolated males did. The remaining seven animals showed the typical precopulatory behaviors of sniffing, nuzzling, or circling the female, but they did not achieve mount or penetration. It is important to note that the isolated males that achieved penetration were able to carry their responding through to the point of ejaculation. This is a typical finding for many kinds of experiments; it suggests that achieving a completed copulation brings behavior under the control of different processes (Beach, 1956).

Valenstein *et al.* (1955) concluded from their results that the effect of isolation was to prevent the effective organization of copulatory behavior during early development. They performed several additional experiments to bolster this conclusion (Valenstein *et al.*, 1955; Valenstein and Goy, 1957). For example, they showed that the effect of raising males in pairs was more variable than that of raising males with females; this was apparently due to the fighting among the males, which may have suppressed later attempts at copulation. They also demonstrated that raising males with spayed females, who were not as receptive as normal females to the male's incipient sexual advances, decreased the percentage of males who later were able to ejaculate with normal females. These results suggested that it was the opportunity actually to practice copulatory responses during the prepubertal period that was important for sexual adequacy in adulthood. However, Gerall (1963) found that if males were raised in individual wire-mesh cages next to other animals, so that they could see, hear, and smell the others but not touch them, then they showed essentially normal copulatory behavior when tested in adulthood. Thus, we may not conclude that actual practice is required (Beach, 1965). Moreover, as was shown in Table 7-2, 30% of the completely isolated animals in Valenstein *et al.*'s experiment showed normal copulatory behavior. Taken all together, these data mean that some male guinea pigs can organize their adult sexual responses adequately even when all contact with other guinea pigs is denied them during development, but that social contact is required to facilitate the organization of these responses in other animals. We still do not know why some males need the social contact while others apparently do not.

A similar state of uncertainty exists in our understanding of the development of sexual behavior in male rats. In one experiment, for example, the proportion of isolated males that copulated normally in adulthood was essentially as great as the proportion of socially-reared animals (62% and 75%, respectively; Beach, 1958), while in another experiment zero out of 16 males raised in isolation were able to mount a female in the normal fashion (Folman and Drori, 1965, Experiment II; see also Gerall, Ward, and Gerall, 1967). We cannot account for this difference between results in a systematic way.

Experiments with pure-bred beagle dogs have also demonstrated that prepubertal social isolation can affect later sexual performance (Beach, 1968). Males that had been raised under conditions of virtually no contact with other animals for the first 14 months of life showed a strong tendency to mount the estrous female inappropriately, i.e., on her side or head. An average of six exposures to a receptive bitch was required before these males achieved their initial intromission. Other males, which had been allowed 15 minutes of daily social contact with each other, achieved their first intromission after an average of 2.4 tests, and a third group of males that had been raised with females in a large open field achieved intromission after an average of 1.5 tests, with no male requiring more than two tests. In general, the males that had the 15 minute periods of daily exposure to other males copulated as rapidly and effectively as the males who grew up in the unhindered heterosexual setting.

Domestic male cats raised in social isolation may or may not exhibit adequate copulatory behavior when they are tested in adulthood. Rosenblatt (1965) reported on four males that had been isolated from ages 2-15 months. Two of the males failed to mate even after repeated opportunities, while the other two mated normally within one hour of the beginning of their first test. An additional pair of males that had been raised in isolation immediately from birth, demonstrated copulatory behavior after several tests but then never copulated again.

Finally, social isolation or restriction for the first six months of life prevents normal copulatory behavior in male rhesus monkeys (Harlow, 1965). This result first came to light when the monkeys were deprived of their mothers and were raised with cloth "surrogate" mothers. Deprivation of maternal care *per se* is not, however, a sufficient condition for eventual copulatory failure, for males that are allowed daily contact with peers during the first six months can later copulate adequately. The sexual inadequacies of the socially isolated monkeys are only one part of a generally pathological behavioral condition.

In summary, we can conclude that social contact before puberty facilitates the organization of normal copulatory behavior in the males of several mammalian species. We may not conclude that this contact is a necessary condition for organization, however, except perhaps for the rhesus monkey, because at least a few males of the species tested copulate normally even after prolonged and essentially complete isolation. Experiments designed to elucidate the differences between individuals responsible for their different resistances to social deprivation are very much needed.

Before concluding this section, we should point out that the initial experiences with copulation themselves have an effect on subsequent copulatory behavior. In the male rat, for example, repeated opportunities to copulate result in generally more efficient behavior: the males show briefer mount, intromission, and ejaculation latencies, and intercopulatory intervals (Dewsbury, 1969).

Hormonal Factors

In an earlier discussion we noted that increasing the amount of androgen in noncopulating male rats was not sufficient to induce complete copulation. This does not mean, of course, that the amount of androgen available to the male is unimportant in determining his copulatory responsiveness. Rather, it serves as a clue that the relation between androgen level and copulatory behavior may be very complicated and partially dependent upon other factors, for example the sexual and social history of the male. The relation is further complicated by the fact that the hormone exerts its effects at numerous locations in the body: the brain, spinal cord, and genital tissue, for example. And finally, the action of androgen at one period of the male's life, e.g., prenatally and immediately postnatally, is different than its action at another period, e.g., puberty. All these factors, and others that could be listed, make a complete discussion of the role of androgen in the determination of copulatory behavior well beyond the scope of this chapter. We will attempt only to sketch some basic findings and highlights of current research. For fuller discussions and bibliographies, the reader is referred to Beach (1967; 1970), Bermant and Davidson (1973), Davidson and Bloch (1969), and Whalen (1968).

We will begin with some background information on the role of androgen in adult behavior. It has been observed in a large number of experiments that the castration of adult males with prior copulatory experience produces a cessation or diminution of subsequent copulatory behavior. The immediacy and strength of the effect varies between both individuals and species. Consider the following examples: hybrid laboratory mice may take as few as three days or as many as sixty days to cease ejaculating following castration; McGill and Tucker (1964) reported a median of twenty-eight days. Davidson (1966) reported that one laboratory rat exhibited ejaculatory responses for 147 days after castration, but in general only 10 to 25% of his males exhibited this response for more than seventy days after the operation. Male hamsters and guinea pigs are also very unlikely to exhibit the complete copulatory sequence for more than ten weeks after castration (Beach and Pauker, 1949; Grunt and Young, 1953). Variability in individual response to castration is increased in dogs and cats. Rosenblatt and Aronson (1958a) observed one cat that exhibited intromission responding for thirty-five weeks after castration and another that ceased intromission after the second postoperative week. Beach (1970) observed one dog that continued to show a completely normal "locking" response (swelling of the penile bulb within the vagina) for twenty-one months after castration, while the performance of other males in this regard deteriorated severely within two to five months. However, it is important to point out that for all these dogs, and for at least some of the males of the other species, castration did not reduce the number of times the males mounted the females and attempted insertion. In general, but with some exceptions, the effect of

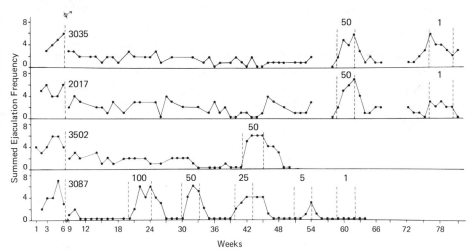

From "Copulatory behavior of the ram, *Ovis aires:* III. Effects of pre- and postpubertal castration and androgen replacement therapy" by M. Clegg, W. Beamer, and G. Bermant, in *Animal Behavior* **17,** 1969.

Figure 7-4 The numbers of ejaculations achieved by rams. Each point is the sum of ejaculations achieved with two consecutively presented ewes. The first dotted line at the left of the figure represents the point of castration. The identification number of each ram is given to the right of that line. Subsequent pairs of dotted lines demarcate the periods of androgen replacement. The amount of androgen given per day, expressed in milligrams, is shown in each case.

castration is seen first on ejaculatory capacity, than on the ability to achieve intromission, and last, if ever, on mounting *per se.*

The decline in copulatory efficiency produced by castration may be remedied by injections of androgen. The effectiveness of androgen therapy is shown graphically in Figure 7-4, taken from the work of Clegg *et al.* (1969) on male sheep. The ejaculatory performance of four rams is shown both before and after castration. Note first that there was substantial variability among the males in response to castration: one ram ceased ejaculatory responding within six weeks, while two others continued responding for approximately a year, although at reduced levels. But in all cases daily injections of 50 mg. of androgen (testosterone propionate) restored copulatory responding to preoperative levels. And when therapy was discontinued the behavior fell rapidly to pretherapy baseline levels. Smaller doses of androgen proved less effective.

Taken all together, the results of studies on castration and androgen replacement in adult males with prior copulatory experience indicate that the presence of androgen facilitates the execution of the complete copulatory sequence; although some males continue to show intromission and ejaculation responses for long periods of time after androgen has been removed, the quality and quantity of these responses is typically diminished. Within limits of age, the behavior can be restored with sufficient androgen replacement. Mounting and thrusting against the female is not so dependent on androgen as is intromission and ejaculation. This finding suggests that the spinal and peripheral components of the copulatory system are more dependent on androgen than are the higher nervous components involved in bringing the male to the female and controlling copulatory attempts.

How important is prior copulatory experience in prolonging copulatory behavior after castration? The answer to this question is clouded somewhat by the different results that have been obtained from different species. Experiments with rats have shown that virgin males castrated as adults did not differ significantly from experienced males in the rates at which their copulatory behavior declined (Bloch and Davidson, 1968). And Hart (1968b), working with beagle dogs, found no differences between virgin and experienced castrates. However, the definition of "virgin" in this experiment needs some qualification, because before they were castrated these males had been allowed to mount and thrust against the female but had been prevented from penetrating her. On the other hand, Rosenblatt and Aronson (1958a) found that male cats who were virgins at the time of castration were markedly less resistant to the effects of castration than experienced males were. It is of course quite possible that these results represent a definite difference between cats and the other species that have been tested; we need more experiments to be sure.

The experiments we have so far discussed all involved castration after puberty. We need next to consider the effects of prepubertal castration, and we have to pay particular attention to the ages of the animals at the time of surgery, because castration during the first few days of life has more profound effects than it does if performed later in the pre-pubertal period.

Beach and Holz (1946) castrated groups of male rats at ages 1, 21, 50, 100, 150, and 350 days. Puberty in the strain used normally occurs at approximately 50 days. The adult animals used in the experiment were virgins. Three months after castration, when the youngest animals could safely be considered adult and all testicular androgen had surely disappeared, the males were tested with receptive females. None of the total of 50 males ejaculated, three achieved intromission, and eight exhibited mounts with thrusting. There was no relationship between age at castration and the level of postcastration behavior. All the males were then given uniform injections of androgen and additional tests for copulatory behavior. Animals in all groups responded to the therapy with an increase in copulatory behavior: every animal in every group achieved at least one intromission during the testing period. The major difference among the groups was in the percentage of animals achieving ejaculation: 40 out of the 41 males castrated between 21 and 350 days of age ejaculated at least once, while only one out of the nine males castrated on the day of birth did so. The day-one castrates also achieved far fewer intromissions during each test.

Failure of day-one castrates to achieve regular intromission and ejaculation could be traced to an inadequate length of the shaft of the penis. The shaft lengths in the day-one group ranged from eight to eleven millimeters, while in the other groups the range was fourteen to nineteen millimeters. Beach and Holz concluded that the short penises of the day-one castrates prevented regular intromission, and thus that adequate stimulation could not be provided for ejaculation. In other respects, i.e., approaching, mounting, and thrusting against the female, day-one castrates were not impaired relative to the other animals. Hart (1968d) found that male rats castrated on day four showed marked deficiencies in intromission and ejaculation when injected with androgen in adulthood, while males castrated on day twelve behaved essentially normally. Thus, Hart's day-four males were similar to the day-one animals observed by Beach and Holz; we may conclude that the presence of androgen is required for at least four, but less than twelve days, if later penis development under the influence of injected androgen is to be complete. Moreover, Hart demonstrated that the spinal reflexes associated with ejaculatory behavior were impaired in the day-four, but not in the day-twelve, castrates. This finding may be interpreted to mean that failure of very young castrates to ejaculate may not be due entirely to insufficient stimulation delivered to the penis because of inadequate intromission behavior. A fuller discussion of these and related findings on spinal functions in sexual behavior is presented in Chapter 6.

The ability of prepubertally castrated male sheep to respond to androgen therapy is presented graphically in Figure 7-5. These five animals were castrated at two weeks of age; daily androgen therapy was begun at age two years, one week before the first point shown in the

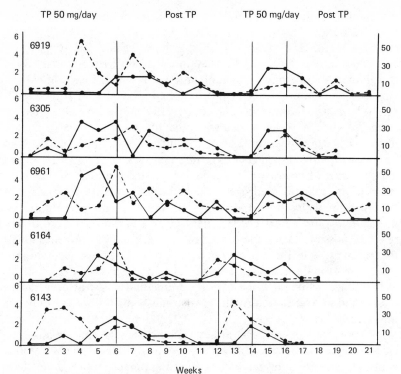

From "Copulatory behavior of the ram, *Ovis aires:* III. Effects of pre- and postpubertal castration and androgen replacement therapy" by M. Clegg, W. Beamer, and G. Bermant, in *Animal Behavior* **17**, 1969.

Figure 7-5 Changes in ejaculation (solid line) and mount (dashed line) frequencies in prepubertally castrated sheep. Several periods of androgen (testosterone propionate) therapy and withdrawal are shown. Injections were begun initially one week before the first test shown.

figure. Before therapy none of the males showed any indication of sexual behavior when placed with a receptive ewe. But by the end of the sixth week of therapy every male had demonstrated at least two ejaculatory responses; the data on numbers of mounts with and without ejaculation are summarized in the left-hand portion of the figure. The figure also shows that the males stopped responding within no more than seven weeks after androgen therapy was discontinued, and that reinstitution of the therapy resulted in a reappearance of the behavior (Clegg *et al.,* 1969). The ejaculatory responses shown by these prepubertal castrates under therapy were well within the normal range.

We have mentioned that prepubertally castrated rats and sheep show very little copulatory behavior in adulthood unless androgen is provided them; this is also true for cats (Rosenblatt and Aronson, 1958b). The case appears somewhat different for dogs, for Beach (1970) found that two prepubertally castrated male beagles showed normal mounting and thrusting behavior even before they were given androgen in adulthood; however, their intromission and ejaculatory behavior was almost completely obliterated. These failures of intromission may have been due to inadequate development of the penis, because the two animals showed erections and ejaculation-like responses when they were stimulated manually.

The small amount of data that exist for nonhuman primates suggest that at least some male monkeys are able to demonstrate normal copulatory responses for years after prepubertal castration (Phoenix, Goy, and Young, 1967).

The Importance of Sensory Integrity

We have mentioned that a male rat cannot perform the deep thrusting response of intromission if his penis has been anesthetized; thus the sensory integrity of the penile receptors is required for the execution of the complete copulatory pattern. The effect of loss of penile sensation is even more profound in male cats. Aronson and Cooper (1968) have reported on the effects of severing the dorsal nerve of the cat's penis, the nerve which transmits information from the tip (glans) of the penis to the spinal cord. All the males were sexually experienced before surgery. The immediate effect of the denervation surgery was a total cessation of intromission. Although all the operated animals continued to mount the females and achieve erections, none was capable of penetration. However, two of the fourteen operated animals recovered the capacity for regular intromission; one after six weeks, the other after eighteen weeks. This was due apparently to the regeneration of the severed dorsal nerves. The remaining animals demonstrated a long-term deterioration of their mounting responses; in particular they tended to fall over on their sides after they had mounted and clasped the female. The amount of time the males spent mounted on the females decreased substantially over the four-year period of observation. Aronson and Cooper concluded that the repeated failures of penetration produced a decline of sexual arousal; the males became sexually disinterested.

In a subsequent experiment, Aronson and Cooper (1969) denervated the penises of prepubertal cats. Three of the eight males failed to show any sexual behavior when tested as adults, and the remainder demonstrated the same abnormalities that the sexually experienced animals had. Several of the males exhibited successful intromission during the four-year testing period. These consequences of anesthetization or denervation stem from the failure of the altered penis to locate the vagina; the detection function of the penis has been hindered.

A number of experiments have been performed to determine the importance of other sensory modalities (vision, audition, olfaction) for the initiation of copulation. We pointed out in an earlier section that the courtship sequences of some birds and fishes depend to a significant degree upon information transmitted across specific sensory channels. Thus it might also be that male mammals deprived of sight or hearing or the sense of smell would not initiate copulation.

Until quite recently it was generally believed that male mammals were not dependent upon input from a single sensory channel for the initiation and maintenance of their copulatory behavior. Several experiments with rats had shown that blinding or deafening the males produced relatively small effects which could be largely overcome by extensive copulatory experience. Similarly, removing the sense of smell produced quantitative deficits in responding (increased intromission and ejaculation latencies) but did not reduce the probabilities of intromission or ejaculation (Bermant and Taylor, 1969). However, it has been recently discovered that both virgin and sexually experienced male hamsters are totally dependent upon their sense of smell for the initiation of copulation (Murphy and Schneider, 1970). We now need experiments with a wide variety of species in order to determine the generality of the hypothesis that male mammals do not rely on information from a single sensory channel for the initiation of copulatory behavior.

The Function of Mating Behavior

If reproduction is to occur the egg must first be fertilized and then situated in a place where it may develop successfully. There are only two alternatives for the location of a developing fertilized egg: either inside or outside an animal's body. If development takes place outside, then fertilization may also be accomplished outside; the courting animals can synchronously shed their gametes into the proper external environment. This is in fact what many species of molluscs, fishes, and other aquatic species do. However, as we will point out with examples below, there have been some very intricate adaptations of the general process of external fertilization.

The terrestrial environment is not well suited to external fertilization and development. The aquatic environment provides a medium through which gametes can travel and also protection from temperature fluctuations, evaporation, and so on; dry land does not have these properties. Hence fertilization is internalized in terrestrial species. And in these cases, if development occurs externally, a special provision is made to protect the embryo from adverse outside influences: it is surrounded by a coating of durable material, i.e., an egg-shell, and enough nutritive material to last until it emerges. Indeed the additional care that adult animals, particularly birds, provide for their developing eggs, by hiding and protecting them from predators, keeping them warm by sitting on them, and so on, is in itself a vitally important behavioral function which is more fully discussed in Chapter 8.

Mammalian species invariably practice internal fertilization and development. Within the class of mammals there is the universal appearance of the male intromittent organ, the phallus or penis. This is not to imply that no other male animals have penises; for in fact the phalluses of many insect species are very completely developed, and are often used for taxonomic purposes. Male fish from species that practice internal fertilization, e.g., guppies, also have phalluses. The primary reproductive function of the penis is obvious enough: it provides a highly directed channel for the introduction of sperm into the reproductive channel of the female. But there is another function of the mammalian penis which also needs to be appreciated: it serves as a sense organ which receives and transmits information to the rest of the nervous system. Ejaculation, the neuromuscular process of sperm and semen deposit, typically occurs only after the penis has been stimulated. As we pointed out above, there are species differences in the relations between stimulation of the penis and the eventual release of the ejaculatory reflex. The penis also serves as a source of essential stimulation to the female. In some species (e.g., rabbit, cat) ovulation can be triggered by penile stimulation. In other species that ovulate spontaneously (e.g., rat, hamster, mouse), fertilization and implantation may depend upon repeated penile insertions (Adler, 1969; Adler and Zoloth, 1970; Diamond, 1970; 1972). Now we will give some examples of the process of external fertilization.

External Fertilization

Straightforward instances of external fertilization may be found among the bivalve molluscs: for example, the North Atlantic American oyster, *Crassostrea virginica*. Mature oysters begin to manufacture gametes during the early spring, and the gametes are spawned during the summer. Of course these animals are stationary, and they cannot rely on typical channels of communication (e.g. visual or auditory) to synchronize the shedding of their gametes. Courtship does not exist for the oyster. How do males and females, living together in an oyster bed, manage to spawn at approximately the same time?

The answer seems to be that the male's spawning response is more rapidly and easily stimulated by environmental changes than the female's is, and that the female spawns most readily if she is chemically stimulated by sperm in the water surrounding her. The male's response consists simply of strong beating movements of the cilia lining the genital ducts which propel the sperm through the epibranchial chamber into the water. The response may be elicited by a number of stimuli, including swift increase in temperature or any of a large number of biological compounds. The latency of the male's response to stimulation is only a few seconds. The response of the female is more complex and protracted. Eggs are expelled into the water as a result of sharp snapping movements of the two valves (shells). The latency of the snapping response to appropriate stimulation is usually between six and twenty minutes. The response cannot be triggered by as many stimuli as the male's response can; the most effective stimulus is a particular chemical fraction of mature sperm cells. The relatively long latency of the female's response is apparently due to its greater complexity and to the necessity that the stimulating chemical penetrate between the female's valves; in fact, it may have to be absorbed into the intestinal tract or gills before it is effective (Galtsoff, 1961). In any case, the differences between the sexes in sensitivity and latency of the spawning response create a high likelihood that eggs will be shed when sperm are in a position to fertilize them.

Species of fishes that rely on external fertilization have evolved highly specific behavioral procedures that insure a successful reproductive outcome. We may take as examples the Siamese fighting fish, the three-spined stickleback, and the bitterling.

The courtship of the Siamese fighting fish (*Betta splendens*), and other anabantid species as well, is concluded when the male wraps his body around the middle of the female, who at that time is swimming upside down. The pressure of the male's contact triggers the release of ova; the male also releases his sperm at about this time. However, the male soon unfolds himself and swims to the downward-floating eggs, which he captures in his mouth (Figure 7-6). He takes the eggs to the

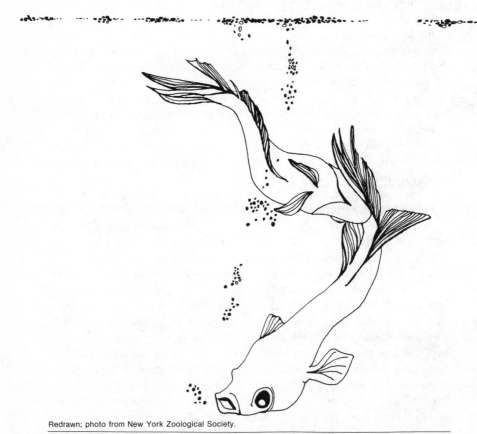

Redrawn; photo from New York Zoological Society.

Figure 7-6 After the male and female fighting fish have spawned, the male gathers eggs in his mouth and places them in the bubble nest he has constructed beneath the water's surface.

nest of air bubbles, lying just beneath the surface of the water, that he had earlier constructed. He attaches the mucus-covered eggs onto the bottom of the nest, where they are insured an adequate oxygen supply for their initial development (Goodrich and Taylor, 1932).

The three-spined stickleback (*Gasterosteus aculeatus*) has solved the problems of external fertilization in a different way. Like the male Betta, the male stickleback builds a next before a female enters his territory. But the stickleback's nest is on the bottom; it is a short tunnel constructed from material in the vicinity and held together by a sticky substance secreted from the male's kidneys. The much-studied zigzag courtship dance of the sticklebacks concludes when the male guides the female into the tunnel nest, from which she protrudes at both ends (see Figure 7-7). When the female has entered the nest the male begins a characteristic movement ("trivering" or "trembling") against her tail; this stimulates her to release her approximately 260 eggs into the nest; after spawning, she exits. The male then enters the nest and deposits his sperm on the

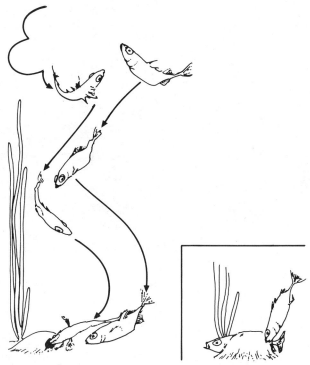

Figure 7-7 Courtship and mating in the Three-Spined Stickleback. The zig-zag maneuvers are completed when the female enters the nest the male has constructed. He pokes her and she releases her eggs, which he can then fertilize.

eggs. Interestingly enough, this is not the end of the male's participation in the reproductive process. After he has fertilized several clutches of eggs, he remains by the nest for about a week and pushes fresh water over the eggs by rhythmically moving his fins. The extent of his fanning activity is determined in part by the amounts of oxygen and carbon dioxide in the immediately surrounding water (Hinde, 1966; Sevenster-Bol, 1962; Tinbergen, 1951; van Iersel, 1953).

One of the most interesting examples of external fertilization occurs in the bitterling (*Rhodeus amarus*). Like both bettas and sticklebacks, the male bitterling defends a territory into which, eventually, a receptive female will come. But unlike the others, the bitterling does not build a nest and then defend and court within the area immediately surrounding it. The focal point of his territory is instead a living mussel (e.g., of genus *Unio*) which has attached itself to the bottom (see Figure 7-8). During courtship the male bitterling directs the female over the gills of the mussel. The female has developed a long tube, the ovipositor, through which her eggs pass from the ovaries to the outside. Courtship is successfully concluded when the female inserts her ovipositor into the gills of

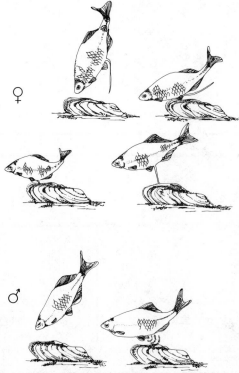

From "An ethological analysis of the reproductive behavior of the bitterling (*Rhodeus amarus bloch*)" by P. R. Wiepkema, in *Archives Neerlandaises de Zoologie* **14**, 1961, 103–109.

Figure 7-8 Mating in the bitterling. Male and female deposit gametes in a mussel.

the mussel and deposits her eggs there. The male then skims over the inhalant siphon of the mussel and deposits sperm. The eggs are fertilized within the body of the mussel. The male bitterling may repeat the procedure with several females, and he subsequently guards the mussel from the use of other males. After about a month, the young fish exit the mussel through its exhalant siphon (Wiepkema, 1961).

Internal Fertilization

Normally, the process of internal fertilization involves the deposit of sperm by the male into the reproductive tract of the female. There the sperm meet the ovulated ova and fertilize them. But there is at least one exception to even this basic biological process, and it occurs in that improbable fish, the sea horse (genus *Hippocampus*). Courtship in sea horses is a complicated affair which concludes when the members of the courting pair appose the genital openings on their ventral surfaces. The female possesses a protruding genital organ (papilla) which she inserts into the male's brood pouch. Having inserted the papilla, the female

Figure 7-9 The female sea horse transfers eggs to the pouch of the male where they are fertilized. When the young are sufficiently mature the father expels them.

From "Physiology of reproduction in molluscs" by P. S. Galtsoff, in *American Zoologist* 1, 1961, 273–289.

Figure 7-10 Mating in the octopus. The male extends his hectocotylus into the mantle cavity of the female.

deposits her eggs. The stimulation associated with these actions triggers the release of sperm by the male into the brood pouch, and the eggs are fertilized. The embryos remain inside the male and grow until they are perfect miniature sea horses; the male becomes considerably distended while harboring them (see Figure 7-9). Then, finally he expels them from the pouch with rhythmical muscular contractions (Jones, 1967; Wendt, 1965).

Both fertilization and substantial development take place within the male sea horse's brood pouch; but the physiological relationship between the father and his offspring during this period is very different from that of a mammalian mother and her young. For the sea horse father does not nurture the fertilized eggs directly from his bloodstream. He simply provides a safe place for them to develop using the nutritional resources available in the egg. The young are hatched within him and then are expelled. This procedure, called *ovoviviparity,* occurs in numerous other species of fishes and reptiles, even when it is the mother who is harboring the developing embryos until they hatch. It represents a kind of biological compromise between the procedures of *oviparity,* in which virtually all development is externalized (e.g., birds), and *viviparity,* in which the young are born alive after developing in direct contact with the bloodstream of the mother (e.g., mammals).

We have mentioned that among the mammals the process of depositing sperm invariably involves the use of the specialized intromittent organ, the penis. Other classes of animals have evolved different anatomical and behavioral mechanisms for the transfer of sperm from the male into the female. One interesting instance occurs among the cephalopod molluscs: octopi, squids, and cuttlefishes. One of the tentacles of the male is specialized as a device for transferring sperm; it is called the *hectocotylus,* which means "arm of a hundred suckers." The male's sperm are packaged in chitinous bags called *spermataphores.* In some species, for example the common octopus *Octopus vulgaris,* the spermataphores reside in a fold of skin along the hectocoylus. But in other species the spermataphores are stored in the male's mantle cavity.

Courtship between cephalopods is protracted and sometimes violent. It is consummated when the male inserts his hectocotylus into the female's mantle cavity and deposits a spermataphore (see Figure 7-10). The membrane surrounding the sperm ruptures, and the sperm become motile when they come into contact with the water. The female then releases eggs. The mixed gametes are expelled and drift to the ground, where the fertilized eggs develop. During the violent thrashing and biting that attend courtship and copulation the male sometimes has his hectocotylus pulled off. Females may be observed swimming alone, bearing the hectocotylus of a dismembered male. This observation, and the rather remarkable semi-independent movement of the hectocotylus in reaching and penetrating the female, led early investigators to postulate several interesting but inaccurate theories of reproduction in cephalopods (Galtsoff, 1961; Lane, 1960).

Still more bizarre (from our point of view) methods of sperm transfer have been evolved by other species of invertebrates (Rothschild, 1956). Each of these methods of internal fertilization is accompanied by other specialized reproductive processes. We will not pursue these examples here but instead will turn to a more complete account of copulation in mammals.

Mammalian Copulation

As we have seen, most of our detailed information about mammalian copulatory behavior is based upon those species which we can observe most easily, the animals of the laboratory, the farm, and to a lesser extent, the zoo. One of the major challenges facing the directors of zoological gardens has been to provide public environments for their animals in which they will mate and successfully rear their young. As late as 1923 experts believed that placing wild animals in captivity made them sterile; since that time we have learned that the failures of reproduction can be traced, not to physiological sterility, but rather to inhibiting effects of the unnatural physical and social environments on courtship and mating. Because the construction of completely natural environments is practically impossible for large animals in a zoo setting, zoo managers and designers have made efforts to isolate the critical environmental features required to facilitate mating in particular species. These efforts have been successful in many instances. For example, between 1951 and 1961 several important zoo specimens mated successfully for the first time: Indian rhinoceros, okapi, gorilla, cheetah, and flamingo. The practical nature of this work has fostered fundamental appreciation of the ways in which animals interact with their environments (Hediger, 1965).

In addition to experiments in the controlled environments of the laboratory, the farm yard, and the zoo, numerous investigations of mammalian copulation have been accomplished under natural field conditions. The difficulties inherent in these investigations are substantial. To begin with, many mammals, for example the rodents, are nocturnal copulators. This fact, coupled with their small size and tendency to remain under cover, makes systematic observation of sexual behavior under natural conditions very difficult. Other species may move over large distances of rugged terrain in brief periods of time, so that the observer is unlikely to be on the spot when copulation occurs. A further difficulty is that many nonprimate species tend to have strictly limited breeding seasons; for most of the year no copulation occurs. Other species may breed during all months of the year, but even then the female's receptivity is limited to relatively brief periods of time. Finally, field observation does not always allow the investigator to identify or mark separate individuals with certainty. As a consequence of these and related difficulties, field investigations have tended to be purely descriptive, with limited or no quantification. Of course there have been some exceptions to this generalization, for example the work of DeVore (1965) on baboons and that of LeBoeuf and Peterson (1969) on elephant seals.

From H. K. Buechner, (1965). "Ceremonial mating behavior in Uganda kob (*Adenota kob thomasi* Neumann)" From *Zeitschrift fur Tierpsychologie* **22**, 209–225. Verlag Paul Parey, Berlin and Hamburg. Original sketches by Dr. Fritz Walther.

Figure 7-11 Mating in the Uganda kob. The male rests lightly on the female's back. Copulation is very rapid.

When observations of several species whether conducted under laboratory, zoo, or field conditions, are combined and organized appropriately, interesting insights may be developed. One such insight has been provided by Hediger (1965), who contrasted the typical copulatory patterns of large herbivores (e.g. antelopes) to those of the large carnivores (e.g. bears and wild cats), and showed that the distinct differences in copulation between these groups are correlated with similar differences in other kinds of behavior. To begin with, hooved animals, in particular the wild antelopes, copulate very rapidly and with minimal physical contact. The brevity of copulation is seen most clearly in the gerenuk: copulation occurs with both members of the pair moving at a trot. The male rises to his hind legs while he is still several yards behind the female, rapidly closes the gap, and inserts and withdraws his penis without touching the female with his forelegs. While less extreme, the copulation of other species is also very rapid (Ewer, 1968; Hediger, 1965). Figure 7-11 shows the very high and light posture the male Uganda kob assumes on the female.

The duration of copulatory contact in the large terrestrial predatory mammals is, by contrast, quite prolonged. Copulation of bears may last for hours, and in wild and domestic canines the male and female remained "tied" together for twenty minutes or more after full insertion, because the base of the male's penis swells up inside the female. In felines the duration of each insertion is not so prolonged, but there are repeated bouts of copulation over a period of several hours. For example, Whalen (1963) found that domestic cats engaged in as many as 13 bouts of completed copulation during a single period of three hours.

Given these obvious differences in copulatory behavior, Hediger went on to show that similar differences could be found in other behavioral categories; his conclusions are shown in Table 7-3.

In addition to the categories described by Hediger, we may also contrast the typical feeding habits of, say, large cats, with those of ante-

Table 7-3 Behavioral Comparisons of Large Carnivores (Bears and Lions) and Herbivores (Antelopes)

Behavioral Category	Bear/Lion	Antelope
Duration of copulation	prolonged	brief
Duration and depth of sleep	prolonged, deep	brief, light
Rate of drinking	slow, particularly in cats	rapid, in brief bouts
Preparation for birth of young	some preparation of nest or birthplace	little or no preparation
Rate of development of young	slow; young, weak and helpless at birth	very fast; young able to walk very soon after birth

Hediger, H. (1965). "Environmental factors influencing the reproduction of zoo animals" Beach, F. A. (ed.), *Sex and Behavior,* © 1965 by John Wiley, p. 319–354.

lopes. After making a kill the cats are likely to lie with it and gorge, while antelopes remain on their feet, move about and graze. Perhaps further related contrasts could also be made, but we already have enough in front of us to see the pattern: the behavioral "styles" of these two divergent groups of animals reflect the selection pressures, or lack thereof, placed on their ancestors. From feeding through copulation and parturition, the behaviors of antelopes are well-suited to the demands placed upon an animal of prey; execution of these behaviors is unlikely to get in the way of escape should the need arise. Understanding this feature of the species' behavioral profiles enriches our understanding of the relations between behavior and evolution.

In summary, it has been our intention in this chapter to describe different kinds of courtship and mating behaviors and to show that these activities can be subjected to two kinds of analyses. On the one hand, it is possible to explore the immediate causal basis of a particular behavior and, on the other hand, to show how the behavior meets an ecological demand confronting the animal. These two approaches together offer a deeper understanding of the nature of reproduction.

8   Parental Behavior

Dale F. Lott

Parental behavior is wonderfully various. There are snakes that sit on their eggs (if it is ever proper to speak of a snake sitting on anything). Some fish transport their young in their mouths. There are birds as well as bees that feed their offspring "milk." There are penguins who spend much of the long Antarctic winter with an egg cradled between foot and breast, incubating it in temperatures that reach sixty degrees below zero. The offspring of cuckoos, cowbirds, and other parasitic brooders are raised by foster parents of another species. There are pelicans who spend hot summer days on the shores of barren lakes in the American desert holding up their wings to cast a protective shadow over their nearly naked young. Mallee-fowl use the heat from decaying vegetation to incubate their eggs. And the maternal koala bear feeds her young a kind of soup which she delivers from her anus at certain times of the day.

All of these parental procedures have served the species which use them so well that they have evolved and survived into our own time; they are successful biological adaptations. Biologically speaking, parental behavior, like all behavior, has only one justification: it increases the probability of the survival of particular genes in the gene pool of the species. The parental behavior exhibited by any individual is successful to the degree that it results in a contribution of its associated genes to the pool of the next generation.

As we shall see more clearly below, "parental behavior" is a category that is defined by the outcome of what certain animals do under certain circumstances; it is a descriptive, functional category of behavior. Our approach in this chapter will be to consider a few instances in detail. We will surrender the opportunity for a kaleidoscopic tour through the diversity of parental behavior to concentrate on careful description and on a discussion of the relation of parental behavior to other prominent features of animal life styles. But first we need to place some limits on the kinds of behavior that we will consider parental.

Parental behavior is so obviously important for the survival of the young of human beings and other familiar species that we might assume it to be universally present among animals. But this is far from true. The range of behavioral response across species to the presence of a fertilized egg, hatchling, or live-borne young is tremendous, ranging from intense and prolonged caretaking to no caretaking at all. Where to draw the line in this range of responses, between those to be called and those not to be called "parental," is to some degree arbitrary. With this in mind, we will adhere to the following working definition of parental behavior: *behavior that contributes directly to increasing the survival probability of fertilized eggs or offspring that have left the female's body.* A few brief examples will help to clarify the classes of responses included in and excluded from this definition.

The reproductive behavior of palolo worms is simplicity itself. For example, the Pacific species, *Eunice viridis,* spends its life hidden in crevices in the reefs surrounding Samoa (Bullough, 1961). On a night in November, timed by the phase of the moon, gamete-bearing rear sections of the worms detach from the rest and swim to the surface where they float until sunrise. Sunrise stimulates the release of sperm and ova (Michelmore, 1964). This system of bringing gametes together is inefficient, but since each individual worm releases its millions of gametes at the same time others are shedding sperm and ova, the probability of fertilization is increased. Furthermore, the risk of predation is such that it is to the animal's advantage to stay in its burrow rather than leave a protected hiding place in order to achieve a more efficient transfer of gametes. The worms' participation in reproduction ends with shedding its gametes; there is nothing in the animals' behavior that can be classified as parental.

Consider next the mating behavior of bullfrogs (*Rana catesbeiana*), which has a positive effect on the probability of egg fertilization (Noble, 1931; Capranica, 1965). Reproduction in this species requires a substantial body of water in which the eggs can develop into tadpoles and the tadpoles into adults. In preparation for mating, the males go to breeding pools where they set up a chorus of their species-specific mating calls. Females are attracted to the pools by the calls. When they reach the pool they may find several species of males, each giving its characteristic call. Females selectively approach the call of their own species, so hybridization is rare. When a female approaches a male, he stops croaking and clasps her around the abdomen with his front legs. His squeezing (*amplexus;* see fig. 6-2) triggers the release of the female's eggs into the water. The male releases his sperm at the same time. The proximity of the two animals during the simultaneous shedding of gametes results in a high rate of fertilization. However, the fertilized ova are then simply left without further attention in an area that is part of the frog's ordinary environment; there is nothing in the adults' behavior that contributes directly to increasing the probability of zygote survival. Thus, by our definition, the bullfrog lacks parental behavior.

There are many creatures who never interact directly with their offspring but whose reproductive behavior nevertheless greatly increases the probability of the survival of their zygotes. An outstanding example is the spawning behavior of the grunion (*Leuresthes tenuis*), a small, silvery fish found along the coast of southern California.

In order to understand how well situated a grunion's eggs are you must recall something about the nature of tides. There are a number of tidal cycles, but only two (daily and biweekly) are really important to this discussion. They are illustrated in Figure 8-1.

The grunion's reproductive cycle is marvelously adapted to the tidal rhythm. A female grunion spawns just after the peak of high tide. When the tide begins to recede from the sandy beach where she is swimming just offshore, the female comes in with a wave and begins to bury herself in the sand, tail first. When she is half or more buried in the sand, one or more males curl around her body and release their sperm. The sperm sink down through the wet sand and fertilize the eggs which the female is releasing from an opening near the base of her tail. The female then struggles to the surface and leaves the fertilized eggs buried some three or four inches down in the wet sand.

The tide continues to recede, so the sand in which the grunion zygotes are buried remains untouched by waves for nearly two weeks, until the tide again reaches its maximum. Meanwhile, the zygotes develop into viable fry which are encased in the membranes which originally surrounded the ova. As the incoming waves begin to pound the sand around the fry, these egg cases rupture and the fry are released into the wet sand. They leave the sand and enter the water within the next few minutes (Walker, 1952).

The responses of the female grunion exemplify behavior that contributes to the probability of the survival of the offspring, even though it consists entirely of leaving the zygote in exceptionally good circumstances. Thus, by our definition, the grunion exhibits parental behavior.

Admittedly the difference between the functional significance of the responses of bullfrog and grunion is not large, but it represents the

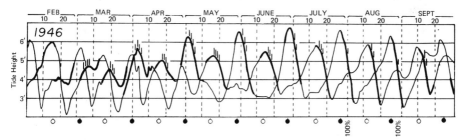

From "A guide to the grunion" by B. W. Walker, in *California Fish and Game* **38**, 1952.

Figure 8-1 Grunion runs observed at La Jolla, California in 1946 are plotted in relation to variations in the heights of high tide. The two series of curves represent the two tides of each day. Heavier lines indicate the tides occurring during darkness. The short vertical lines above the tide curves indicate the occurrence of grunion runs. The phases of the moon are indicated at the bottom of each graph: a solid circle indicates a new moon, and a hollow circle a full moon. All data are based on observations made at Scripps Beach in La Jolla, with data for the time and height of tides taken from records of the tide-recording machine maintained for the Coast and Geodetic Survey on Scripps Pier.

threshold of parental behavior. The borderline is not precise, but it will do for our purposes. The examples we will go on to consider in detail will not pose any problem of definition. We will be concerned directly with a description of what the animals do and an analysis of why they do it. This "why" is really two questions, and will require two rather different answers.

First, we will consider the *adaptive significance* of the particular forms of parental behavior found in the different species. We will see that the details of parental responding in particular species are often elegantly related to the demands placed on the animals by their environment and their basic biological role (niche). This analysis will allow us to appreciate why natural selection has led to the evolution of particular behavioral forms, and how behavior contributes to the species' distinctive style of life.

Second, we will consider the *causal mechanisms and pathways* that lead to the exhibition of species-typical parental behavior. We do not suppose that any nonhuman animal is controlled in his parental behavior by an intellectual appreciation of its adaptive significance. For example, birds do not incubate eggs in order to make them hatch. Their behavior is controlled in the immediate sense by combinations of the internal and external events to which their nervous systems are exposed.

Three classes of events are of particular importance. First, changes in the absolute and relative levels of various hormones in the bloodstream have great impact on the initiation, maintenance, and cessation of parental behavior. Second, certain stimuli may elicit and/or reinforce the exhibition of parental behavior of particular species. Often, but not always, these stimuli are associated with the presence of the offspring. Some of these stimuli are effective only if the nervous system has been already "primed" by hormonal action; others actually trigger the release of hormones and thereby lead to the production of behavior. Third, events occurring at one point in an animal's life may affect its behavior at a much later time; hence the effects of earlier experiences are often key factors in understanding the control of parental behavior. These effects include, but are not limited to, those normally considered under the heading of "learning."

In summary, the following three questions will be uppermost in mind as we examine some instances of parental behavior: 1) What is the behavior? 2) What is its adaptive value? 3) What mechanisms control it? Information about any one species is rarely complete enough to answer all these questions completely. Consequently, the emphasis will shift somewhat as we move from species to species.

Insects: Cicada Killers and Honey Bees

There are more species of insects than there are of all other animal species taken together. Although the number of insect species whose behavior

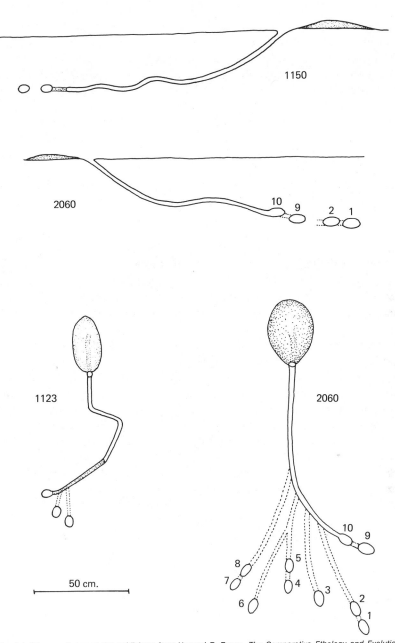

Reprinted by permission of the publishers from Howard E. Evans, *The Comparative Ethology and Evolution of the Sand Wasps.* Cambridge, Mass.: Harvard University Press. Copyright, 1966, by the President and Fellows of Harvard College.

Figure 8-2 Three incompleted nests of the cicada killer, *Sphecius speciosus.* Number 1150 is a profile of a nest from Arkansas County, Arkansas. Number 1123 shows the plan of a nest from Pittsford, New York. Both profile and plan are shown for number 2060, a nest from Ithaca, New York, though some cells and cell burrows are omitted from the profile. Burrows indicated by broken lines had been filled, and were impossible to trace exactly.

has been studied is very small in proportion to the total number of species, the sample is large enough for us to be sure that insects exhibit tremendous variety in their behavioral adaptations. Hence, it becomes impossible to give a comprehensive survey of parental behavior in insects, or even to select instances that typify the class. But we cannot ignore some basic lessons that insects teach us about parental behavior; here we will discuss one species of wasp and the common honey bee.

Cicada Killers

Among the hundreds of species of wasps which capture prey to feed their young, one of the best studied is the sand wasp, *Sphecius speciosus,* commonly known as the cicada killer. The behavior of these remarkable creatures has recently been summarized by one of their most noted students, Howard E. Evans (1966).

When the female cicada killer emerges as an adult from her cocoon, she finds herself in a dark, underground world. She is in one of the nest cells located at one end of a thirty- to forty-inch-long branching tunnel that extends under the surface of the ground at a depth of about four inches.

The female makes her way to the surface of the ground and emerges for the first time into the summer sunlight. She soon flies, and her first flight is almost surely interrupted by the males which have emerged from the tunnel before her and are now defending small territories around the emergence holes. The males are vigorously pursuing most of the insects and a few of the small birds which pass through their territories. The female, who seems to be sexually receptive upon emergence, is soon mated.

She probably never breeds again; rather she uses the sperm collected from her first mating for the remaining thirty days or so of her life. Soon she begins to dig a burrow like the one from which she emerged. She is a strong, persistent digger and completes the burrow in a few hours.

The female obtains the energy for this heavy work from a diet of nectar from flowering plants and some occasional plant sap. But when the nest has been completed with large cells at the end of each branch of its tunnel, the female suddenly exhibits a new appetite: she begins a search for cicadas.

She appears not to conduct her search at random. She goes to likely places, primarily trees, and flies around them in ways that maximize her chances of encountering cicadas. She flies within a few inches of the limbs and trunk and moves along them systematically. When she discovers a cicada she quickly attacks and injects her prey with venom from her stinger. The venom immobilizes the prey but does not kill it: the cicada will remain paralyzed, in a state of suspended animation, for at least a week.

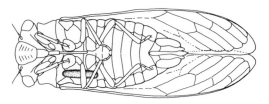

Figure 8-3 An adult female cicada is shown bearing the egg of the cicada killer, *Sphecius speciosus,* on her ventral surface.

The wasp, a very capable flyer, grasps the cicada firmly and flies to her burrow. The effort of this flight can be judged from the fact that a cicada usually weighs four to six times more than the wasp.

When the cicada killer arrives at her burrow she lands, drags the cicada into one of the nest chambers, and lays a single egg on it. This behavior seems to be quite stereotyped. The cicada is always placed back down and head foremost into the cell, and the egg is always deposited in a particular part of the cicada's ventral surface, as shown in Figure 8-3. When the cicada and egg have been deposited, the female seals the cell with a wall of dirt and never opens it again.

The egg hatches in a few days, and the larva feeds on the fresh, relaxed, unresisting cicada. Thus the new cicada killer grows until it is ready to spin its cocoon and pupate.

The behavior of this wasp illustrates some interesting points. The female's parental behavior involves gathering food that is very different from what she herself eats. This development of appetites specific to the needs of the offspring is an intriguing feature of the parental behavior of numerous other species as well. Why does the female wasp behave this way?

In part, the answer depends on the meaning of "why" in the question. If we mean to ask what physiological or stimulus factors are acting on the female to control her behavior, the answer is that we don't know yet. But if we mean to ask about the evolutionary history of this relationship between wasp and cicada, then some insight is available. These wasps, while largely nectar feeders as adults, were derived evolutionarily from wasps that were carnivores both as adults and larvae. The larvae of the modern wasps have retained the ancestral nourishment requirement, while the adults have shifted to a new ecological niche in which nectar is the primary source of sugar and pollen is the primary source of protein.

The retention of the ancestral trait in the larvae requires the care-providing adult to seek out kinds of food that it does not eat itself. The same behavioral requirement is observed in many bird species, also descended from carnivores, in which the adults eat seeds but must gather enormous numbers of insects to feed their fledglings.

Honey Bees

The female honey bee (*Apis mellifera*) is related to the provisioning wasp but shows quite a different pattern of parental behavior. In fact, the term "parental" is a bit of a misnomer in the case of honey bees: although the developing larvae receive a great deal of care, they are never cared for by their parents. Every honey bee worker's parents are the colony's queen and a drone with which the queen mated during the first few days of her adulthood. However, every bee larva is cared for by a worker rather than by the queen or a drone.

The queen bee stores the sperm she collected during her mating flight and uses them to fertilize most of the eggs which her ovaries produce during the rest of her reproductive life. She deposits her eggs in hexagonal cells of wax that the worker bees have constructed. The fertilized eggs all develop into workers (females). The queen also deposits some unfertilized eggs which develop, parthenogenetically, into drones (males).

During the first ten days or so of the larva's life in the cell, it is visited perhaps two thousand times by workers functioning as "nurse bees." On most of these visits the larva is merely inspected or the cell cleaned, but probably between one hundred and two hundred visits include feeding. At first the larva is fed a protein-rich substance called *bee's milk* which the nurse bees manufacture from pollen. Later the nurse bees shift most of the developing larvae to a diet which includes much less protein.

Eight or nine days after the fertilized eggs have been deposited, the nurse bees seal the cells with wax. The larvae pupate and then emerge as adult worker bees about ten days later. However, there is a special case in which this typical procedure is not followed. If for any reason the queen leaves the colony, the workers will often begin to feed developing worker larvae as they would queen larvae; with their rich diet extended these larvae grow so rapidly that they soon outgrow their cells. Worker bees must then enlarge the cells and, eventually, the extra-large larvae are sealed into their enlarged cells and allowed to pupate. They will eventually emerge as new queens (Butler, 1955). The results of the worker's response to the larvae in the queenless colony demonstrates three important facts: 1) All larvae from fertilized eggs have the potential to become queens. (Unfertilized larvae always become drones.) 2) The difference between queens and workers is a matter of individual development, which in turn is completely determined by the diet they are fed. 3) Bees respond to the absence of the colony's queen by producing one or several new queens to replace her. Why do workers change their caretaking behavior when the queen is removed from the colony?

The adaptive significance of the changed behavior is obvious: a colony without queens cannot reproduce its members, for the queen is the only reproductively capable female in the colony. The workers do not know this in any conscious or rational sense. Rather, changes within the worker bees, and in the stimuli they encounter in the queenless colony, initiate the special behavior that leads to the production of new queens. Experiments have shown a healthy queen continuously produces a material called "queen substance" which is transmitted from bee to bee throughout the hive. The substance originates in glands in the mouth parts of the queen. The workers closest to the queen acquire it from her as they stroke her body while she moves over the comb depositing eggs. These bees pass it along, from mouth to mouth, to other bees as they come into contact with them, until it is present in every part of the hive. The queen substance, which has been chemically identified as 9-oxodec-trans-2-enoic acid, thus may stimulate workers which never have direct contact with the queen. It turns out that one of the effects of the substance is to modify the nervous system of the workers so as to inhibit the "queen-producing" behavior of those who are working as nurse bees. When the queen is absent, the inhibiting effect of the substance decays, and queen-producing behavior is released (Butler, 1955).

There are some fascinating contrasts between the parental behavior of honey bees and cicada killers. First is the difference between the solutions reached by the two species for the problem of providing a protein-rich diet for the developing young. The cicada killer provides meat while the honey bee provides transformed pollen. These are two solutions that have been evolved to meet the same ecological problem. Both species are nectar feeders which have evolved from carnivores. However the development of the larvae is enhanced by a high-protein diet (because large amounts of protein are essential for the construction of cell membranes and other cellular structures and in the building of enzymes which catalyze all biochemical reactions). Thus cicada killers and honey bees which have in one way or other provided high-protein foods for their young have enjoyed a selective advantage.

A second interesting difference is that while in the wasp parental care is provided by the individual's parent, in the honey bee it is provided by the individual's sisters or half-sisters. Thus the evolutionary selection of these behaviors has been different. In the case of the wasp, genes relating to parental behavior are selected directly. In the honey bee, genes are selected in the same way, and just as rigorously, but the selection occurs when the genes are expressed in the parent's offspring rather than in the parents themselves. This indirect action of natural selection is also evident in a good many other social insects. Since only a few of the individuals in these species participate in reproduction, selection acts on the characteristics that are expressed in the nonreproducing members of the species. One can view the workers as no more than an extension of the queen herself. Thus queens are selected on the basis of the kind of workers they produce. Queens with workers which provide competent parental care have surely enjoyed a reproductive advantage in the past.

Birds: Mallee-Fowl and Doves

To anyone interested in animals and their behavior, Australia is a paradise. The radiation of its marsupials into a continent with widely varied ecological opportunities has produced such truly exotic creatures as the koala bear, wombat, and kangaroo. The reproductive behavior of these animals is as intriguing as their other qualities, yet we will pass them by to study an ordinary-looking bird from which we can learn some even more remarkable things.

Mallee-Fowl

In southern Australia there is a large semi-arid region called the *Mallee*. The land is not productive. The soil is dry with patches of nearly pure sand. The summer is hot, the winter cold, and the rainfall, which occurs

From *The Mallee-Fowl* by H. J. Frith. London: Angus and Robertson, 1962.

Figure 8-4 A photograph of the Mallee-fowl.

only in the spring, scant. After the spring rains herbs grow, flowers bloom, and seeds develop, but all wither in the heat of the dry summer. The main plant life in the Mallee is Mallee bush, a combination of several varieties of eucalyptus which grow neither very tall nor very thick as individual plants, but in some places form a moderately dense, tangled growth overhead with fairly open ground below.

In this demanding land lives a dull-colored bird about twice the size of a pheasant. This is the Mallee-fowl, and the male of the species is one of the world's most remarkable parents. His behavior has been patiently and ingeniously studied by H. J. Frith (1962).

The male is engaged in the work of reproduction for eleven months of the year. During all this time he remains in his territory, an area of Mallee scrub at least 300 yards across which he defends against intruders. Near the center of the territory, located in an unshaded opening in the bush, is a mound of sand two to three feet high and ten to fifteen across. The male leaves this mound only to rest in nearby shade during the hot summer days, to feed for an hour or two each day, and to roost close by in the Mallee scrub during the night. He occasionally gives his territorial call while on his mound, and there even his mate's approach is greeted with a mild threat. Often he is hard at work on the mound, digging a large cavity in the top of it with quick strokes of his powerful legs or filling in the cavity and heaping sand on top of it. This work is hard but vital, for the male is incubating eggs. To understand how, one must know the history and construction of the mound.

The Mallee-fowl began work on the mound during the winter when there was a light scattering of leaves and twigs on the ground. He dug a hole a foot or so below ground level probably in the site of an old mound. When finished it looked like a very small volcano, with a cone-shaped cavity of dirt excavated by thousands of strong sweeps of the bird's feet. Into this hole the male scratched leaves and twigs he had gathered from as far away as fifty yards. This material was swept into the hole and piled inside until the heap it formed rose to ground level or a little above.

At this point the male stopped construction until the spring rains wet the material in the hole. Then he dug a depression into the top of the wet pile and filled it with a mixture of finer organic material and sand. This done he piled about two more feet of sand on top; the mound was then complete. The male had been working about four months. See Figure 8-5.

With construction completed, the male's behavior changes. Almost daily he comes to the mound and digs a cone-shaped cavity in the top down to the depression filled with fine organic material and sand. He then thrusts his open bill into this material, stopping just short of his eyes; he is testing the temperature in the portion of the mound that will become the egg chamber. When he has completed the test, he rebuilds the top of the mound.

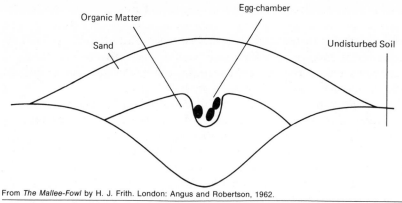

From *The Mallee-Fowl* by H. J. Frith. London: Angus and Robertson, 1962.

Figure 8-5 Diagram showing cross-section of a Mallee-fowl mound.

Summer is coming, but the increasing heat of the sun has little effect on the temperature of the egg chamber. At this time of year its temperature is determined primarily by the heat generated from the fermentation of the vegetable matter that the male has built under and around it. The sand on top allows the heat to accumulate and the temperature to rise. Finally, on a day when the temperature has risen high enough (somewhere between 90 and 96 degrees, depending on the individual male), the male allows his female to enter the excavated mound and lay a fertilized egg in the carefully prepared egg chamber. Any attempts by the female to enter the mound, before it reaches the proper temperature, are vigorously thwarted by the male.

The egg itself is relatively enormous; it weighs a full tenth as much as the female herself (the significance of this will become clear later). After the female has laid the egg and left the mound, the male refills the egg chamber and rebuilds the mound to its cone shape. For the remaining four or five months of the laying season, i.e. throughout the hottest months of the year, the female comes back every ten to fourteen days and lays another egg.

Every two or three days during the early part of the laying season the male opens the mound early in the morning and tests the temperature of the egg chamber. At this time of year the heat from the rotting vegetation has often made the temperature too high. If so, the male leaves the mound open until some heat has dissipated and then rebuilds it. For the rest of the spring, the male's behavior serves primarily to keep the decaying humus from overheating the developing eggs. The thick layer of soil insulates the eggs from the sun's heat during the day and prevents loss of heat from the decaying humus during the night. However, as the summer progresses, the heating potential of the decaying humus begins to decline, and the soil over the egg chamber now insulates it from the only remaining heat source, the sun.

At this point the male fowl's behavior changes once again. He no longer opens the egg chamber briefly in the morning before the heat of the sun can reach the eggs. Instead he waits until ten or eleven A.M. and then opens the egg chamber wide, leaving only a few inches of sand directly over the eggs. Solar heat penetrates to the eggs and warms the soil around them, thus providing the heat for incubation. The male spreads the sand he has taken from above the egg chamber out on the ground nearby, stirs it about and, as each layer is warmed, kicks it back onto the mound. By four P.M. he has rebuilt the mound with warm sand. This heat lasts until the next day, when the process is repeated. Thus, the male maintains the incubation temperature until autumn, when solar heat becomes insufficient and the egg chamber cools beyond correction. Then he abandons the mound.

Meanwhile, what has happened to the eggs? Most of them hatch after fifty to sixty-five days; the timing depends partly on the temperature of the egg chamber. The chick, after breaking out of the egg shell, finds itself buried in as much as two feet of sand. Now the huge egg pays off. It has made possible a well-developed, self-sufficient chick. Eyes still closed, barely breathing, it begins forcing its way slowly upward through the sand. Some chicks die before they reach the surface, but most emerge from the mound after several hours of effort. Then they open their eyes for the first time and look around, panting. Soon they free themselves entirely from the sand and stagger and stumble down the slope of the mound to the shelter of a nearby bush where they rest. Within a few hours they can run swiftly, and within twenty-four hours they can fly.

It is well that they can. The male who has tended them has worked very hard at behavior which contributes directly to the probability of their survival as embryos, but neither he nor his mate gives any care to the hatchlings. Moreover, the male shows no solicitude toward the eggs themselves on the relatively rare occasions when he can see them. Frith repeatedly observed that when birds opened the nest chamber to lay more eggs, they were quite indifferent to the fate of the eggs already laid. These eggs were often scratched out of the chamber just as a rock or any other obstruction would have been. Other eggs were broken. The hatchlings were treated with equal indifference. Despite many, many hours of observation, Frith never observed any direct social interaction between an adult Mallee-fowl and a chick. However, since the chicks spend a long time digging their way out of the egg chamber and the males frequently dig into it, they must encounter one another at least occasionally. Frith observed one of these encounters. The male, scratching the sand out of the cavity of his mound, uncovered a struggling chick with one sweep of his powerful foot. The father did not miss a stroke; the next sweep caught the chick and pitched it out with the rest of the debris. The chick scrambled away as soon as it was able, and that was that.

But if the Mallee-fowl is not "taking care" of its eggs and chicks, then what is he doing? *Apparently he is controlling the temperature of the egg chamber, no more.* This is not very sentimental, but it is quite sufficient.

This bird's unusual behavior raises at least two interesting questions. First, given that the Mallee-fowl changes his "incubation strategies" as the heat of the organic material in the mound waxes then wanes, how does he determine the appropriate behavior for the particular time of the year? And second, why does the male expend such enormous amounts of energy to obtain the heats of fermentation and the sun? Why doesn't he just sit on the eggs to provide heat from his own body, like so many other birds do?

Temperature Control Studies

The male's shift from a strategy in which his digging serves to keep the chamber temperature down, to a strategy in which the digging and refilling with sun-warmed sand keeps the temperature up, is truly an amazing behavioral adaptation. Frith was fascinated by it, and he proposed several possible mechanisms.

Normally, the male's behavioral shift is correlated with both a decreasing temperature inside the mound and an increasing temperature outside the mound. Hence the bird could be responding to either or both of these environmental cues, or to related ones, such as a change in the fermentation activity of the humus. The shift also occurs after the bird has been working on his mound for some months. Perhaps the mechanism that controls the shift is correlated with the passage of time, or a change in season, or a change in the bird's physiology related to fatigue produced by his long period of mound tending.

In order to separate these alternative possibilities, Frith performed some experiments.

Artificially increased egg-chamber temperature. The first set of experiments was designed to examine the bird's reaction to increased mound temperature at various times of the year. An electric heating element was placed on the mound during the building period, and the male incorporated it into the mound's structure. An underground cable led from the heating element to an electric generator some 300 yards away. With this device the experimenters were able to raise the internal temperature of the mound by as much as twenty degrees in an hour. The bird's reaction to a high mound temperature was tested first in the spring, when the normal source of heat is the fermenting humus.

The bird reacted appropriately to these circumstances. What the mound needed was dissipation of internal heat; and indeed the male opened it early in the morning and permitted it to cool. If the temperature was particularly high on a certain visit, he would increase his visits from once every two or three days to once every day, cooling the egg chamber each time until the experimenters allowed its temperature to return to normal. Thus, the male's response to the kind of temperature fluctuation that might normally have occurred at this time of year, i.e., an increase, was effective in dealing with the experimentally induced fluctuation.

During the summer, however, when the primary source of heat comes not from the decaying vegetation but from the heat of the sun, the bird's behavior became markedly inappropriate. Now when the egg-chamber temperature was raised by the heating coil, the male did not leave the mound open for cooling as he had done in the spring, but rather he filled it up and piled even more sand high on top of it. Normally, at this time of year, this procedure would have shielded the chamber from the sun's heat, but in these experimental circumstances it trapped the heat coming from below the chamber even more effectively; the bird was tricked into behaving inappropriately.

These results demonstrate that the bird's behavior is not controlled by the immediate effects of his efforts to cool or heat the mound. They suggest that the bird employs strategies which are usually effective during a particular season, but that he is unable to employ alternative strategies when the need arises.

Two hypotheses remain: 1) It may be that the normal shift in incubation strategy is triggered by some relationship between ground and air temperatures. 2) It may be that after spending a certain length of time behaving according to one pattern, the bird shifts to another regardless of environmental stimuli.

The next experiment made it possible to decide between the two hypotheses. This time the mound was artificially heated during the autumn, when the declining heat of the sun is the only source of heat for incubation. Remember that ordinarily the bird opens the nest late in the morning and spends the rest of the day refilling it with hot sand. But now, when the mound was artificially heated, the bird behaved as if it were spring: he merely cooled the mound and rapidly refilled it.

The key to understanding this result lies in seeing that when the mound is artificially heated in the autumn, the relationship of air to mound temperatures is the same as in the spring. The air temperatures are the same in both seasons, and the heat of fermentation in the spring is equivalent to the artificially produced heat in the fall. Therefore, it seems likely that the bird's choice of response to egg-chamber temperature is determined by the relationship between the air temperature and the mound temperature.

Artificially decreased egg-chamber temperature. An experiment was designed to test the bird's reaction to unexpectedly low temperatures. Early one spring the experimenters removed all the organic material from an actively fermenting mound and replaced it with cold sand. Naturally, the egg-chamber temperature fell abruptly. The male's behavior changed just as abruptly; he began immediately to reduce the top of the mound and expose the egg chamber to the sun under only a few inches of soil. He worked the removed soil until it had warmed in the sun, then he returned it to the mound. In other words, he shifted to his autumn strategy. He continued this behavior until the hottest part of the summer, when the sun's heat raised the internal temperature to about ninety degrees. At this point he shifted to summertime behavior, piling the soil high. While this would normally prevent the sun from overheating the egg chamber (remember that humus heat would be declining at this point), in this case it insulated the egg chamber from the only source of heat; hence the egg chamber temperature dropped. The bird immediately shifted once again to the autumn pattern, but soon gave it up and abandoned the mound.

Is the Mound Necessary?

But surely the most pressing question has yet to be discussed. Why does a Mallee-fowl do so much hard work? The species would seem to have many alternatives. The simplest would be to lay a batch of smaller eggs, incubate them by sitting quietly over and on them, keep track of the young for a few days, and then be done. But, given the evolutionary history of the bird, this is not a real alternative for the Mallee-fowl. The Mallee-fowl is one of thirteen members of the family *Megapodiidae* (mighty-foot) distributed in Australia and Southeast Asia. The Mallee-fowl is the most markedly different from the likely ancestral form, which probably resembled the modern Jungle Fowl, a common bird along the sandy beaches on the northern coast of the most tropical part of Australia.

The Jungle Fowl's incubation behavior is quite variable. In some areas it consists in simply opening a hole in the beach, dropping an egg in, covering it up, and leaving everything else to the sun and to chance. Some beaches are occasionally completely inundated by the tide and the resultant soaking dramatically reduces the probability of survival of the young. In these areas the Jungle Fowl have evolved a pattern of digging tunnels into dunes or sandy sea cliffs that are exposed to the sun. The eggs are laid at the end of the tunnel. The tunnel is filled with sand and the sun provides heat for incubation. Where dunes or cliffs are not available, some Jungle Fowl build mounds of sand into which the incubation tunnels are dug.

When the chicks hatch, they simply struggle to the surface and dart away into the jungle which borders the beach. There they live until they return as adults to the beach to lay their eggs and let them be incubated in the sand. All these patterns appear to save energy and reduce predator pressure, when they are compared with the behavior of most other birds. Therefore, we can appreciate how this behavior might have evolved.

It is an easy step from a Jungle Fowl's building mounds of sand to the behavior of the related Brush Turkey which lives inland and along the eastern coast of Australia (Fleay, 1937). There the dense overhead foliage prevents direct use of the sun's heat in incubation, but extremely lush plant growth allows the males to build a mound entirely of organic material. In the rainy season, this mound quickly becomes a compost heap. The male tests the mound's temperature regularly with heat receptors in its beak, and when the decaying vegetation is producing enough heat for incubation, he permits the female to dig tunnels into the heap and lay eggs at the end of them. The male works the compost heap part of the time and probably effects some modification of internal temperature. Unfortunately, we know very little about this, but we do know that in the end a chick is hatched from most of the eggs. The Brush Turkey works harder than the Jungle Fowl, but its behavior still seems relatively efficient.

Now just as the Brush Turkey probably evolved from an ancestral population of a Jungle Fowl-like bird, so the Mallee-fowl is probably descended from an ancestral population of birds similar to Brush Turkeys (Ashby, 1922). The area in Australia where the Brush Turkey lives was once moist and heavily forested. As it dried out the Brush Turkey pattern was found wanting; in order to survive in the rather barren circumstances under which the Mallee-Fowl now live, it was necessary to add several embellishments to the basic pattern of building a mound and dropping an egg in it. The result of these embellishments is that the Mallee-fowl is surely one of the hardest working of all bird parents.

But for the Mallee-fowl there was no alternative. Through the radiation from the Jungle Fowl-like ancestor, the Mallee-fowl had certain basic behavioral equipment which could not be radically altered but only gradually modified by the accumulation of small differences. The environment changed gradually, making small elaborations of the compost-pile effective and advantageous, then favoring more small elaborations and selecting the birds that produced them. Thus the evolution of the Mallee-fowl limits the options of living birds. However, it is worth noting that, although the behavior seems tremendously inefficient, it works; male birds do successfully hatch out the eggs.

Doves

It's hard to imagine a park without pigeons. There is some magic in the puffed chest, waddling stride, hushed flight, and beautiful cooing calls that fascinates and charms us at first sight, and the more we learn about these birds the more fascinating they become.

Pigeons and doves make up the family *Columbidae,* of which there are more than two-hundred and fifty living species. Members of the family exist in every continent except Antarctica and have radiated into many habitats. The smallest is a dove, about the size of a sparrow, that lives in northern Mexico; and the largest is an Asian pigeon about the size of a pheasant. A number of species in the family have adapted to human habitation, and several have become domesticated.

The ring dove (*Streptopelia risoria*), a native of North Africa, has lived around and with humans for centuries. It is often represented in

Egyptian art, and Aristotle discussed the bird and its behavior. Like the other members of the family, ring doves are vegetarian. Their food consists largely of seeds which they swallow whole. The seeds are first stored in the bird's crop (an enlargement of its esophagus), and later passed through the stomach (Heinroth and Heinroth, 1958).

The crop makes possible one of the unique aspects of the parental behavior of the *Columbidae*. Many parental birds use their crop to store food gathered for their offspring, then regurgitate the contents of the crop back at the nest. Doves and pigeons feed their young cells from the lining of the crop itself.

We will say more about this remarkable adaptation later. But first, one of the lessons that the ring dove has taught us is that sometimes reproductive behavior is composed of parts which are so interrelated that often one cannot really understand parental behavior without first knowing some of the general features of reproduction.

Ring doves become sexually mature at about six months of age. A typical reproductive cycle of a pair of ring doves breeding in a laboratory is described by Lehrman (1965). Suppose the birds have been exposed to a day-night cycle of fourteen hours light and ten hours dark. Now they are placed together in a cage containing some four cubic feet of space, with a supply of seeds, water, and grit. There is also some nesting material, for example tree needles or alfalfa, and a place to build a nest, for example a glass bowl. The male approaches the female almost immediately. He orients toward her, tilts his body forward until his back is parallel to the floor, and then hops toward her while giving a call that sounds a little like "hnh hnh hnh." The female generally retreats from this charge, and the male soon shifts to a new behavior sequence in which he bows very deeply and coos in synchrony with the bowing. The female seems passive as she moves around and watches the male as he continues to orient his bow-coo toward her.

Within a few days (occasionally a few minutes), the male's bow-cooing behavior begins to alternate with another response at the place that will eventually become the nest site. Here, head down and wings flipping slightly out from his sides, he emits a somewhat different cooing call. This behavior pattern is called the "nest-coo." The female frequently responds by approaching the male. Copulation occurs often during this courtship sequence.

Nest building begins after three or four days of courtship. Typically, it is the female who builds the nest and the male who carries material to her, but these roles may be alternated or reversed. The first of two eggs is laid seven or eight days after the beginning of courtship. About forty hours later, the second egg is laid and the pair begins to incubate the clutch of eggs.

The female does most of the incubation, sitting on the eggs for about eighteen out of every twenty-four hours. The male sits the remaining six hours. By this time courtship is almost nonexistent, having been replaced by incubation. The eggs hatch after about fourteen days. The squabs (as the young are called) are naked, blind, clumsy, and weak. Without intensive care they could not survive. It is nearly impossible for humans to raise them even with the aid of electric incubators and specially formulated diets.

The squabs are soon hungry and push their heads up high with their mouths open. The parents feed each squab by regurgitating the contents of their crops into its mouth. At first this consists of "crop milk," a thin, whitish liquid with whitish lumps in it. It looks very much like curds and whey, but is the sloughed-off lining of the ring dove's crop. During the last few days of incubation, the epithelial cells lining the crop begin to proliferate very rapidly and to slough off, forming masses of cells floating in watery liquid. The change in the lining of the crop is caused by an increase in the circulating level of the pituitary hormone prolactin.

From "Interaction between internal and external environments in the regulation of the reproductive cycle of the ring dove" by D. S. Lehrman, in F. A. Beach (Ed.), *Sex and Behavior.* Copyright 1965 by John Wiley & Sons.

Figure 8-6 Initial courtship behavior of the male ring dove. Above: bowing; below: cooing.

Production and feeding of crop milk is the pigeon solution to the ecological problem common to many herbivores: how to provide a protein-rich diet for the rapid, early development of offspring. Many other birds that are primarily vegetarians as adults feed their offspring a diet that is eighty to ninety percent animal matter, generally insects and worms.

The young ring doves are warmed by their parents and fed, first crop milk, then a mixture of crop milk and seeds (Wortis, 1969). By the time the parents stop feeding them, twenty-one days after hatching, the young birds are as large or larger than the parents. As the feeding declines and the squabs near their full development and ultimate independence, the male resumes courtship. Thus a new reproductive cycle is initiated at the time of weaning or shortly thereafter.

For this complicated behavior sequence to succeed, the two birds must coordinate their responses very closely. It is easy to demonstrate that the behavioral sequence exhibited by each bird is dependent upon the presence of the other. Birds left alone in the same conditions will not go through any part of this sequence. For example, the male has a very marked effect on the occurrence of two aspects of parental behavior: incubation and feeding of the squabs.

Initiation of Incubation

Eggs are normally laid some six or seven days after the experienced pair is introduced into the cage. Both birds immediately take turns incubating them. Of course, this does not demonstrate that either of the birds has influenced the behavior of the other. In the natural reproductive cycle, ring doves sit on eggs whenever they are available. It could be that the birds are always ready to incubate eggs, but have no opportunity to show this readiness until they are laid. Lehrman (1965) has provided an experimental analysis of the events in a normal cycle. Instead of waiting until the eggs were laid in the course of a normal reproductive cycle, he introduced a completed nest with two eggs in it at various points in the cycle and tested the birds' readiness to incubate. In the first experiment the nest and eggs were placed in the cage at the same time that the male and female were put together. None of these birds incubated the eggs before the fifth day after introduction and most did not until the sixth or seventh day. The ring doves were not ready to sit on eggs until they had been stimulated by a mate for a few days.

The social stimulus of the mate is obviously important; are other stimuli involved as well? Recall that every situation in which the birds have incubated up to this time has included a nest bowl and some nest-building material (in the experimental situation most recently described, that material was already built into a nest). Does the nest-building material itself influence incubation?

To answer this question, several pairs of birds were placed in cages without a nest bowl or nesting material. Six days later completed nests with two eggs were introduced into the cages along with empty nest bowls and loose nesting material. (Remember that most of the birds that had been given a completed nest with eggs on the first day began to incubate by the sixth day.) Only a few birds incubated on the day that the nest bowl was introduced. The rest spent the day in vigorous nest building. All of these birds incubated on the following day. In the absence of nesting material, therefore, pure social stimulation prepared most of the birds for nest building but not for incubation. Incubation required stimulation from the mate plus feedback from the activity of building.

Physiological Mechanisms

These experiments have told us something about the external stimuli that initiate incubation behavior. How do these external stimuli cause behavioral changes? What physiological changes do they produce in the bird? What is the mechanism by which the behavior is controlled?

A ready and reasonable hypothesis is that reproductive behavior is under the control of the reproductive hormones, the supply of which fluctuates predictably and systematically with stages in the reproductive cycle. Since it is now clear that external stimuli control the behavior of the sexually experienced ring dove, is it not likely that these stimuli are operating through changes in the hormonal condition of the birds?

A mated female undergoes changes in her reproductive apparatus. The *oviduct,* a tube down which the fertilized ovum travels picking up nutrients and finally an egg shell, develops rapidly in the days immediately preceding egg laying. Full development of the oviduct in female ring doves depends on the combined effect of the two ovarian hormones, estrogen and progesterone (Lehrman and Brody, 1957). This means that the levels of these hormones increase sharply about the time that doves are ready to incubate, suggesting that they may be critical for this behavior.

Lehrman kept a number of birds in individual cages and increased the level of these two hormones circulating in the blood by intramuscular injections. One group received only the neutral medium in which the hormones were dissolved and not the hormones themselves, another group received progesterone, and a third group received estrogen. After seven days of daily injections each bird was put into a cage with a mate and a nest bowl with eggs. All the birds in this experiment had already been through at least one successful reproductive cycle.

The birds that had received only the neutral medium courted, built nests, laid eggs, and incubated in the normal sequence. In other words, these injections had no effect on the birds' behavior. However, all the birds that had received progesterone incubated within a matter of hours. They did not exhibit the behavior that usually precedes incubation. On the other hand, most of the birds injected with estrogen spent the first two days building nests and sat on eggs on the third or fourth day.

These results suggested that incubation is controlled by progesterone and nest-building behavior is controlled by estrogen. During a normal cycle, the hormone-behavior interactions probably operate like this: the presence of the mate and the subsequent courtship stimulate the production of estrogen. Estrogen induces nest building. Nest building stimulates increased progesterone secretion, which in turn induces incubation. The behavior of the estrogen-injected birds supports this hypothesis, as does the behavior of the birds, described above, that had been paired in a cage without nesting material for six days. When material became available they built nests frantically for a day before they began to incubate. Probably the day of nest building produced enough progesterone to cause incubation.

The Hormone-Nervous System Interface

The next problem is to determine how progesterone has this effect. To change behavior, a hormone must affect the nervous system. One way it might do this is to act directly on the activity of a small, localized set of nerve cells in the brain (see discussion of central motive state in Chapter 5). Komisaruk (1967) tested this hypothesis by placing very small amounts of progesterone into various brain areas. To accurately assess the consequences of this procedure, however, he first needed to know the effect of circulating progesterone on the behavior of the particular doves he was studying. He tested one-hundred and sixty-eight ring doves, approximately half males and half females, by injecting them with progesterone every day for seven days and then giving them a mate and a nest with eggs. He found that seventy-two percent of the birds incubated the eggs after this treatment.

Of course these particular birds might have incubated without progesterone. Therefore, they were isolated to allow the effects of the progesterone injections to wear off, and then tested again for incubation behavior. Only seven of the hundred and twenty-two subjects that had incubated when injected with progesterone also incubated without it. The remaining hundred and fifteen ring doves were the subject for the rest of the experiment.

The next step was to supply minute amounts of progesterone to specific areas of the brain. Komisaruk achieved this by filling the end of a very small tube (.009 inches in diameter) with crystalline progesterone, then permanently inserting the tube into various parts of the brain; exact locations were selected in advance and verified after the experiment. After these implants had been in place for seven days, the birds were again tested for incubation.

The behavioral responses varied dramatically with the region into which the progesterone had been placed. Forty-four percent of the birds with implants in the preoptic area incubated, as did twenty-nine percent of those with lateral forebrain tract implants. Implants anywhere else in the brain were never associated with an incidence of incubation higher than four percent. This suggests that when progesterone levels are high in the blood, the hormone affects specific groups of neurons in the brain leading to incubation behavior.

During the normal reproductive cycles adults will of course feed their own young; less obviously they will also feed any other squabs placed in their next. However, Lehrman (1955), and Lott and Comerford (1968) have shown that when mature birds are not moving through the reproductive cycle they will not feed squabs. Therefore, some changes in the bird correlated with events in its reproductive cycle must underlie the occurrence of feeding.

Crop Development

A necessary condition for successful feeding is the proliferation of cells within the crop to produce crop milk. This development normally occurs late in the incubation stage of the normal reproductive cycle, and Friedman (1966) has shown that it is triggered by the social stimuli normally present at that time. Beginning early in the incubation stage of a normal cycle, she separated males from their mates and nests with a glass partition. The males could watch the females incubating eggs but could not themselves incubate. Some crop engorgement occurred in all these males. However, only those males that had had an opportunity to incubate briefly prior to separation showed full crop development. Therefore, full crop development is dependent upon the stimuli received from the activity of incubation as well as those received from the mate. These stimuli increase the levels of circulating prolactin, which stimulates the development of crop milk.

Feeding

Given an adequate supply of crop milk, why does the dove begin to feed its young? This behavior occurs in a situation in which there is a high level of circulating prolactin and a resulting engorgement of the crop with crop milk. One might hypothesize that either or both of these two circumstances control the onset of feeding.

Lehrman (1955) studied the reaction of adult male ring doves to squabs after developing adults' crops with prolactin injections. He found an interesting relationship. The adult males with previous breeding experience, including successful rearing of one set of squabs, fed the squabs that were presented to them. On the other hand, inexperienced males, though treated with the same hormonal regimen and experiencing the same degree of crop development, did not feed squabs. This suggests that a prolactin-developed crop is a necessary but not sufficient condition for feeding, and that to produce feeding both experience and crop development are necessary.

However this does not account for the fact that inexperienced dove parents will feed their first set of squabs. Lehrman observed new parents and developed the following hypothesis: the parent, which had been sitting on the eggs, ends up sitting on the squab when the egg hatches.

The squab has a strong tendency to lift its head and thrust upwards with its beak. If these thrusts upward happen to correspond to a moment when the adult is looking down and has its mouth open the squab's bill will enter the adult's throat and have the same effect as fingers down your throat—regurgitation. Since this regurgitation results in the squab being fed, and the parent's engorged and possibly uncomfortable crop being emptied both the squab and the parent are reinforced and the rate of this behavior goes up sharply.

Once the parent has learned a response that empties its crop, it will feed young whenever its crop is engorged. On the other hand, if it lacks that experience, crop engorgement alone will not lead to feeding squabs who are placed in a nest to which the adult has no attachment. This accounts for the difference between experienced and inexperienced ring doves reacting to crop engorgement when the engorgement is due to hormone injections, rather than the natural reproductive cycle.

This line of argument is so persuasive that it was startling when two studies reported that crop development (and presumably prolactin, since crop development is a highly sensitive indicator of the presence of prolactin) is not necessary to initiate feeding.

In the finest analytic style, these two studies eliminated a number of the conditions forming the usual context of feeding behavior, thus substantially reducing the number of variables to which it could be attributed. The normal context of feeding behavior includes a full crop, and a recent history of some fourteen days of incubation, preceded by a period of nest building, courtship, and copulation. Klinghammer and Hess (1964) and Hansen (1966a) eliminated a substantial part of this normal background by introducing young squabs to parents early in the incubation stage of their first reproductive cycle. At this early stage there is little or no engorgement of the crop and therefore only a small amount of circulating prolactin. Nonetheless, regurgitation feeding motions were observed in both studies, though the incidence was fairly low.

These studies demonstrated that neither high levels of prolactin nor crop engorgement are necessary conditions for the occasional occurrence of feeding behavior. However, while the design of these studies allowed them to reveal a great deal about which variables are not essential to the occurrence of the behavior, they could contribute only indirectly to knowledge about which variables *are* essential.

In an attempt to fill this gap Lott and Comerford (1968) tested the following hypothesis: progesterone, which is available early in the incubation stage, has the effect of attracting the mature ring dove to the offspring and inducing some components of feeding behavior. Prolactin and an engorged crop should increase the tendency of the inexperienced parent to feed its young.

Their subjects were reproductively inexperienced males treated in one of four ways, 1) injected with progesterone by itself, 2) injected with prolactin by itself, 3) injected with progesterone and prolactin in combination, 4) injected with media for these hormones.

At the end of the injection series each bird was placed in a large cage with a one- to three-day-old squab and observed for a total of seven hours out of the next fifty. The seven hours were distributed identically for all birds. The results can be summarized as follows: birds injected with progesterone alone frequently sat but rarely fed; birds injected with prolactin only neither sat nor fed; birds injected with both progesterone and prolactin both sat and fed; and birds injected with media never sat and very rarely fed (Table 8-1). These data generally support the hypothesis that Lott and Comerford had generated, but do not fully explain parental behavior. Even the most effective hormonal regimen did not produce parental behavior in all, or even most, of the subjects. Nonetheless, these data do show that the response of regurgitation feeding in this species is enhanced by a combination of progesterone and prolactin, and that in inexperienced birds feeding is very unlikely to occur in the absence of either of those two hormones.

But while this seems to be true of inexperienced birds, Lehrman (1955) had shown that it was not true of experienced birds. Experienced birds fed when injected with only prolactin. Apparently experience can substitute for progesterone in motivating this behavior.

Mammals: Ungulates, Rodents, and Primates

Bison

I don't know anyone who goes cow watching. And the truth is, anyone who did could describe ninety percent of a cow's behavior by saying "she eats grass and sleeps." But the other ten percent contains some very specialized behaviors that permit cows and other large ungulates (hooved animals) to thrive on grass or browse. Parental behavior is one of these

Table 8-1

	Treatment			
	Progesterone	Prolactin	Progesterone & Prolactin	Control
Number injected	20[1]	12[2]	12	18[3]
Number brooding	10	0	4	0
Brooding incidents	41	0	12	0
Time[4]	13:51	0:00	12:27	0:00
Number feeding	2	0	6	1
Feed squabs incidents	3	0	67	1
Time	0:03	0:00	1:04	0:01

From Lott, Dale F., and Comerford, Sherna (1968), "Hormonal initiation of parental behavior in inexperienced ring doves" *Zeitschrift fur Tierpsychologie* **25**, 71–75.

[1] Includes 14 birds injected with progesterone alone and 6 injected with progesterone plus saline.

[2] Includes 6 birds injected with prolactin alone and 6 with prolactin plus sesame oil.

[3] Includes 6 birds injected with sesame oil, 6 injected with saline and 6 injected with both.

[4] Time is total time for the group in hours and minutes.

adaptations whose ecological significance becomes clear when we observe a wild grazing species, the bison. There are not many bison watchers either. I recall the horrified response of a zoology student listening to a proposal to study the behavior of the American bison: "They're nothing but hump-backed cows!" True enough, but if we skip the grazing and sleeping they are very exciting animals. They make a good starting point in studying ungulates because we may still observe their behavior in the environmental context in which it evolved and is still adaptive. Indeed, the bison is one of the largest and most successful grazing animals in all biological history.

For most grass-eating animals, including the bison, parental behavior means maternal behavior. The male participates little if at all in the protection, feeding, training, or other care of the young. The popular picture of the giant buffalo bull defending his herd from ravenous wolves in a howling Montana blizzard or under a burning Texas sun is a tribute to the imagination and anthropomorphism of some writers, but it does not describe any behavior that occurs in the bison. The mature bull is not really a member of the same social group as the cows and calves. He spends only the breeding season with the herd; this amounts to about three weeks out of the year. The rest of the time he remains solitary or in the company of other bulls.

On the other hand, the female and subadult bison are relatively gregarious. Cows spend most of their lives in small groups of twenty to twenty-five individuals. These groups are made up of bulls up to the age of four years, and mature and immature cows. Members of this little society graze together, go to water together, and lie down in the heat of the day together (McHugh, 1958; Shackleton, 1968). The one time in an adult female's life when she is not gregarious is when the birth of her calf is imminent. A few hours before parturition, most cows move at least a small distance away from the rest of the herd. The calf is born after a labor of two to five hours (Egerton, 1962).

At birth and for the first few minutes thereafter, the calf is basically a wet, struggling heap. The mother attends closely to it, licking and frequently eating the placental membranes, then licking the calf until the amnionic fluid is removed from its hair. The calf makes initial efforts to rise about fifteen or twenty minutes after birth. It is unsteady and occasionally falls during the first half hour, but while on its feet it moves to its mother and orients to her belly and legs. The calf's movements at this time tend to keep its head under its mother's belly with the muzzle elevated and rubbing against her underside. The mother's movements during this period often bring the calf closer to the udders. She may do this by pushing the calf back toward the udders with her head or by moving her body with respect to the calf's so that its head is closer to her udders.

As the first days of the calf's life pass, the mother continues to attend him closely. She licks him often, both while he is suckling and while he is not. She often follows and calls after the calf with a nasal grunt. The calf is usually close to the mother at this age, frequently joining her as she walks, seldom straying far away. Sometimes he interrupts her movement by running up beside her and swinging his head and neck across her forelegs. The mother stops and the calf usually moves immediately to the udders and begins to suckle. The calf is not attracted only to the mother; he is also attracted to other females and, in fact, almost any large moving object. Calves often approach and follow other cows, young males, and even a man on a horse.

Since nearly all bison in natural or seminatural state have calves within a few weeks of one another, there are a number of other calves in the herd. They are likely to approach the new mother and attempt to suckle. The mother rejects these calves by butting them, kicking at them, or sweeping them away with the side of her head. She permits only her own calf to approach and suckle.

Thus the mother's care being given exclusively to her own calf is due to the active rejection of strange calves by the cows, rather than to a selective approach to their own mothers by the calves. It is easy to anticipate what the fate of an orphan calf will be. He will approach all the large objects in his environment, including many lactating cows. Almost certainly he will be rejected by all the cows and eventually will starve to death.

It would be an error to interpret rejection of an orphan calf as evidence of "poor mothering," for just the reverse is true. I have already said that the only biological justification for parental behavior is to enhance the probability of one's own genes surviving in the gene pool. Therefore, energy expended on young other than one's own, at least in the absence of a social system which will result in reciprocity of such care, is biologically maladaptive. The cow has only so much energy to spend in defense and grooming of her offspring, and she produces only enough milk to feed one buffalo calf. If she were to expend any of this energy caring for any other cow's calf, she would be decreasing her own chances of genetic survival. Therefore, her particular ecological circumstances, as an animal living in a group where other calves are certain to seek her nourishment and care, require the female to show a high degree of discrimination between her own calf and others.

But why do bison live in groups, given the problems it creates for females with offspring? Judging from the high degree of gregariousness of female bison (and many other herd animals), social cohesion is extremely important, probably because it contributes to protection from predators. However, during the birth of her calf the female usually leaves the herd and its protection just at the time that she and her offspring are most vulnerable to predators. In leaving the herd she is apparently driven by some urge ("instinct" if you will) which implements an even greater adaptive necessity. This is the necessity to establish a close bond with, and an unequivocal identification of, her own offspring. The advantage gained from separation must historically have outweighed the increased danger. To be a good mother, she must overcome her normal gregariousness in order to learn an evolutionarily vital fact: which calf is hers.

Exactly what that experience consists in and how it works is not known for very many wild species; however, it has been successfully studied in a domestic ungulate, the farm goat.

Farm Goat

The maternal behavior of domestic goats seems very little different from that of their wild counterparts and is generally similar to the maternal behavior of bison (Hersher, Richmond, and Moore, 1963a). The doe isolates herself from the herd shortly before parturition, although, as in the case of the bison, separation is not invariable nor is it necessarily complete. The doe, like the bison cow, cleans her newborn and is closely associated with it for some time before she returns to the flock. Like the bison, goats respond to the acoustic signals, in this case bleats, from the young. Mothers often approach their kids when the kids bleat, but they may also approach alien young who are bleating. This suggests that in this species auditory signals are not the cues used by the female to identify their young. Visual cues are appparently used to some extent because the female approaches most rapidly and shows the most acceptance of kids who look like her own. Sometimes a kid is allowed to approach close enough to attempt suckling. However, the female always seems to make a last olfactory check of the kid before she will nurse it. Usually the kid is smelled in the anal region. On the basis of the stimuli thus perceived, the female rejects an alien kid even if it looks very much like her own.

In goats, as in bison, the exclusive relationship between a mother and her offspring is maintained by active rejection of alien young. But how does the mother tell them apart? One hypothesis is that the female is attracted to her own offspring because in some way it resembles her. Since there are a number of indicators that the final recognition of the young is olfactory, it could be that the female recognizes some genetically determined aspect of her own odor in the young. Thus, the mere fact that the young is genetically the female's offspring would account for the exclusive care that she gives the young. But this hypothesis seems improbable. For one thing it would obviate the most likely reason for self-isolation during parturition. Moreover, nearly all mothers can be induced to adopt kids born to other females and give them perfectly normal care if presented with a substitute for their own newborn. While adoption can be produced at various stages in the reproductive cycle (Hersher et al., 1963b), females are most receptive to a foster kid immediately after the birth of their own (Collias, 1956).

Klopfer, Adams, and Klopfer (1964) designed a study to examine closely the period of high susceptibility to maternal attachment. Kids from one group of mothers were taken immediately upon their emergence from the birth canal, thus preventing any contact whatsoever between the mother and infant. Kids in another group were left with the mother for five minutes after birth, and then removed from her. During these five minutes the mother behaved normally, licking off most of the amnionic fluids. The duration of separation was varied in both groups: some kids were replaced with their mothers after one hour, some after two hours, and the rest after three hours. Nearly all the mothers whose young had been taken from them immediately upon birth rejected their kids regardless of the duration of separation. By contrast, nearly all the mothers whose young had been left with them for five minutes accepted their kids when they were returned, again regardless of the length of separation.

These results suggest that there is a very short critical period (see Chapter 4) during which the maternal bond is normally established. The exact length of the critical period was not measured in this study, but it is clear that one hour after parturition is too late to develop a normal maternal bond.

It is unusual for a durable and discriminating relationship between a mammalian mother and her offspring to be established during such a brief period. Klopfer and Gamble (1966) were interested in the possible mechanisms responsible for this phenomenon. They hypothesized that the attachment of the mother to the kid might be developed through some very powerful reinforcement that the mother received from the kid during the immediate post-parturition contact. For example the amnionic fluids might contain some labile odorous substance which reinforces the mother's maternal responses. If so, smelling it on a kid at any time could establish a strong bond. Normally, according to this hypothesis, the only time the mother has an opportunity to smell this substance on an infant is during birth and immediately after. At this time the rewarding substance would be associated with the unique odor of the particular infant, and smelling it would produce both the bond and later recognition by smell.

They tested their hypothesis by depriving some females of their sense of smell either during or immediately after parturition. They argued that if the reinforcing stimulus is olfactory, the females that had no opportunity to smell the reinforcing substance would reject all young. On the other hand, a female with normal olfactory sensitivity during parturition might later accept all young if her sense of smell were temporarily disabled, because she would not be able to discriminate her own from the others.

Cocaine was applied to the nostrils of a group of pregnant females whose parturition was imminent; this procedure temporarily made them anosmic. The females were allowed five minutes of contact with their offspring immediately after parturition, then they were separated and the young were returned three hours later to test whether the females would accept or reject them. Females from another group were allowed to deliver their kids and attend to them for five minutes before the kids were removed. The kids were replaced with the mothers three hours later. Just before replacement the mothers were made anosmic.

The results of the experiment were surprising. Females who were deprived of their ability to smell during parturition and hence were predicted by Klopfer and Gamble to reject all young, accepted not only their own but alien young when tested three hours later. Apparently, the five minute exposure to the newborn animal had established maternal behavior, but not being able to smell the kid eliminated the ability to distinguish between it and strange kids. Thus, it appears that while the discrimination of individual kids is based on an olfactory cue, the attachment or the readiness to exhibit maternal behavior is not based on an olfactory reward.

The females who were normal during and just after parturition, but anosmic three hours later, were tested with their own young only. Four out of seven of these females rejected their own young. This result is difficult to interpret with certainty, but it suggests that early contact with the newborn kid may cause mothers to be maternal, but only to kids they can positively identify as having been there at parturition.

Taken together, the above results indicate that the manifestation of maternal behavior in the doe is dependent upon immediate contact with a kid, but that no reinforcement from the kid is necessary for the mother to form a bond with that kid. Thus, it seems likely that there is a critical period of very short duration during which mere exposure to the kid's unique odor produces a durable attachment of the mother to her offspring. The analogy to imprinting in young animals is extremely close— except, of course, that the mother "imprints" rather than the offspring. To test this hypothesis, Klopfer and Klopfer (1968) substituted a foster kid, from whom all amnionic fluids had been removed, for the mother's own kid immediately after parturition. Although some of the foster kids were almost a month old when they were first placed with the newly parturient mother, nearly all the mothers accepted these young. The hypothesis was supported.

Rats

Some species have come to be so closely associated with man that man must be included as part of their natural habitat. *Rattus norvegicus,* often called the Norway rat, is associated with human habitation over much of the world. Although the species does occasionally live elsewhere, areas modified by man are, at this point in time, its natural habitat. Despite the long and close association between rats and ourselves, we have only a few descriptions of their behavior outside of laboratory conditions. The best sources of information are Barnett (1963), who has drawn together much of the existing information on the Norway rat in nature, and Calhoun (1963), who has published a very full description of a semi-natural rat colony.

Norway rats are born in an underground dormitory called a burrow. The burrow is a hollowed-out space with tunnels leading to it from several different directions. Some burrows are no more than long tunnels with occasional widenings, while others feature a single, rather large central area. The rats living in the burrow are a social group. They spend a good bit of their time together, and during certain periods in their lives they may collect in particular areas of the burrow and sleep piled on top of one another.

However, there are other parts of the burrow which are like tiny personal territories a few inches across; in such territories individuals build their nests and hoard their food. Each such area is, in a certain sense, owned by an individual rat and will be defended by it. Females approaching the end of their pregnancy spend increasing amounts of time in their territories and become increasingly possessive about it. At this time a female expands and elaborates the nest in which she will care for her pups. Our knowledge of the maternal behavior of the wild Norway rat is not detailed; however, the species includes a number of domesticated strains (Barnett, 1963), and the behavior of some of these strains has been extensively studied (Rosenblatt and Lehrman, 1963). There appear to be few differences between the maternal behavior of the wild Norway rat and its laboratory strains.

All aspects of life proceed at a rapid rate in most rodents. The total gestation period of the Norway rat is only about twenty-one days. On the twenty-first or twenty-second day of her pregnancy, the female delivers a litter of some eight or ten pups. As the pups emerge from the female's vagina, they are cleaned and licked. Often, especially in the case of primiparous females, the pups are at first scattered around the nest area, but within a few hours the mother has gathered them closely together in the nest. There the mother crouches over the litter in a posture which brings her mammary glands close to them. They begin suckling immediately and continue vigorously for some time. Their suckling is sometimes interrupted by the mother, who leaves the nursing posture and licks the pups. This licking is not random, but is usually begun and always most heavily concentrated in the region of the anus and genitalia. Anal-genital licking is important to the pups because it stimulates elimination of urine and feces, which would not occur without it.

The mother will faithfully attend to the pups, retrieve them to the nest, nurse, lick, and generally clean them for about twenty days, when the pups are finally weaned. As weaning approaches, the female spends less and less of her time with the pups. Her efforts to keep them in the nest end soon after the pups are capable of active locomotion.

Mothers defend their pups aggressively. Even laboratory rats, certainly one of the most placid of all living rodents, are relatively dangerous while defending pups. But female rats are not very selective about the young on whom they lavish their intense and vigorous care. Nearly all female rats will readily accept foster pups.

This behavior is sharply different from the behavior of bison and goats. We recall that maternal behavior directed toward offspring other than one's own does not contribute to the preservation of one's genes in the gene pool and will be selected against. In our discussion of ungulates the analysis of the mechanism that achieves rejection of alien young was our main focus. Rats seem to be flying in the face of this biological demand. But this is not so surprising when one thinks of the ecological differences between the rats and ungulates. The ungulates live in a dense social milieu without territories to separate the mothers. Besides, the young are highly precocious and remain dependent for a long period during which they are quite mobile. Young rats, on the other hand, are essentially immobile and their mothers have individual nests. This means that the mother can locate her own young simply by going to her own nest. Therefore, there has been little selective pressure to reject alien young since the chance of encountering foreign pups in nature is probably too small to favor rats with the mechanisms for exclusive care.

Motivation

Given that these ecological and evolutionary considerations help us understand a mother rat's failure to reject offspring other than her own, we are still left with a very important question: why does she take care of any young at all? Prior to motherhood her behavior was devoted entirely to her own well-being. She ate, drank, slept, groomed herself, and engaged in social interaction. Motherhood has transformed her life into a demanding routine of defending, nursing, nest building, retrieving, and licking. And she is not an unwilling slave to her pups, but an eager one. She will risk her life in their defense. If she is separated from them she struggles desperately to rejoin them. Some powerful force has altered her behavior. What has produced this change? What motivates her?

This question has stimulated a lot of research. The basic strategy of this research has been to identify the changes in the female's environment or her physiological state (or both) that are correlated with the onset of her parental behavior, and to try to find which changes truly cause it.

Environmental Changes as Motivation

One change is in the stimuli to which the female is exposed. Before her delivery there are no pups around; afterward there are several. Perhaps this stimulus change causes the change in her behavior. This could work if the pups simply give the female a stimulus for an always ready maternal response.

This hypothesis has been tested a good many times by placing some pups in with a virgin female to see if she will care for them, for example, by retrieving them if they are placed some distance from the nest. Wiesner and Sheard (1933), Rosenblatt and Lehrman (1963), and others all report that most virgin females do not behave maternally (although some fifteen to twenty percent do) within the first few hours of exposure to pups. Some students, e.g., Leblond (1940) and Rosenblatt (1967), have shown that most females will begin to mother pups after several days of exposure to them. This finding is interesting in its own right, and it will influence our later thinking about the mechanism controlling maternal behavior in parturient animals. But it does not explain the motivation of maternal behavior as it normally occurs. Mothers begin to care for their pups within a few hours after birth. Whatever mechanism controls the behavior, it operates more rapidly than mere exposure to the young. Clearly, the mere presence of pups is not a sufficient condition for the rat's normal maternal behavior.

Hormonal Effects on Maternal Motivation

For many years we have known that hormones profoundly affect the sexual behavior of both males and females in many species (see Chapter 7). Many hormonal changes take place during pregnancy, parturition, and lactation; perhaps one or a combination of them produces maternal motivation.

Wiesner and Sheard (1933), working before the development of modern endocrinology, prepared crude extracts from the relevant endocrine glands and injected them into virgin rats in an effort to stimulate parental behavior; their results were unconvincing. However, with new techniques, Riddle, Lahr, and Bates (1942) reported a set of experiments that for many years seemed conclusive. They injected each of several different groups of virgin females with one of several different hormones. They reported that either of two reproductive hormones, prolactin (from the anterior pituitary) and progesterone (from the ovaries), initiated maternal behavior in some eighty percent of virgin females.

Unfortunately, their testing technique sometimes involved long exposure to pups. Because long exposure in itself can produce maternal behavior, it is not too surprising that later studies all failed to find evidence that either prolactin or progesterone initiated maternal behavior in virgins (Lott, 1962; Lott and Fuchs, 1962; Beach and Wilson, 1963). This work suggests that Riddle *et al.*'s effect was related to exposure to pups rather than hormone injections.

But still more recent studies have offered good evidence that there must be some hormonal influence in the initiation of maternal behavior. Lott and Rosenblatt (1969) investigated the development of the readiness to behave maternally during pregnancy. Pregnancies were interrupted by Caesarean section at various points and the subjects tested for retrieving foster pups. Females whose fetuses and placentas were removed on the thirteenth day after conception were no readier to be maternal than were still pregnant females. But by the sixteenth day after conception, the surgical termination of pregnancy produced females that were much readier to retrieve than were the control (still pregnant) females. This suggests, although it does not prove, that the hormonal state of the female is important in the initiation of maternal behavior since that is the characteristic of a female which is most profoundly altered by Caesarean section.

Terkel and Rosenblatt (1968) have offered still more convincing evidence for the hormonal initiation of maternal behavior. They devised an apparatus which made it possible to transfer blood from an actively maternal female to a virgin in a neighboring cage. Mixing of these blood supplies stimulated maternal behavior in the virgin female. Because the greatest difference in the blood of these two kinds of females is in the levels and ratios of reproductive hormones, this study strongly indicates that the hormonal state of the mother initiates her maternal behavior.

Unfortunately, neither of these studies indicates which of the many temporary hormonal states occurring at this time has this behavioral effect, but we do know that it is short-lived. Terkel and Rosenblatt (1972) transferred blood from mothers to virgins 1) 24 hours before birth, 2) 30 minutes after birth, or 3) 24 hours after birth. Only the transfer occurring 30 minutes after birth was effective in initiating maternal behavior.

In summary then, the analysis to date seems to show that the motivation of maternal behavior in the female rat is very much influenced by her hormonal state. We do not know yet precisely what substances are involved, nor where or how they act.

Primates

The mammalian order *Primates* includes (among other animals) the monkeys, apes, and man. The word "monkey" is likely to conjure up a picture of one of those long-limbed, long-tailed, timid trapeze artists who are common on the "monkey islands" of zoos and parks. In nature, these animals are tree-dwellers, exploiting both the food resources and protection that trees offer. Their bodies are light, their limbs are long, and their tails prehensile. They are well equipped for their reaching, grasping way of life. They are often mild-tempered and friendly, and some of them make good pets.

Like any other adaptation, tree-dwelling leaves certain resources unexploited. There is often food on the forest floor, and some primate species have moved down to eat it. The rhesus macaque (*Macaca mulatta*)

From *Primate Behavior: Field Studies of Monkeys and Apes* edited by Irven DeVore. Copyright © 1965 by Holt, Rinehart, and Winston, Inc. Reproduced by permission of Holt, Rinehart and Winston.

Figure 8-7 A six-year-old male rhesus stands with his tail high.

is a vegetarian species living on the forest floor of northern India. This habitat makes different demands than the arboreal environment; hence a rhesus macaque bears little resemblance to our typical stereotype of a "monkey." The rhesus looks more like what might happen if you crossed a monkey with a terrier.

Full-grown males weigh thirty pounds or more, with deep chests and short, powerful limbs that often carry the monkey pacing along on all fours like a dog. Their tails are quite short. The males, in particular, are armed with a set of very impressive canine teeth and the will to use them. They are tough, aggressive, resourceful animals. They have adapted so well to man's modification of the environment that their numbers may be larger in urban India than in the forest (Southwick, Beg, and Siddiqi, 1965). They are one of the few nonhuman primates whose parental behavior has been both described in natural settings and analyzed in the laboratory. For this reason they can teach us something about at least one solution to the problems many primates share in rearing their young.

On the forest floor the rhesus lives as a member of a group averaging about fifty individuals of all ages and both sexes (Southwick *et al.,* 1965);

From *Primate Behavior: Field Studies of Monkeys and Apes* edited by Irven DeVore. Copyright © 1965 by Holt, Rinehart, and Winston, Inc. Reproduced by permission of Holt, Rinehart and Winston.

Figure 8-8   A female rhesus with her two-week-old male infant.

the group forms a stable and lasting social organization. Males occasionally move from one group to another (Lindburg, 1969), but in general the group membership of males is stable, and that of females is strikingly so.

The social system is structured as a dominance hierarchy (rigid for males, less so for females) enforced, in the long run, by the long canines and short tempers of the most dominant animals and in the short run by vocal and gestural signals that substitute for fighting. In addition to the dominance hierarchy within each group there is also a dominance relationship between groups, so that one group will give way to another when they come together at favored local areas for feeding and sleeping. The day-to-day life of the rhesus involves many social interactions and requires many social skills.

Mature females are usually found in small subgroups made up of mothers and their sons and daughters of previous years. Since rhesus females usually give birth once a year, the group around one old female may include several subadult sons, both subadult and adult daughters, and some grandchildren. These subgroups spend a great deal of time together, peacefully grooming one another, caring for the infants communally, lying or sitting quietly, and feeding side by side. From time to time, especially during a period of sexual receptivity, the adult females will approach males and invite copulation. The receptive female may form a temporary consort relationship with a dominant male and move away from her customary social group to spend some time alone with him in mutual grooming and copulation.

The infant rhesus is relatively helpless, unable to feed, defend, locomote, or keep itself warm. All these problems are solved by its mother, who brings it to her breast immediately after birth and holds it there. For the first few weeks of its life the infant will be in continuous contact with her. It feeds at her breast and sleeps in her arms; and when it moves it rides, first under her belly and later on her back. When the mother is at rest the infant is cuddled, groomed, nursed, and, from time to time, engaged in gentle play.

At first the mother is the infant's only social contact. But from the beginning she has many offers of help, and soon she begins to accept some of them. Rhesus infants exert a powerful attraction to females of all ages. From the moment of their birth they are literally surrounded by willing babysitters. In the wild state the females closest to the mother are her own daughters, so the infant's first babysitters are usually its older sisters. Older brothers may take some interest in an infant but their interest declines with age, so by the time they are adults they rarely show much interest in infants, or even tolerance for them. An exception is the protection any adult male (and indeed any adult rhesus) gives when an infant is endangered. Humans handling an infant have been injured and sometimes killed by a combined attack of nearby adults (Singh, 1969).

At first the infant's weakness and the mother's firm restraint keep it always by her side. Usually the infant is simply held, sometimes by its short tail, but the female can also signal a wandering infant to return by means of a particular facial expression ("silly grin") or the "affectional present" (pronounced pre*sent*) (Hansen, 1966b), a posture indistinguishable from that which the female assumes to invite copulation or signal submission. It is here termed an affectional rather than a sexual present because of the function it serves in this context.

After a few weeks, the infant is able and permitted to move around on its own. The infant then begins to play with others its own age as well as older sisters and brothers. This play includes a lot of "rough and tumble" contact and some components of the aggressive and sexual behavior of adults. By now the infant has become increasingly active and aggressive in all its social interactions. Often, when its aggression is directed at the older animals around it, particularly the mother, older animals respond with punishment.

At the initiative of both the mother and infant, contacts between the two decline rapidly during the first three months. At the end of that time the infant no longer suckles (though in moments of stress it may take a nipple into its mouth), seldom clings to the mother's ventral surface, and in general has much less physical contact with her (Harlow, Harlow, and Hansen, 1963). But the mother remains the touchpoint for the infant's actions for months, and, to a lesser extent, for years to come. Moreover, when the infant leaves the mother, however long the duration of its separation, it remains physically and socially close to other group members, both younger and older.

We see then that social behavior for a rhesus is intense and continuous, and that for the first several months of life it centers almost totally around the mother. Why should this be so? One assumes that devoted prolonged maternal care increases the probability that the cared-for baby will itself reach maturity, reproduce, and pass on its mother's genes. But primates are very unusual in the amount of time and energy they spend on the care of babies and juveniles. Why don't they behave like other animals? Perhaps because the exceedingly complex nature of adult primate social interactions favors the babies given the time and care to learn how to behave appropriately in a demanding social environment.

Until the baby reproduces, the parent has not really reproduced. Therefore reproductive behavior is not successful until it leads to replacement. This criterion holds for the reproductive behavior of all species, but in many species the social interaction necessary to prepare the infant for successful reproductive behavior is terminated shortly after birth. In most primate species, the rhesus included, this is not the case. The rhesus needs many social skills as an adult and has few as an infant. There is good evidence that to acquire them it needs the right experiences and social interactions.

What facts support this claim? Part of the answer is both self-evident and trivial. Without food, warmth, and protection, the infant would soon die. But it is possible to meet these purely physical needs without ever exposing an infant to another member of its own species. If this is done, will the outcome still be a powerful dominant male or a functional female? In other words, will one of the parents have been replaced?

A number of studies have used the basic strategy of creating a physical environment adequate to the infant's needs while withholding social experience for some period of time (see also Chapter 4). Physically normal rhesus monkeys can be produced that way, but they sometimes show a number of behavioral abnormalities that make proper social functioning impossible. The kind and degree of effect depend on the timing and duration of the isolation. Various studies have concentrated on the effects of isolation on a variety of social behaviors. Here we will consider the effects of social isolation on the maternal behavior of the resulting adult.

Two important studies are similar enough to be considered together. Seay, Alexander, and Harlow (1964) and Arling and Harlow (1967) studied the maternal behavior of seven adult females who had been socially isolated for at least the first eighteen months of their lives. They were observed with their offspring regularly for at least the first 180 days after parturition and their maternal behavior was compared to that of normally raised mothers giving birth at about the same time.

All the isolate-reared mothers (motherless mothers) were abnormal in their behavior. Although the details of the behavioral abnormalities were highly variable, even in this small number of animals, the behavior of each female took one of two principal forms: indifference or abusiveness.

Some mothers avoided the infants by moving away at their approach. Some reacted violently: these mothers beat the infants brutally, banged them against the walls and floor of the cage, and even threw them to the floor of the cage and jumped up and down on them. Strangely, the main effect of this treatment on the behavior of the infants was to increase their approaches to their mothers.

Other mothers were just as indifferent or abusive for some time after the birth of the infant, but then they shifted, sometimes abruptly, to a more normal pattern of cuddling, protective behavior. We cannot tell from the studies if this change was influenced by the persistence of the approaching offspring. Still other mothers were both affectionate and abusive by turns. In fact the most abusive mother in one study (Seay et al., 1964) was also the most affectionate, hugging her infant one minute and beating it the next.

These infants survived maternal neglect and hostility because the experimenters saw to their basic physical needs. They have since grown up and their social behavior, as subadults, has been studied. They were found to be more aggressive and less competent in social communication than normally reared animals.

Motherless mothers are not completely unchangeable. Subsequent experience can alter their maternal behavior. Normal rhesus females are rougher with their second infant than they are with their first, but the opposite is true for motherless mothers. In these cases the mother's punishing and rejection behaviors are much closer to normal (Mitchell, 1970). Mitchell believes, however, that this change is due more to the passage of time than to the experience the mother may have had with her first infant. His interpretation is supported by the fact that the older the isolate-reared mother is when she *first* gives birth to a baby the less likely she is to beat, bite, and abuse her baby (see Chapter 4).

In several species we have seen strong evidence that the motivation for maternal behavior has an important hormonal component. That hormones are involved in the rhesus is suggested by the difference between the behaviors of males and females toward infants (see below). But several other lines of evidence suggest that the hormonal state of the female is of little consequence in motivating her maternal behavior. First, there is the powerful attraction that newborn infants exert on every female in the rhesus population. A second line of evidence stems from the observation that when several adult females, some completely out of the reproductive cycle, were placed alone with infants, they quickly began to care for them (Hansen, 1966b). The evidence on this point is not yet strongly developed, but so far it seems that any normal female will mother any available infant. In nature, of course, infants other than the mother's own are rarely available, except for short babysitting stints.

There seem to be at least two evolutionary puzzles involved in the parental behavior of the rhesus. The first is that females other than the infant's biological mother devote considerable energy and expose themselves to risk in caring for and protecting the infant. It is necessary to put aside the human bias to regard these as being indications of being a "good mother" and try to understand why it is that they are good mothers to infants who do not carry any of their genes into the next generation. The question is quite simply this: how can selection act favorably on behavior which requires energy and produces risk but does not increase the individual's contribution to the gene pool? It is important to recall that in nature this behavior is nearly always exhibited by either the biological mother or her daughters of previous years. We can draw an analogy between the behavior of the mother's daughters toward her infants and the behavior of the worker bees toward the offspring of their colony. A character of providing infant care expressed in the older daughters can be advantageous to the mother in terms of increasing the probability that her genotype will contribute to the next generation. This makes it analogous to the characters of the queen bee. Some of these characters are expressed by individuals who do not themselves contribute directly to the gene pool, but rather who contribute indirectly by making the next generation possible.

The second evolutionary puzzle concerns the generally low level of male rhesus interest in infants. All anthropomorphism about good and bad fathers aside, it seems there would surely be some evolutionary advantage to caring for one's own offspring. The crucial fact in understanding why they do not is that there is only low probability that the infant they expend energy on would carry their genes.

Rhesus do not form single-pair bonds; this substantially reduces father-infant contact. Moreover, the males may shift from troop to troop (Lindburg, 1969), further reducing the probability of correspondence of biological and social relationships in male-infant interaction. Care given to an unrelated infant would be wasted in the thrifty economy of natural selection. Under these circumstances one might well expect selection against male involvement in infant care and for paternal indifference. That some kind of balance may be struck is suggested by the fact that males occasionally protect infants.

This suggests that the intensity of parental interest of the male rhesus is a compromise: a mix of indifference and attention that has come to characterize the males of this species because it optimizes their contribution to the gene pool of the next generation.

In this the parental behavior of the rhesus is exactly like that of all the other species we have considered. Whether exotic or commonplace, whether in bird or beast, parental behavior is an adaptation that has been shaped by the past to optimize the number of successful offspring in the present, and hence the number of like-behaving individuals in the future.

The transformation of mechanisms through which this is achieved must be marvelous, if only we knew more about them. The mechanisms controlling the dramatically different behavior of ring doves and Mallee-fowl, for example, must be fashioned from about the same original raw material: the basic nervous system, reproductive hormones, and hormone cycles of birds. Yet how different must be the action and interaction of these basics in Mallee-fowl and ring dove. And how delightful and enriching it would be to know what those differences are.

Parental behavior, both as an adaptation and as a set of mechanisms, is under ever more active investigation, and every year several gaps in our knowledge are filled; perhaps this one will be filled next year . . . or the year after.

9 The Comparative Psychology of Learning: The Selective Association Principle and Some Problems with "General" Laws of Learning

Robert C. Bolles

There are several ways to arrange a learning situation so that an animal will have difficulty learning anything. We can expect poor learning if we require an animal to learn a response it is physically incapable of making, or even one it rarely makes. But although the study of different animals' motor abilities is an important part of animal psychology, this chapter is concerned with a more curious phenomenon, namely, that we may find poor learning in situations where the animal is quite able to make the required response, and where the response may be quite easy to learn in another context. The familiar bar-press response situation provides an example. It is easy to train a rat to press a bar for food, but much more difficult and sometimes impossible to train it to press a bar to avoid shock.

We can expect poor learning if we require an animal to respond to stimuli which it cannot readily detect. But again, although the study of animals' sensory capacity is an interesting and important part of animal psychology, the present chapter is concerned with the more puzzling fact that sometimes an animal can easily discriminate a certain stimulus and yet have considerable difficulty learning to respond to it in a particular manner.

There is another obvious reason why learning can be difficult or

Research Supported by grant GB-20801 from the National Science Foundation.

impossible to obtain: an animal may be required to learn a stimulus-response association which exceeds its capacity to associate. Psychologists have always been interested in what kind of problems a particular species of animal can and cannot learn to solve; and although we can no longer entertain the idea that there is a single scale of animal intelligence (Hodos and Campbell, 1969), some useful comparative generalizations have been established. Thus, we may think of primates as being more intelligent than insects in one sense, because they can learn a wider variety of associations. However, even this well-established generalization does not always hold. There is the case of the digger wasp which looks into each of several nests in the morning, and, upon returning with food in the afternoon, deposits in each nest a number of morsels of food which is appropriate for the number and size of young that were there hours earlier. Experiments have shown that the digger wasp accomplishes this precise rationing after a delay of several hours on the basis of the conditions that prevailed in the morning rather than the conditions in the nests at the time of feeding. If primates, even the very intelligent chimpanzee, are tested in almost any sort of delayed responding task, they cannot learn if they have to delay their response more than a few minutes. We have to conclude that while the chimpanzee is "intelligent" in the sense that it is capable of learning a wide variety of different things, at least one "stupid" animal, the digger wasp, is superior in at least one situation. As another example of paradoxical learning ability, it has been found that human subjects with all their vast intellect are inferior to the lowly laboratory rat when it comes to learning mazes.

The main argument of this chapter is that sometimes learning is much poorer than we would expect from our knowledge of an animal's sensory, motor and associative abilities. Then, as if to balance these instances of poor learning, sometimes an animal learns much more readily than we should expect from the general laws of learning. Indeed, it can be argued that our general laws of learning have very little generality (Seligman, 1970).[1] The ability to learn is not a general ability, it is highly selective. It should be assumed from the start that a given animal will be easily able to learn some things and will have difficulty learning others. This idea, which I will call the *Selective Association Principle,* is illustrated by the following experiment.

A rat is confined for thirty minutes to a small, plain box. During this interval a painfully loud noise (95db) comes on whenever the animal makes a particular response. The instant the response occurs the experimenter turns on the noise so as to *punish* it. Suppose the experimenter chooses to punish a response which frequently occurs when the rat is exploring a new situation, such as rearing up on the hind legs. To control

---

[1] After this was written I discovered that soon to be published was an excellent review of comparative data which provides strong support for the Selective Association Principle: S. J. Shettleworth, "Constraints on learning." In D. S. Lehrman, R. A. Hinde, and E. Shaw, (Eds.), *Advances in the Study of Behavior, 4.* New York: Academic Press, 1972.

for the possible emotional effects of the noise, control rats are tested in exactly the same way, i.e., they get the loud noise at the same time as the experimental animals regardless of their behavior at the time. (These animals are called "yoked controls," because the times at which they are shocked are completely determined by, or "yoked to" the behavior of the experimental animals.)

The results from such an experiment (Figure 9-1) indicate that the loud noise is not an effective punisher. Rearing gradually drops out as exploration dissipates, but there is no difference between the two groups in the rate at which it drops out.

Suppose we take another response which occurs when rats explore, namely, poking the nose into a one-inch hole in the wall of the box. Again, for the experimental animals the loud noise comes on to punish the nose-poking response, and for yoked control animals the noise is presented at the same time regardless of what the animal is doing. Now, the noise rapidly eliminates the nose-poking response; it is an effective punisher for this response (see the data in Figure 9-2).

These results are difficult for the traditional learning theorist to explain. How can learning occur very rapidly under one set of conditions, and not at all under very similar conditions? The subjects were the same albino rats; the stimulus was the same aversive loud noise (the fact that animals sometimes learn justifies the assertion that the noise is aversive);

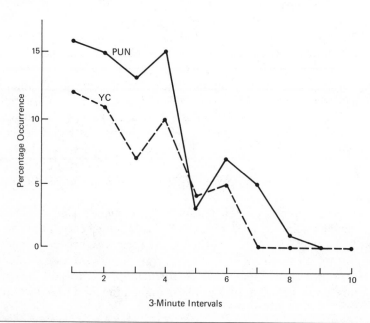

Figure 9-1 The percentage of occurrence of an exploratory response, rearing up on the hind feet, during successive three-minute intervals for rats punished for this response, and for yoked controls.

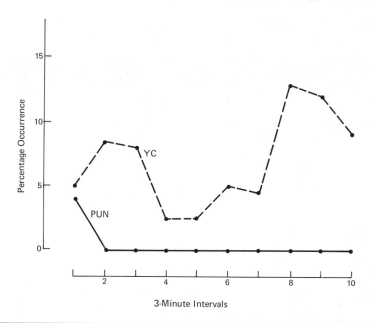

Figure 9-2 The percentage of occurrence of an exploratory response, poking nose in a hole, during successive three-minute intervals for rats punished for this response, and for yoked controls.

the same punishment contingency was used in both cases; and both the nose-poking and rearing responses had considerable initial strength. But one response was subject to learning while the other was not. Current theories of learning suggest no reason why learning should occur in one case and not in the other. While the Selective Association Principle also provides no explanation of the difference in learning, it does alert us to the existence of such differences, and it prepares us to expect them and to accept them as a normal part of animal learning rather than as anomalous failures of universal laws of learning.

In the following pages we will look at some of the history of the selective association idea; a little historical perspective makes it easier to understand the resistance that has developed to the idea. Then we will look at some experiments that demonstrate the selectivity of association. These include the food aversion phenomenon that has been called the Garcia Effect, the recently discovered auto-shaping phenomenon, and some data from my own laboratory on selective association in avoidance learning. The chapter will conclude with a few speculations about the selectivity that may exist in human learning.

Theoretical Background

One reads in vain through the writings of the great behavior theorists such as Hull and Skinner for any hint of a selective association principle.

Ironically, Thorndike (1911), the first learning theorist, did recognize the selectivity of association. But psychologists became enthralled with Thorndike's primary laws of learning and ignored his secondary laws, which embodied the selective association idea. Thorndike's primary laws of learning were the Law of Reinforcement and the Law of Exercise (or what we might call *practice*). According to Thorndike, learning occurs readily if reinforcement is given, but it also might occur in the absence of reinforcement, if the S-R association is repeated sufficiently often.

In addition to these primary laws, Thorndike proposed several secondary laws which suggested that the animal should 1) be relevantly motivated, 2) be prepared to learn, and 3) pay attention to the relevant cues.

Thorndike was an astute observer of behavior and worked on a variety of research problems with both animal and human subjects. He was well aware of the complexity of living organisms and did not want to oversimplify the learning process. His secondary laws of learning added the necessary richness to his theory, but they tended to make the theory rather loose. That Thorndike could see in the great complexity and diversity of learning a primary law of reinforcement and could attribute so much importance to it reveals his great genius.

Later theorists, especially Hull, tried to develop much tighter theories. Hull lost some of the richness of Thorndike's formulation but his theory was superior by being more closely knit, more explicit, and more formal. Hull developed mathematical formulations to describe the strength of a learned response. Only those factors that could be placed in the equation were considered to be determinants of behavior. In Hull's system all response learning was based on his version of Thorndike's primary law of reinforcement. There was no place for Thorndike's secondary laws—no place for the idea that some S-R associations are learned while others are not. Instead Hull asserted at one point (1952) that the speed with which an association is formed depends only upon the number of times it has been reinforced.

Hull's quantitative approach to the experimental study of behavior appeared to be a great step forward at the time it was presented. Because his equations were stated explicitly, it was possible to test all of his assertions. Unfortunately, the results of most tests have been negative, and we have finally been forced to conclude that virtually all of Hull's mathematical model of animal behavior was inadequate in one way or another.

Another unfortunate aspect of the research stemming from Hull's theory (which includes most of the animal research done during the 1940's and 1950's) is that it was restricted almost entirely to hungry rats. Moreover, it was restricted to hungry rats with a simple response to learn—turning left in a T-maze, going toward a black card, or pressing a bar. Hull stated in his last work (1952) that his theory had become so specialized that it only applied to the hungry rat in certain kinds of tasks. It did not necessarily apply to other organisms, to other motivational states, or to other tasks.

In the 1960's the leadership in animal learning theory and research passed from Hull and his students to Skinner and his followers. At this point a peculiar twist of fate helped keep the idea of selective association buried. The typical operant researcher is not very interested in how rapidly behavior can be established and stabilized on a reinforcement schedule. If he has sufficient diligence and can ultimately get the animal to respond at a certain rate in a certain situation, then he is satisfied. The operant psychologist, like the Hullian before him, has enormous confidence in the law of reinforcement; it is almost sufficient to explain all learning. As more emphasis is placed upon the primary law of reinforcement, less importance is attached to the secondary factors that concerned Thorndike. If an animal learns under one set of conditions, that learning demonstrates that the reinforcer was effective and that the animal is able to learn that response.

As in Hullian research, most of the work done in operant laboratories involves few species of animals and few responses. Rats are taught to press bars and pigeons are taught to peck keys to get food. The law of reinforcement has been verified again and again in these situations. Isn't it surprising, though, that no operant researcher reported pigeons pecking a key to avoid shock until as late as 1969? When Hineline and Rachlin (1969) finally reported establishing such behavior, their description of the study indicated that they had a great deal of difficulty. They found it necessary to give extensive training with a number of unusual procedures to get the behavior established. We must wonder about their conclusion that since pigeons can peck a key to avoid shock the law of reinforcement has been reaffirmed once again. The failure of other experimenters to train key-pecking avoidance behavior with standard procedures is not seen as particularly interesting or important. I would suggest a different conclusion: pigeons learn to peck a key to avoid shock only with difficulty and under quite special conditions. This learning contrasts dramatically with the rapid learning of the same response to obtain food.

The purpose of this brief theoretical background is not to place blame anywhere, but rather to describe some of the trends in theoretical thinking and research that have drawn attention from the selectivity of association. Thorndike's secondary laws of selective association were overlooked because so many animal psychologists were determined to find a universal law of learning. This determination was based upon the belief that there was just one kind of learning and that earnest studying of the hungry rat in a bar-press or runway task would reveal it. The mechanisms that were to explain all learning were to be universal, like the physicists' law of gravity or the chemists' law of valence bonding. This was an elegant working assumption.

The trouble is that this simple, elegant assumption of a universal law of learning is wrong. It is not consistent with the facts. It is peculiar that there has been so much reluctance to admit that the assumption was wrong, and so much effort directed toward demonstrating how right it was. Enormous numbers of rats have been run in alleys and Skinner boxes for food reinforcement, apparently in the hope that even if the law of reinforcement could not account for all learning, at least it could account for most of the learning reported in the published experimental literature!

In the last few years a number of psychologists have begun to notice failures of the law of reinforcement. Some have tried to resurrect Thorndike's old law of preparedness (Seligman, 1970), while others have proposed that there are special cases where the ordinary rules of learning don't apply. It would appear that the law of reinforcement has all of the empirical support it warrants, and that now it is time to look at other factors that affect learning and performance so that we may get a better perspective of the place of learning in animal behavior. Let us look first at a variety of learning that occurs much *more* readily than learning theorists could have predicted.

The Garcia Effect

John Garcia was one of a group of animal psychologists who worked in the 1950's with the effects of radiation (gamma rays and X-rays). This early work has been reviewed by Garcia, Kimeldorf, and Hunt (1961), who found that fairly low doses of radiation, much lower than would produce any apparent radiation sickness, would cause rats to avoid a food substance that had been associated with the illness. In a typical experiment, rats were placed in a box and confined there for several hours, during which time they were subjected to radiation of about 35R (R, the unit of radiation, is for Roetgen, the discoverer of X-rays. The rat shows signs of illness at 400R, and its life is in jeopardy at about 1000R.) Then, sometimes during the confinement and irradiation period, the animal was given a saccharin solution to drink. The presence of radiation did not appear to inhibit drinking, so there seemed to be nothing aversive about the radiation itself. But some days later when the animals were given the opportunity to drink saccharin they avoided it, or drank much less than control animals receiving no radiation in the original situation.

Garcia showed that radiation had this effect even when it could not be directly sensed (there is evidence that animals can detect gamma radiation through their sense of smell, but Garcia controlled for this possibility). Garcia and his co-workers next conducted a series of experiments to show that this aversion to saccharin was a learned phenomenon based upon classical conditioning. The taste of saccharin can be viewed as the conditioned stimulus, and the slight digestive disturbance produced by the radiation can be viewed as the unconditioned stimulus. According to their 1961 paper, the CS is paired with the US over a period of time and then subsequently, when the saccharin is presented in the test some days later, it produces some of the same distressed digestive reactions that the original radiation had produced. It was this conditioned reaction that led to aversion in the subsequent drinking test. Garcia had already shown that if radiation preceded the ingestion of saccharin, there was little or no aversion to it, which was consistent with the general rule of classical conditioning that the CS has to precede the US. These findings were noted at the time with some interest, but they really only provide a background for the Selective Association Principle. The importance of the aversion phenomenon was not fully appreciated until a few years later when Garcia and Koelling (1966) reported an experiment that led to a completely different interpretation of the data.

Garcia and Koelling (1966) trained one group of rats to drink bright, noisy water. This was accomplished by an electronic circuit arranged so that whenever the animal's tongue made contact with the fluid a light flashed and a clicker clicked. Another group of rats was given tasty water, i.e., water containing saccharin. Half of each group was then given painful electric shock to the feet following drinking, while the other half was made ill by the administration of either X-rays or lithium chloride (a fairly effective poison). Subsequently, all animals were tested by being given either bright, noisy water or tasty water—whichever they had received prior to the aversive experience. The results indicated that poisoning inhibited drinking the tasty water but not bright, noisy water. By contrast, foot shock produced an inhibition of drinking for the animals that received the bright, noisy water but not those that received the tasty water. In summary, then, it appears that rats that are given foot shock will develop an aversion to bright, noisy water but that rats that are made ill will develop an aversion to water with a characteristic taste.

This last experiment is conceptually different in several ways from Garcia's earlier experiments. Its emphasis is upon the response and the consequence of the response. Garcia was no longer talking about "pairing" the taste of food with illness, but about illness as a *consequence of drinking*. In other words, Garcia shifted theoretically from classical conditioning to reinforcement learning. There is a stimulus situation (water with particular characteristics), the animal makes a certain response (drinking), and there is a consequence (the animal becomes ill). These are the conditions that define an instrumental learning (or learning-by-punishment) procedure. However, Garcia and Koelling's procedure differs from the ordinary punishment-learning situation in three ways. The first is that a number of punishments are usually necessary to reduce response strength, whereas Garcia and Koelling used only three daily illness sessions. Subsequently investigators have obtained the same effect using only a single illness. The second consideration is that, ordinarily, if punishment is to be effective in weakening a response, it must occur immediately after the response. If punishment is delayed one minute after a bar press response, generalized fear may produce some suppression of bar pressing, but the response itself will not be weakened. The animal only becomes upset when punishment is postponed. The child who is punished when father comes home from work because he did something naughty early in the day is a classic case of ineffective punishment. The child may end up more respectful of his father, but no less inclined to be naughty when father is away. In the case of the Garcia Effect, however, punishment in the form of illness can be delayed for hours and yet it remains effective in producing a learned aversion. The third unusual feature of the study is the specificity of the learning. Long-delayed punishment in the form of illness produces an aversion to tasty water but not to bright, noisy water.

In summary, the Garcia Effect is the name we give to Garcia's discovery that illness (or more correctly, digestive disturbance) as a punisher is 1) fantastically effective in terms of how infrequently it needs to be repeated and how long it can be delayed, and 2) remarkably restricted in the sense that it affects only one S-R association, i.e., taste-consumption.

The Garcia Effect has been shown to depend upon appropriate experimental variables. For example, the learned aversion is stronger if larger dosages of X-irradiation are used (Revusky, 1968; Garcia, et al., 1955). Garcia, Ervin, and Koelling (1967) found that the aversion to food increased with larger injections of nitrogen mustard (which induces illness). A variety of different agents can be used to produce illness. Intragastric injections of saline, insulin, glucagon, and a number of other agents all produce the same effect. So do intravenous infusions of isotonic saline or forced ingestion of lithium chloride. The more frequently illness follows the ingestion of a novel food, the stronger is the resulting aversion. A variety of novel foods have been used as the test substance, and there seems to be no limit to the kind of flavor to which aversion can be induced. It is important to note however, that for optimum effects the food should be novel, or relatively unfamiliar, at the time that illness is induced. Aversion to drinking water or to lab chow can be produced, but the effects in these cases are apparently much smaller and harder to get than those obtained with novel flavors (see Revusky and Bedarf, 1967; Kalat and Rozin, 1970). Kalat and Rozin had also shown that there is a recency effect, i.e., the sooner illness follows the ingestion of a novel food, the stronger will be the aversion to it. It is presumably possible to increase the novelty or increase the recency and get any combination of these two effects. As far as the maximum delay of illness is concerned, the record seems to have been established by Smith and Roll (1967) who induced illness with X-rays 0, $\frac{1}{2}$, 1, 2, 3, 6, and 12 hours after ingestion of saccharin and found nearly total supression of drinking at all intervals up to 12 hours, and some suppression for the 12-hour group.

Nearly all of the research on the Garcia effect has been done with the laboratory rat. Indeed, the rat has been used so extensively that we may begin to wonder if perhaps the effect is restricted to the rat. But similar results have been obtained with cats and mice (Kimeldorf,

Garcia, and Rubadeau, 1960) and with monkeys (Harlow, 1964). The effect therefore appears to have some generality across species. However, we might expect the rat to be able to learn food aversions better than most species because it is a generalized eater, i.e., it eats all kinds of food. Moreover, the rat is a scavenger, an opportunist: it eats when food is available. Packs of rats are constantly on the search for new food supplies, and since the rat normally depends upon a highly diversified diet, it needs special protection against novel foods which are harmful to it.

Different animals show different kinds of food aversions. Wilcoxon, Dragoin, and Kral (1971) compared food aversions in rats and Japanese quail. Both rats and quail were trained to drink a specially prepared fluid which was both sweet (saccharin had been added to it) and blue in color. Following an illness induced by X-irradiation, the animals were tested with solutions which were either blue or sweet. The rats avoided only the sweet-tasting test solution while the quail avoided only the blue fluid. Although this study is clever in concept, it presents some problems of interpretation, since rats are color blind and may have had no sensory basis for responding to the color of the test substance. On the other hand, many grain-eating birds have a poorly developed sense of taste. Therefore the quail that Wilcoxon *et al.* tested, which is a grain eater, may not have been able to detect the taste of the test substance. Thus Wilcoxon *et al.* may have only discovered something about the sensory capacity of their subjects and nothing that really bears on selective associations. Differences in sensory equipment impose limitations upon what can be learned just as motor limitations prevent us from teaching the rat to fly or the bird to run on all fours.

On the other hand, there are important and interesting differences in what sensory dimension an animal *uses* from among those that are available to it. It is well known that birds, like men, rely upon vision; they depend upon, pay attention to, and readily make associations with visual inputs, even when alternative sensory channels could be used. We ignore the world of smells around us (or try to), and so think of ourselves as being primarily visual and only secondarily olfactory. But a blind man is perfectly well able to respond to odors once he comes to depend upon them and pay attention to them. Man, of course, enjoys a great deal of flexibility in the kinds of stimuli and responses that he can associate, and it is likely that other creatures would not be able to make the same extensive use of a secondary sensory dimension if deprived of their primary or favorite sense, even after extended training with the secondary sensory modality. If associations are selective, then most animals would have relatively little flexibility of this sort; this is one implication of the Selective Association Principle.

The main point of the bright, noisy water experiment is that Garcia and Koelling were working well within the rat's sensory capacity, since their animals learned to respond to both bright, noisy stimuli and taste stimuli. But these stimuli entered selectively into association with particular responses. The taste of food controlled aversion only when illness was used as a punishment, and external stimuli controlled avoidance behavior only when an external stimulus, shock, was used as a punisher.

To return to the food aversion study with Japanese quail, if it is true that this particular species could not taste the training or test fluids, then the study is not pertinent to the Selective Association Principle. But there are other possibilities. One is that the quail does have a sense of taste but does not use taste cues (i.e., taste does not enter into association with behavior) when visual cues are available. Another possibility is that quail can use taste cues but that they do not readily enter into association with feeding behavior. If either of these alternative possibilities is correct, then the data are relevant to the Selective Association Principle.

Capretta has reported a pair of studies that provide somewhat clearer evidence for the selectivity of associations. In the first study (Capretta, 1961) he found that chickens would not avoid food with a particular taste that was paired with shock, but they would refuse to eat a food of a given visual appearance which was paired with shock. Here again there are several possible interpretations. But in a second study (Capretta and Moore, 1970) chickens were required to discriminate between the tastes of sweet and normal foods. They did learn this discrimination, and even showed an initial preference for one taste over another, but only under conditions where both of the test foods were visually different from the ordinary food. It is as if a bird looks at things it eats but does not ordinarily taste them, even though it is capable of being trained to do so. Summarizing the comparison of rats and birds, it appears that foods that have been associated with illness *taste* bad to rats, but *look* bad to birds.

Let us consider next Alcock's (1970) study with black-capped chickadees. He first trained them to turn over an empty sunflower seed to reveal food underneath. Then the chickadees were divided into three groups and given a series of presentations with either nothing under the seed, a bad-tasting (salty) mealworm, or a bad-tasting and illness-producing mealworm (injected with quinine sulphate which tastes bitter and is a powerful emetic). Alcock found that the birds given the empty seeds learned to avoid them on the test trials and that on a retention test given some days later the birds recovered only a slight tendency to approach the empty seeds. Similarly, birds given emetic quinine-treated mealworms quickly learned to avoid them and showed a permanent aversion to the seeds hiding this "prey." However, the birds presented with bad-tasting (salty) insects under the seeds attacked the seeds in the retention test despite having learned to reject them during the initial training trials. In other words, bad taste alone was not sufficient to produce permanent rejection of a possible food item. Alcock argues that this persistence is in keeping with the chickadees' feeding habits: the chickadee is a broad-ranging feeder and since the long-term consequences of eating unpleasant-tasting food is still beneficial, it is to the animal's advantage to consume such food if it has few alternatives. In summary, the most durable learning found by Alcock involved the chickadee's avoidance of a particular visual stimulus, an object which, if eaten, would lead to illness. There are two important biological consequences of such avoidance behavior. One is that a predator learns not to attack certain stimuli. The second is that there will be an enormous advantage in the evolution of the prey animal either to develop poisonous properties of its own, as the monarch butterfly has done, or to develop stimulus characteristics which make it closely resemble such prey, as a number of other butterflies have done in mimicking the monarch.

The Misbehavior of Organisms

"The Misbehavior of Organisms" is the title of a paper written in 1961 by Keller and Marion Breland. The authors were among the first to call attention to the selective association idea. The Brelands had been working for some years training animals to perform for commercial displays, tourist attractions, movies, and television. They had used traditional Skinnerian methods, gradually shaping successive approximations to the desired sequences of behavior. Their commercial animal-training program had been quite successful and many of the better trained animal acts seen on the popular media were trained by the Brelands. However, their 1961 paper was concerned with some curious failures of operant training techniques.

The Brelands describe one instance where the observer sees a door open and a chicken walk out and go over to a round rotating platform. The chicken climbs up on it and scratches vigorously while going around and around. Then after fifteen seconds, an automatic feeder operates in the retaining compartment. The chicken abandons the platform and enters the compartment to eat. Thus the audience sees the chicken come out, turn on the juke box, and dance. The curious thing about this behavior is how it is acquired. All that the chicken has to do to get fed is to stand on the circular platform for fifteen seconds. None of the scratching and walking around is necessary for reinforcement; this behavior is just waste motion. In the course of a three-hour daily performance the bird produces thousands of unnecessary scratching and scraping responses.

An operant theorist could readily explain the occurrence and persistence of this extra behavior by noting that reinforcement automatically follows whatever kind of behavior occurs during the fifteen-second wait for food. The food reinforcement which follows the behavior will strengthen it. It does not matter that its occurrence is unnecessary for reinforcement. What does matter, according to reinforcement theory, is that the superstitious behavior occurs and is reinforced and is therefore more likely to occur again so that it becomes firmly established over an extensive period of training. Although the Brelands admit that this behavior looks like it might be superstitious, and could be explained in this way, they proceed to indicate other failures in animal training that are not so readily explained.

In one case, they attempted to train a raccoon to pick up coins and deposit them in a small metal box. Their raccoon was a very tame and eager subject, and they had no difficulty training it to pick up the coin. The trouble started when the metal box was introduced and the animal required to drop the coin into it. Here the raccoon, normally an intelligent and adaptable animal, seemed unable to let go of the coin. He would rub it up against the inside of the box and pull it back out again. A worse problem arose when the animal was put on a ratio of two coins per reinforcement. The raccoon would not let go of the coins but spent all its time rubbing them together and dipping them in and out of the box. This behavior became so entrenched and persistent that the Brelands abandoned the whole enterprise.

In another case chickens were trained to play baseball. The chicken pulled a wire loop which released an electrically-operated bat which knocked a small ball across a playing field. The idea was that if the ball got past the miniature players and hit the back fence, the chicken was automatically reinforced with food, and could run in the general direction of first base to get it. In spite of the simple design of the experiment and the straightforward reinforcement contingency, everything went haywire because the chicken became wildly excited when the ball started to move. It would jump onto the playing field, knock the ball onto the floor, and chase it in every direction. This so disrupted the ball game and the chicken's base-running behavior that no reinforcement was ever received. Here, chasing the ball was never reinforced, so there is no possibility of arguing for superstitious reinforcement as there was in the case of the chicken scratching on the platform.

The Brelands describe another instance in which a pig was supposed to learn to pick up a large wooden coin and deposit it in a piggy bank (a task similar to that on which the raccoon had been so promising and had developed into such a dismal failure). Pigs are intelligent animals, they have good appetites, and they have no trouble learning a variety of behavioral tasks. In this particular situation the pig had no trouble at the start, but after a while it became slower and slower in picking up the coins and taking them over to the piggy bank. The problem was that instead of carrying the coin in its mouth and depositing it neatly in the piggy bank, it would repeatedly drop the coin on the floor, root it, drop it again, root it along the way, pick it up, throw it in the air, drop it, root it some more, etc. Continued training only made the situation worse. Ultimately the pig project had to be abandoned.

Notice that the basic dilemma in all these cases was that instead of the reinforced response becoming more and more efficient and all of the waste motion and inefficient responses dropping out, just the opposite happened. The animal became so involved in its extraneous behavior patterns or so wrapped up in fiddling with the manipulanda, coins or whatever, that it ceased to receive any reinforcement. We appear to have a failure of all of the operant conditioning principles that have been used so effectively in getting pigeons to peck keys and rats to press levers.

But there is also something lawful in the misbehavior of these animals. In each case the intruding behavior, i.e., the behavior which occurs instead of the desired operant, is characteristic of the particular species, and in each case the undesired response is one which ordinarily precedes eating. Pigs get food by rooting in the ground. Chickens obtain food by scratching the ground and by pecking. In this connection, the Brelands observed that those animals which did not develop the pattern of dancing on the platform developed persistent pecking. In the case of the raccoon, the intruding behavior is a rudimentary kind of food-washing; dipping and rubbing food is a consistent part of the raccoon's preingestion behavior. If raccoons are given sugar cubes they wash them and scrub them and rub them together until they dissolve. The raccoon will literally wash its food away.

We can draw the conclusion that whereas it is possible to teach an animal some responses to obtain food, we have to be very careful about the animal's innate food-getting responses. These other responses are likely to prevail *in spite of* the existing reinforcement contingency. Moreover we may wonder about the familiar cases of pigeons pecking keys and rats pressing levers. Are these responses really pure operants, selected at random from the animal's repertoire, or are they special cases, because they are so similar to the animal's natural food-getting responses? It seems doubtful that much learning is really involved in pecking a key for food, at least as far as the *response* is concerned. All the animal has to learn is to connect its natural food-getting response to particular stimuli, namely, the stimulus configuration presented by the Skinner box, or, in some cases, the occurrence of a cue light or some other stimulus such as the lapse of time or the number of preceding responses.

Support for such an interpretation has come from a study by Brown and Jenkins (1968) of the "auto-shaping" response in pigeons. As all students know who have been through an introductory laboratory in animal psychology, an experimenter is first supposed to "magazine train" the animal and then gradually shape up the desired key-pecking behavior. Obtaining food (primary reinforcement) and the click of the magazine (secondary reinforcement) are supposed to shape up successive approximations to the final key-peck response. However, what Brown and Jenkins did was simply to present the animal with food every sixty seconds—and that is all they did. What happens on this free-food schedule is that the chicken pecks at the food when it is available and pecks everywhere else when it is not. In a very short period of time, comparable to the time it takes to shape the response in the customary fashion, the animal quits pecking at the walls of the box and limits its pecking to the key. One reason is that the key is brightly illuminated and flashes off and on. We may assume a lighted key is perceptually "more interesting" than the dark metal walls. It also helps to have the key near the food magazine so that the pecking that occurs in the general direction of the magazine is likely to land on the key. In short, it is not necessary to shape the response. If the animal is given enough food so that it becomes excited and its appetite is aroused, and if the animal is provided with something interesting on the wall to peck at, such as an illuminated key, the response will be rapidly "learned."

The traditional explanation of these results is that the operant rate of the response is increased by making the key perceptually interesting, so that reinforcement will be more likely to occur and the response will be strengthened more quickly. This argument is based, of course, on the assumption that reinforcement is necessary for learning to occur. But what if this assumption is false? Reinforcement, as it is ordinarily conceived, may have nothing to do with strengthening the response. In this situation reinforcement may serve only to increase the animal's appetite, get it excited, and increase its general level of activity, and the key-peck response follows because the animal is excited. This interpretation is suggested by the recent study of Williams and Williams (1969). They repeated the Brown and Jenkins experiment under conditions where food was presented periodically but a key peck *delayed* food presentation. Under these conditions some birds persisted in pecking the key at a very high rate even though such responding virtually eliminated subsequent reinforcement! It appears, then, as though the reinforcement contingency itself has relatively little to do with auto-shaping the pigeon's key-peck response. Perhaps all that reinforcement does is provide incentive motivation which excites the animal, tells it when to become active and when to seek food, and innate food-getting response tendencies take it from there. Perhaps all of the marvelous sensory-motor coordinations that have been demonstrated in the Skinner box—color discrimination, pattern discrimination, scalloping on different kinds of schedules—all of these factors merely show the different kinds of sensory control of incentive motivation that are possible. For example, when a cue for no food is presented, e.g., when a long ratio schedule is indicated, the animal loses its appetite. Its appetite is restored by the presentation of appropriate stimuli signaling that food is near at hand.

I am suggesting that what is learned in an operant situation is the association of incentive motivation (usually hunger) with the particular stimulus details of the experiment. The instrumental (or operant) response follows. As Mowrer (1960) has observed, we do not have to teach the rat to press the bar, it already knows how to press the bar. What we have to do is to teach it to *want* to press the bar. We do this by signaling that food will come when the bar is pressed.

Not only does the rat come into the Skinner-box situation with the appropriate response already in its repertoire, this response is already relevant to eating. At least, the paws play an important part in eating. The rat also typically eats in a sitting position with its weight on its hind feet holding food with its front paws, a couple of inches off the ground— just where the bar is usually located. I suspect it is not a coincidence that a rat pressing the bar in a Skinner box looks very much like a rat eating. A Brown and Jenkins type of auto-shaping experiment can be done with the rat, too; I have done it in my own laboratory.

On the other hand, the rat may have more flexibility in its food-getting behavior than the pigeon or some of the other animals that gave the Brelands so much trouble. For one thing the rat is a scavenger and a generalized eater, i.e., it eats all kinds of different foods in all sorts of different situations. Probably most of the flexibility, i.e., learning capacity, of the rat is available for food-getting behavior; it may show more plasticity and less selective association in getting food than many other types of animals. Even so, there probably are bounds to what the rat can learn in getting food. For example, the enormous facility rats have for running mazes and runways to get to food suggests that running forward is a part of the rat's natural food-getting behavior just as sitting on its haunches and holding food in its paws is a natural part of its consummatory behavior. Would the rat learn the instrumental response if an experimental situation were arranged so that it had to *back up* to get to the goal box? Suppose we arranged the food cup so that it emerged from the floor as the rat backed up over it, so that the last response the animal made before eating was backing up. Would the rat learn to press a bar if it had to be hanging by its heels from the ceiling before a bar press counted? In short, there is every reason to suppose that like any other animal there are restraints upon what the rat can learn to get food, even though the rat is less severely restrained than animals which are more specialized eaters.

The guinea pig provides an interesting comparison with the rat because it doesn't use its paws to eat. The guinea pig has terrible table manners: it picks up a big chunk of food with its mouth, bites off a piece to chew, and lets the rest fall on the ground. Can it learn to press a bar with its paws for food reinforcement? No, according to George Collier (personal communication); although there is a reasonable operant rate of paw-pressing, i.e., the response does occur from time to time, the alleged reinforcing power of contingent food doesn't seem to strengthen it. The guinea pig can, however, readily learn a face-press response.

Defensive Behavior

Defensive behavior provides another line of evidence for the Selective Association Principle. By "defensive" behavior I mean all of those behavior systems that keep animals alive in the face of physical dangers, predators, etc. Problems of survival have an urgent, top-priority quality that the animal must solve, and learning through reinforcement is a luxury which the animal probably cannot afford.

Consider the following experiment: a rat is put in a box and after five seconds is given an electric shock. What will the rat do the next time it is put in the box? Its behavior will be governed almost entirely by one simple procedural detail: whether there is a lid on the box. If the box is closed so that there is no obvious way out, the animal will freeze and it will continue to freeze for some hours afterward (Blanchard, Dielman, and Blanchard, 1968). But if the box is left open, the rat will promptly jump out (Maatsch, 1959). One important conclusion can be drawn from this simple-minded experiment, namely, that the rat's reaction to shock depends not just upon the shock, but upon other factors in the environment as well. If the environment provides a way out, the rat will take it; if not, it will freeze.

A survey of different avoidance-learning situations reveals, at one extreme, Maatsch's jump-out box and other run-away situations in which an avoidance response is learned in one or two trials; and at the other extreme situations such as the Skinner box, where an avoidance response may be very difficult for the rat to learn. Typically the rat has five seconds to press the bar to prevent an automatically scheduled shock from occurring. After a few *hundred* trials under these conditions, some animals will avoid some of the time, but some of them will not be avoiding at all. A number of experimenters have reported being unable to get any learning of the bar-press avoidance response. In one study (D'Amato and Schiff, 1964), twenty-one of twenty-four rats failed to learn the response after thousands of trials. Some of D'Amato and Schiff's animals made hundreds of bar presses over the course of training, so there was a reasonable operant rate; but most of their animals showed no learning. Here we have a serious dilemma for reinforcement theory: a response occurs, the presumed reinforcer is applied consistently whenever it does, and yet the response is not strengthened.

The following experiment also demonstrates that operant rate in and of itself fails to tell us very much about how rapidly an avoidance response will be learned. Three groups of animals were required to learn different avoidance responses in a running wheel. One group was required to run; any movement of the wheel greater than a quarter turn was effective in avoiding shock and also effective in escaping shock following a failure to avoid. The second group was required to turn around in the wheel, i.e., to do an about-face without rotating the wheel itself. The third group was required merely to stand up on its hind legs. In all cases there was a warning stimulus which preceded the scheduled shock by ten seconds, and the appropriate avoidance response terminated the warning stimulus. The experimental conditions were the same for all three groups; the only differences among them were the different response requirements. The results from this study (Bolles, 1969) are shown in Figure 9-3. Notice that the three learning curves started at about the same point. Actually, the

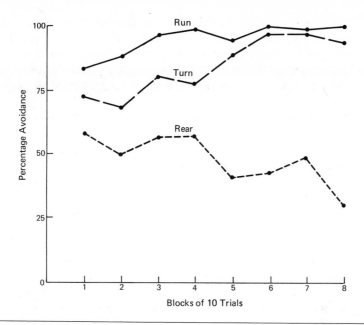

Figure 9-3 Three different degrees of acquisition for rats learning to avoid shock by making three different responses.

three responses were chosen from among the responses in the rat's repertoire so that initially they would all occur between forty percent and fifty percent of the time. But one response, running, was rapidly learned and performed at a high level, while the second response, turning, was learned somewhat more slowly, and the third response was not learned at all. The strength of the standing-up response actually seemed to decline over trials, and there is no indication that any of the animals required to stand up would have learned to do so if given further training.

There are two parts to the selection association argument. Sometimes, as in the case of the Garcia Effect, learning occurs under conditions where reinforcement theory requires it to be impossible. Other times, as in the avoidance learning experiment just described, learning does not occur under conditions where reinforcement theory requires that it should. Both kinds of discrepancies are embarrassing for reinforcement theory and both require us to introduce new principles. For the Garcia Effect our new principle was that animals have channels in which learning can proceed rapidly, as soon as the opportunity for learning is presented. An analogy is the dry riverbed which becomes a river if there is a cloudburst or other sudden source of water. We can predict before the water comes where it must go. In the failures of appetitive learning described by the Brelands, and in the failures of avoidance learning, we find that there are channels in which learning seems to be virtually impossible. To extend our analogy, these failures are like high places in the terrain. We can predict that when water does come, it will spare the hill tops: They will stay relatively dry.

The Selective Association Principle asserts that whether a given species will learn a particular response in a particular situation is determined by special channels in which learning is possible. Attempts to obtain learning outside of these channels are very likely to be unsuccessful— regardless of how earnestly the traditional rules of reinforcement are applied.

One explanatory system has come to prevail in the avoidance area during the thirty years or so in which psychologists have been doing avoidance experiments. It goes as follows: there are stimuli which regularly precede shock, and these stimuli are assumed, therefore, to become aversive. Then, just as shock is a primary reinforcer whose termination is a primary reinforcer, so any stimulus that is consistently associated with shock will become aversive, and its termination will be a secondary reinforcer. Sometimes these stimuli are explicitly presented by the experimenter. For example, there may be a ten-second buzzer or flashing light before shock onset, and the avoidance response in addition to avoiding shock also terminates this signal if it occurs during the ten seconds. Then if learning occurs, it is easy to point to the termination of this warning stimulus as the reinforcement for the avoidance response. (When there is no learning, it is necessary to introduce the idea of competing responses or some other kind of mechanism that interferes with the "normal" learning process.) Avoidance learning can also be obtained in the absence of any external CS (Sidman, 1953). The argument in this case is that there are proprioceptive events occurring within the animal which constitute the effective warning stimulus, and that the termination of these stimuli provide the reinforcement. For example, it can be argued that if the animal sits motionlessly, then the proprioceptive feedback from this behavior is invariably paired with shock, so this feedback becomes aversive. Then after a number of trials, if the animal quits sitting still, it will be reinforced for doing so by the termination of this aversive feedback.

The only complication in this well-accepted account of avoidance learning is the concept of fear. Some theorists (e.g., Mowrer, 1960) argue that it is not the termination of the stimulus, but rather the reduction or termination of the fear that has become conditioned to the stimulus, that constitutes reinforcement. In either case, however, there is a stimulus, either an observable external one or a hypothetical internal one, which regularly precedes shock, the termination of which can be said to reinforce the avoidance response. Considerable evidence against this traditional interpretation has accumulated in the last few years, but the best evidence is that which we have already discussed, namely that some responses cannot be readily learned in the avoidance situation.

How can we account for the fact that in one situation a rat can learn an avoidance response in just a few trials, whereas in another situation the response may not be learned after thousands of trials? The Selective Association Principle starts us off with the assumption that an animal can readily learn to make some responses but not others. To oversimplify the argument somewhat, we can say that jumping out of a painful situation is a natural defensive response for the rat, whereas pressing the bar is not. If we put our theories of reinforcement aside for a moment, and think just in terms of natural behavior of real animals, the remarkable thing is that a rat can *ever* learn to press a bar to avoid shock! That such learning is slow and uncertain, and frequently fails to occur, should not really surprise us.

Elsewhere (Bolles, 1970) I have suggested a different interpretation of avoidance learning which is better able to account for these difficulties. The basic idea is that the experimental procedure that defines avoidance learning is almost identical to the procedure that defines punishment learning. In avoidance, if the animal makes the appropriate response, shock is not presented. In punishment, if the animal makes an unappropriate response, shock is presented. Compare the two procedures in the bar-press situation. If the animal presses the bar before the shock comes on, the shock is withheld—that is avoidance. If the rat freezes before the shock comes on, freezing behavior is punished. If the rat tries to climb the walls of the box in the attempt to get out, climbing behavior is punished. Indeed, all behavior is punished except the avoidance response.

The trick in the avoidance situation is to punish all of the wrong responses so that the right response will occur. When we think of all of the responses in the rat's repertoire, then we see that this would certainly be a clever trick indeed. It is at this point that the Selective Association Principle comes to our aid. After the rat has received a few shocks in the situation its repertoire becomes severely limited. The rat will display only a small number of species-specific defense reactions. It may freeze (freezing may be very useful for a small animal in the wild) or attempt to flee from the situation. In the jump-out box or in a simple one-way box where the rat can leave the danger area and go to a safe area, learning is very rapid. It is only necessary to punish freezing a few times and the rat will run away. It is also very easy to train the rat not to move at all. Here it is only necessary to punish fleeing behavior a few times and the rat will quickly learn to freeze.

The bar-press situation is quite different. The rat cannot flee; it can make some attempts—scramble at the walls, scratch at the floor, pick at the corners—but it cannot get out, so this behavior will be rapidly suppressed by punishment. The other species-specific defensive behavior is freezing. If the rat freezes in a distant corner of the box, that behavior is consistently punished and cannot long persist because the shock will break it up. There is, however, one place where the rat can freeze, and that is standing immediately in front of the bar and holding onto it. When shock comes on, the rat can give a reflexive lurch and quickly terminate the shock. Because our automatic apparatus fails to distinguish such reflexive responses from genuine operant presses of the bar, they all get counted as responses. Bolles and McGillis (1968) have shown that within forty trials the rat will freeze on the bar and turn off shock in approximately .05 seconds after it comes on. The speed of this response suggests that it is reflexive and not an operant. What the animal learns, then, is to freeze. It learns to freeze because all other behavior is more severely punished and because freezing (while holding onto the bar) provides a very good although not optimum solution to the problem. Moreover, this behavior limits the total amount of shock received, so that fear can begin to dissipate and the animal can begin to return to the normal state in which its response repertoire includes many responses other than freezing and running away.

As this happens other kinds of reinforcement mechanisms, such as the warning-stimulus termination that reinforcement theorists have emphasized, may begin to operate. But notice that learning under these conditions has to be relatively slow and unpredictable. It is slow because it takes a number of trials before the animal will freeze on the bar, more trials before fear begins to dissipate, and then more trials before an operant bar-press is finally acquired. It is unpredictable because the animal must remain on the bar in order to keep the shocks brief, but at the same time it has to produce genuinely operant bar-presses if it is ever to become proficient at the response. These contradictory requirements make the acquisition of operant bar-pressing impossible to predict.

Notice that I have made some concession to reinforcement theory. Something like progressive shaping of the response can occur and certainly rats do occasionally really learn the bar-press avoidance response. But this learning must necessarily be slow and uncertain because it is inherently so inconsistent with the rat's natural defensive behavior.

If we look at avoidance learning in other common laboratory situations, such as the running wheel and the shuttle box, we find that rats ordinarily learn to run in the wheel in thirty or forty trials, and that they become quite proficient at the task. Supposedly, the running wheel provides some support for the natural getting-away reaction. The rat is allowed to make the right kind of movement (continuous running), and it does change its immediate environmental cues. But since the animal is not changing its location in space, there is some ambiguity about whether the rat is really running away in the running wheel or whether it is just going through the motions. Thus, learning occurs fairly rapidly, but not as rapidly as in the one-way or jump-out situations, where it is more obvious that the rat is fleeing. The most common avoidance learning situation is the shuttle box. Here the rat typically avoids ninety percent of the time after about one hundred trials. Again, the rat can get away from immediate danger by going to the other side of the shuttle box, but it cannot really get away, because on each trial it must return to the other end of the apparatus—which it just escaped on the preceding trial. So the shuttle box is another ambiguous situation. It neither clearly prevents flight behavior nor clearly permits flight. There is a corresponding rate of learning: faster than in the Skinner box but slower than in the one-way apparatus.

In summary, in all of the common avoidance-learning situations there is no need to invoke the principle of reinforcement except perhaps for the Skinner box, which gives us, ironically, the poorest learning. Most avoidance learning seems to occur because other behavior is punished.

So far I have discussed only the rat. Part of this specialization is due to the scarcity of good data from other species. We know little about the defensive repertoires of different animals, or about the relative ease with which they can learn different kinds of avoidance responses. We do know something about dogs, mostly because of the work done by Solomon and his students. We know, for example that dogs are very proficient at avoidance learning in the shuttle box. They are much better than rats in this situation. But consider the natural defensive behavior of the dog. It is a large, aggressive carnivore; it is generally able to take care of itself when confronted by a natural hazard. When a dog encounters a large predator, such as a mountain lion or a bear, it doesn't scamper for cover like the rat, it simply steps to one side and takes a new look at the situation. Perhaps what the dog is doing in the shuttle box is stepping to one side when confronted with potential shock. If so, then its superior performance in the shuttle box is a result of this relatively specific kind of defensive reaction.

I have gone into some detail on the problem of avoidance learning in order to illustrate two main points. One is that there are now some alternatives to the traditional reinforcement interpretation of avoidance learning. Increments in response strength do not have to be attributed to reinforcement. It is gratuitous to assume that if a response gets stronger there has to be a reinforcer. It is possible to have learning through punishment, and it is also possible to have an increase in response strength which depends upon changes in incentive motivation rather than response learning. Incentive motivation is based upon experience, to be sure, but the learning mechanisms are quite different from those ordinarily assumed to underly operant learning. The second reason we have discussed avoidance learning is to emphasize the idea that some responses are easier to learn than others, and to show that how fast a particular S-R connection will be learned depends upon what S is, what R is, and what species the animal is.

Other Limitations upon Learning

Once we accept the idea that there are inherent restraints upon what an animal can learn, we can see evidence of these restraints everywhere in the animal literature. Konorski (1967) has described a series of discrimination studies with dogs which clearly demonstrates selective associations.

In one experiment dogs were trained on signal either to go to food or to sit still. One set of signals varied in quality: bubbling water versus a ticking metronome. The dogs had no difficulty learning this discrimination, but they found it much more difficult to solve the same problem when the cues consisted of sounds above them versus sounds coming from below them. Are the latter stimuli inherently more difficult to discriminate? Not at all. When other dogs were trained to go either left or right for food, they learned this much more readily when cues were up versus down sounds than when the cues were bubbling versus ticking. The conclusion that some kinds of responses can be associated with some stimuli more readily than with other stimuli has been supported by further reports from the same laboratory (e.g., Dobrzecka, Szwejkowska, and Konorski, 1966; Konorski, 1967, Chapter 10).

In our own laboratory we have found that rats do not learn to anticipate electric shocks which are presented every day at the same time. We found that they show no anticipatory escape responses in the hours just before the scheduled shock. We also looked at physiological measures of the animals and discovered that there was no rise in body temperature or other index of arousal just before shock. By contrast, if we give hungry rats food every day at the same time, in just a few days they show clear anticipation of the meal (Bolles and Stokes, 1965). The anticipation may take the form of increased running in activity wheels, or pressing of a bar, or just heightened general activity. In short, the rat can anticipate a daily meal but it cannot anticipate a daily shock. The principles of learning that might be applicable in the one case clearly do not pertain to the other case. For example, anticipation of food indicates that rats can tell time, in some sense, so this cannot be the reason they failed to anticipate shock. The rat's nervous system apparently just does not permit it to anticipate shock on a twenty-four-hour schedule. If we look at rats in nature we might expect that food may occur on a regular twenty-four-hour-basis. Most animals in nature have a preferred time of day to seek food and water. But danger can occur at any time, and an animal has to be alert for dangers that occur without warning. Perhaps then no provision needs to be made in the structure of the nervous system to anticipate predation, attack, pain, electric shock, or similar events that occur on as unnatural a schedule as every twenty-four hours.

Imprinting in birds has been ably surveyed (e.g., Hess, 1964), so I need say no more about it than to point out that it is an instance of learning, because it depends upon experience, but that it is an unusual type of learning because 1) it occurs only during a "critical period" in the young birds, 2) it is generally irreversible, i.e., it doesn't extinguish, 3) it doesn't seem to involve any obvious primary reinforcement, and 4) it involves specific stimuli (ones that move) and a specific response (following). Thus, imprinting resembles the Garcia Effect in that it permits learning to occur under conditions where the reinforcement theorist would not expect it, but it is very sharply limited to certain S-R associations.

Another study from our laboratory has to do with the failure of learning in the case of sexual behavior (Bolles, Rapp, and White, 1968). It is well known that if the opportunity to copulate is made contingent upon a particular instrumental response, pressing a bar, running in an alley, or whatever, the male rat readily learns the response to get to the female. Is the same true of the female rat? First of all, we may wonder if it is necessary for the female rat to be able to profit from such experience. If the male can find her, does she need to find him? On the other hand if we watch a female rat in heat, she is certainly very responsive to the male. Every time he approaches she runs off—but not too far—and engages in enticing behavior which appears to excite the male and encourage his advances. The female certainly seems to know what she is doing, and all of her postural adjustments and locomotion are what we ordinarily think of as voluntary or operant behavior.

To answer this question we required females to run an alley to get to sexually aggressive male rats. We were concerned with whether learning would occur in this situation. Although running did speed up on successive trials (see Figure 9-4), the performance of the experimental rats was no better than that of control rats who were also in heat, but who ran to males that were not sexually aggressive. The reinforcement for the controls was presumably social contact with another animal, and that factor alone seems to be sufficient to account for all of the learning that was evidenced by any of the animals. Although other investigators have

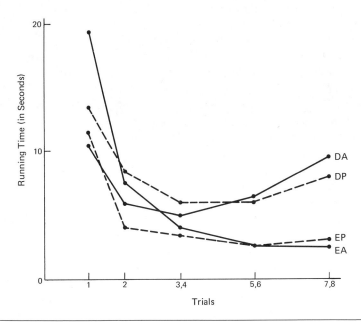

Figure 9-4 Running time of female rats running to either sexually aggressive (*A*) or passive (*P*) male rats while either in estrus (*E*) or diestrus (*D*).

come to other conclusions (Bermant and Westbrook, 1966) our data suggest that copulation is not reinforcing for the female rat. Again, we can think of the efficient use of an animal's limited learning ability: if the male is as proficient at finding females as he appears to be, then there is nothing for the female to do beyond responding appropriately once he's found her. And of course there is no reason to suppose that even if the female rat fails to find sexual behavior reinforcing, that the same will be true of other species. Indeed, there is a great deal of lore from the farm (as well as from our own culture) to indicate that some females are very much motivated for and reinforced by sexual behavior.

Conclusions and Implications

Some implications should already be apparent; one is that it is not safe to generalize across species. If we want to know about the eating behavior, or the defensive behavior, or some other behavior of a particular animal, it is necessary to observe that animal, because it will almost certainly involve particularities.

Another conclusion is that psychology has put too much emphasis upon finding universal laws of learning. It simply is not true that we can select any response from the animal's repertoire and build it up to any strength we wish by the appropriate application of a reinforcer. This assertion is based, for the most part, upon a vast number of experiments in which rats press a bar and pigeons peck a key for food. The ability to learn is a useful biological adaptation to be sure, but the solution of many critical biological problems does not require learning (e.g., we do not have to learn to breathe). Moreover, learning is very narrowly restricted in many cases (e.g., imprinting in birds). If the requirements of a learning experiment prescribe that learning should follow an animal's natural response tendencies, then that behavior is likely to be found in great strength and in very few trials. But if the task imposes an unnatural response requirement upon a particular animal then we may get no learning at all. Organisms have a limited flexibility in their nervous systems, and it may only be available for certain kinds of behavior. It may be that learning for the rat is primarily committed to food and how to obtain it. In the case of the male rat, he may also learn about females and how to get to them. But much of the rest of the rat's behavior is not subject to learning. It may be only in the primates, and particularly in the human, that we have the full flexibility of the organism's nervous system available for any kind of learning.

This last conjecture raises a further question: Are there restraints on the human too? Is the Selective Association Principle applicable to humans? There is an increasing number of psychologists and other scientists who are convinced that it is, but the evidence is far from conclusive. One thing we know about humans is that they are inordinately interested in other humans. We are very social animals, and it is perhaps not accidental that people can learn proper names much more rapidly, in longer lists, and with fewer errors than they can learn lists of nonsense syllables or lists of other meaningful words, even when these are equated for associative value. We are probably also inordinately good at discriminating human faces. If we were required to discriminate other geometric forms which differed to the same extent of contour and shape that faces differ we would be greatly confused. A person with very high cheek bones and a person with flat cheek bones, which we have no trouble discriminating, may have cheek bones which differ by only a fraction of a millimeter, yet we see these subtle differences and pay a great deal of attention to them.

Similarly, we have a remarkable ability to perceive other human voices. The differences in the physical stimulus between one person speaking in one tone of voice and another person speaking in a different tone of voice, are infinitesimal. Transcribed onto an oscilloscope we can see no difference at all, but with the ear as an analyzer, even the ear of a backward child, the speakers are discriminated and the different tones of voice are discriminated, and what the speaker said is perceived. The great orchestra conductor can perform the same sort of miraculous analysis of his orchestra. He can tell who is off key, who is out of tempo, etc., when the ordinary listener hears nothing but the big sound. This is not to say that such analytic capacity is not learned or does not depend upon learning; it obviously does. The point is simply that the human finds it peculiarly easy to detect auditory differences in other human's voices and to learn the significance of these differences, and yet requires immensely more specialized training to make discriminations in a different modality that are objectively no more difficult.

There has also been a great deal of discussion in recent years about whether humans have the rules of language built into their nervous systems. Chomsky, Lenneberg, and others have emphasized that the grammatical rules for the use of language are rapidly acquired even though these rules are so complex that they cannot be properly taught (Dale, 1972). It requires the most skilled linguists even to formulate the rules, and yet the two-year-old child with a vocabulary of a few hundred words is already using them as he builds new sentences. The argument is basically that the rules of language are learned before there is the opportunity for them to be learned. Therefore, the human nervous system has to be laid out in such a way that the rules are already there in some sense. Either that, or else the human is peculiarly able to learn the rules of language.[2]

These questions all require much more substantial empirical investigation, particularly if we hope to include the human among the animals

---

[2] See Chapter 10 for further discussion of this issue [Ed.].

we wish to study. We cannot profitably get involved in the issue of the basic nature of man: whether he is aggressive; whether he is a social being; whether he is a natural-born talker; whether he is good or evil. But we do have to accept the idea that we are not neutral; none of us is born as a *tabula rasa*. We can no longer accept, as psychologists have for a generation or more, that all behavior is learned. We now know that much behavior is not learned, and much that is learned is learned in special ways following special principles. We should now know that there is no universal law of reinforcement. And we should now know that if we want to know if a particular instance of behavior is learned, we should approach the question by finding out more about that particular situation, more about that particular behavior, and more about the particular species of animal involved.

10

Animal Communication: An Overview and Conceptual Analysis

Jarvis Bastian and Gordon Bermant

It is appropriate that we conclude this book with a chapter on communication, because a proper appreciation of the topic depends upon a prior understanding of many other facets of animal behavior. In this chapter we rely on the student's knowledge of much that has already been presented, and we trust that he will be willing to follow us through the theoretical argument that is presented here.

Let us begin by contrasting the idea of communication with other, more obvious categories of animal behavior. Obviously, animals hit, bite, or otherwise touch each other; they also mount, follow, chase, and otherwise interact, in ways that are often peculiar to their species, genus, or higher taxonomic grouping. We can group whole sets of these behaviors into functional classes: courtship and mating behaviors, parental behaviors, and so on. But these functional classes are categories of our own convenience, and their boundaries are not defined precisely.

The concept or category of communication is not equivalent to the descriptive behavioral categories (hitting, biting, etc.), nor is it a simple member of the set of functional categories (courtship, mating, parental behavior, etc.). It is a more abstract category than either of the other two, and it includes some portions of both of them. Some of the events that occur during courtship may legitimately be considered communication, as may some of the events that occur during copulation or parental behavior. The concept of communication includes something about the relationships between or among animals; it is a relational concept. Intuitively, it also seems to have something to do with *information and its transfer from one organism to another.* This intuition, however, can lead

Figure 10-1 A simple representation of a communication system.

us astray in studying animal communication if we rely too heavily on the model of human communication which underlies it. It will help to have a clear view of this model, so that we may acquire a good estimate of its strengths and weaknesses.

A General Model of a Communication System

An idealized communication system has five components, which may be described as in Figure 10-1. The five components are a *source*, a *transmitter*, a *channel*, a *receiver*, and a *destination*. When we think of human communication, for example speech, we can without difficulty fill in the specific components: the source is one person, the transmitter is that person's vocal apparatus, the channel is the air, the receiver is another person's auditory apparatus, and the destination is the other person. This seems simple enough, as does the idea that what is transmitted from one person to another is a *message*, of a unit of communication that may in one way or another be measured in terms of its information.

However, this is not so simple as it seems. To begin with, the idea of destination implies that the source person is beaming or directing his message to some particular other person or persons. Now if I happen to be the source person in question, it is generally true that I have a particular destination in mind for my messages; in other words, I *intend* that my message should reach this or that person or class of persons. But when I am trying to figure out another person's "communication behavior," I don't know the identity of his destinations in the same way that I know my own. I may, of course, infer the destination of his message by observing the behavior of all those persons who are physically able to receive it, e.g., all those in earshot. Moreover, I may ask the person to whom he is speaking. But, if for whatever reason, I can't ask him or he can't tell me, then I can specify the destination of his message only after the fact. And, interestingly, there is no necessary relation between *my* designation of the destination and his designation of it. For example, he may intend to communicate with Mr. X (and succeed or fail), but I may describe his destination to be Mr. Y, because I observe Mr. Y to behave in a way that gives me confidence that he received and understood the message. Of course the sender might object that he wasn't *trying* to communicate with Mr. Y, but that would only be some additional information about the communication situation, not a denial that communication had taken place. Unintentional communication is still communication.

With regard to understanding communication among nonhuman animals, it is best to approach the animals with the realization that in general they do not tell us, directly, the destinations of whatever signals they emit; for the time being, we will assume that all communication among nonhumans is nonintentional. With this view firmly in mind, we can see that the study of animal communication involves our attempt to structure the complexities of animal interactions according to the generalized scheme of a communication system. We are using an abstract idea in order to organize our data on animal social interactions in ways that are intellectually powerful and useful to us.

To say that animal communication is nonintentional is not to say that it is purposeless or functionless. Intentions are not the same things as purposes or functions. On the contrary, animal communications serve very precise and important functions in the regulation of animal communities. But the nature of these functions or purposes needs to be discerned from a populational or evolutionary perspective, as well as from the perspective of the individual animal emitting or receiving a signal. As before, we need to modify our typical approach to human communication if we are to have a clear view of nonhuman communication.

Our intuitive understanding of human communication will sometimes give us initial important insights into animal communication. For example, consider the observations of Jane van Lawick-Goodall on the social interaction of chimpanzees living free in the Gombe Stream Reserve of Tanzania (van Lawick-Goodall, 1968). Under certain circumstances one animal will reach out and make slight, very rapid patting movements on another chimpanzee's face, head, or body. This interaction often occurs when the animal being patted is socially subordinate to the patting animal and after the subordinate animal has lost an aggressive or hostile encounter. The subordinate animal may be whining or screaming loudly, and it is at this point that the dominant chimpanzee pats the subordinate. Apparently as a result of being patted, the subordinate animal quiets down. Patting motions are also directed at infant animals.

How may we characterize "the message" that has been transmitted in this interaction? Van Lawick-Goodall (1968) uses the word "reassurance": patting is an example of "reassurance behavior." And in fact, given the social contexts in which chimpanzee patting occurs, and also given our human experiences with patting and being patted, this way of talking about what the apes are doing makes a certain amount of sense. There is no question here of whether the patting ape wants or intends to reassure the subordinate; rather, the functional consequences of the act are the same as we expect when we engage in reassurance ourselves.

The importance of this generalization from our own experience and vocabulary is highlighted by the fact that on some occasions similar to the one described, the dominant chimp does not pat the subordinate;

instead, the two animals stand facing away from each other and press their rumps together (Figure 10-2). Unlike patting, rump pressing is not an intuitively recognizable gesture of reassurance among humans. Consider for a moment the hypothetical situation in which rump pressing were the only behavior that occurred in this social context. A well-trained observer might still come to the conclusion that this interaction served the functions normally included within the idea of reassurance; but our confidence in the validity of the conclusion, our "feel" for it if you like, would not be so great as it is when we assume rump pressing to be equivalent to patting. In general, communication among animals is more like rump pressing than like patting: it does not have obvious parallels in human behavior.

Functions of Animal Communication

Understanding the advantages and disadvantages of using our common model of communication in the study of animal behavior is not the end of the problem; it is the beginning. Granted the usefulness of the cautious application of terms taken from ordinary human experience, we still need to inquire into why this sort of communication comes to be practiced within the animal community under consideration. We recognize at once that explaining the actions of the animals by reference to "human nature" or "chimpanzee nature" constitutes no explanation at all. We want to know about the functions of an act. It is not enough to label patting or rump pressing as reassurance behavior; we need to know why chimpanzees practice reassurance behavior.

As usual, when we ask the question "why" about some aspect of animal behavior, two distinguishable but related questions are implied. To begin with, we are asking about the immediate causes of the behavior in the individuals that display it; ideally our answers are found in rigorous behavioral and physiological terms combined with descriptions of relevant factors in the animal's environment. Complete descriptions of this sort

From "The behavior of free-living chimpanzees in the Gombe Stream Reserve" by J. van Lawick-Goodall, in *Animal Behavior Monographs* 1, 1968.

Figure 10-2 Rump pressing ("reassurance behavior") in the chimpanzee.

are extraordinarily difficult to accomplish, so we are often forced to accept a kind of psychological shorthand, in terms of "drives," "motives," "intentions," etc., which provides an approximation to an ideal answer.

As interesting and important as the investigation of immediate causes is, there is another approach to answering our "why" question that is equally important, if not more so, in coming to grips with the critical issues involved in understanding animal communication. The focus of this approach is not on an individual animal but rather on a population of animals. This is the focus that is demanded if we are to understand the reasons for the existence of communication systems among virtually all animal species and for the forms that these systems take. As we shall see, the need for communication among members of a species increases as the social roles enacted by the members become more diversified and complex. Diversification, in turn, represents a class of maneuvers in the evolutionary process. The evolutionary process operates at the level of populations. In the next section we will take some time to sharpen our sensitivities to a populational way of thinking in order to prepare, hopefully, for a sound understanding of what animal communication is about.

A Populational Way of Thinking

We will focus here on the fundamental biological concept of *adaptation.* This word takes on several different but related meanings in different contexts. To begin with, it can refer to an advantageous, temporary change in the functioning of an individual in response to a changing environment. For example, "dark adaptation" refers to the increasing sensitivity of our eyes under conditions of dim illumination. Second, adaptation can refer to a permanent structural feature of an organism that equips it to function effectively in a particular environment. Obvious examples are the beak shapes of the various Galapagos finches (see p. 312 and also Chapter 2). In both of these usages, the emphasis is on characteristics of the individual.

Adaptation also refers to characteristics of populations. For example, it may mean the proportion of individuals in the population that possesses a trait that is advantageous in a particular environment: the larger the proportion, the better adapted is the population to that environment. Adaptation may also refer to the ability of a population to change across generations in response to an environmental change. In this sense adaptation has to do not only with the phenotypic characteristics present in a population at a given time, but also with the population's genetic reserve.

The importance of a populational perspective for an understanding of biological adaptation is clarified by relating adaptation to evolution. Wallace and Srb (1961) put this relationship nicely by saying that adaptation is "an *unavoidable evolutionary change* within populations of living things." Adaptations are the evidence of evolution. However, not all evolution leads to adaptation; many evolutionary changes are insufficient or maladaptive in the face of new environmental challenges, and they result in extinction.

Adaptations of populations to their environment may be analyzed at all levels of biological organization, from the molecular to the social. Whatever the level chosen for analysis, one principle remains invariant: changes in a particular trait will be constrained by the characteristics of the other components in the system of which that trait is a part. In order to be adaptive, a change in one part of a biological system must enhance, or at least not weaken, the working of the whole system. Changes in more than one part of a system must be *co-adapted* if the net result of the . change is to be adaptive. For example, the evolution of the chemistry of hormones must take into account the evolution of the tissues which are the hormones' targets (Barrington, 1971). The endocrine system is one example of a biological communication system. In general, for all biological communication systems, adaptive changes in one component will be constrained by the potentials for change in the other components.

Presumably, every adaptation observable at a behavioral level of analysis has some correlate or counterpart at a physiological level, although the reverse need not be true. Some behavioral adaptations, for example those involved with basic feeding patterns, occur uniformly in the members of a species and are relatively independent of the species' type or degree of social organization. An extreme example of such an adaptation is found in the members of one species of the Galapagos finches, which have exploited the availability of grubs in the bark of trees by acquiring the ability to pry the grubs out by holding small sticks in their beaks. Other behavioral adaptations are more closely related to the social system in which the animal lives; their significance can be comprehended only by reference to their effects on the behavior of other animals.

Animal communication systems are clearly adaptations of this latter type. In order to account for their evolution, therefore, we must make primary reference to their significance for the population. For example, changes in the characteristics of the "sender" of a signal may increase the likelihood of the "receiver's" reproductive success without changing that likelihood in the sender. The persistence of this change in the population can be accounted for only by reference to the genetics of the population. Klopfer has dealt with this problem in his discussion of altruistic behavior in Chapter 2 (p. 69). The reader may find it useful to review his argument at this point.

A defining feature of *social adaptations,* not all of which are behavioral, is that the selective pressure operating on them arises as a function of the social environment. The social organization of a population during one generation is the current state of social adaptation in that population; it is also the set of constraints that limit the possibilities of further adaptations in succeeding generations. With reference to animal communication, we can appreciate that communication systems operate not only in regard to transmitting messages about the physical environment (for example, the location of food sources) but also in regard to regulating important features of the social environment. The patting and rump-pressing behaviors of chimpanzees, mentioned earlier, are clear examples of socially adaptive behavioral interchanges.

From this perspective it may be seen that communication among members of a population becomes more important as the degree of their social organization becomes more complex. In order to understand in greater detail what this communication will be about, it is necessary to have a closer look at some basic features of social organization.

Social Organization

Fundamentally, what we mean by social organization is a division of labor among the members of a population, given that the basic "task" of the population is to maintain itself through reproduction. Any heritable feature of the organization that facilitates reproduction will tend to be maintained in the population, i.e., it is an adaptation. Some features of the organization will be concerned directly and obviously with repro-duction; the process of sexual reproduction is the most obvious and basic example. Sexual reproduction involves the division of reproductive labor into two genders, male and female. The requirement that a gamete from one gender fuse with a gamete from the other forces the members of a population to achieve at least minimal behavioral coordination at the time of reproduction. Given the absolutely critical importance of success-ful coordination, it is not surprising that many of the clearest examples of animal communication deal directly with the regulation of behaviors surrounding sexual reproduction, i.e., courtship and mating behaviors. Some examples have already been presented in Chapter 7; additional examples appear later in this chapter.

For most animal species the basic division of reproductive labor into two genders is accompanied by additional behavioral differences between the genders; we may refer to behaviors that are characteristic of a gender as *gender roles*. Some gender roles are closely tied to reproductive physi-ology: for example, only female mammals nurse infants. Other gender roles are less immediately related to reproduction: for example, male songbirds initially establish the territories that will eventually become their breeding grounds, and male baboons lead the troop as it forages for food. Gender roles account for a major portion of social organization in all sexually reproducing species.

A second major determinant of social organization is the social roles correlated with age. Infants, juveniles, young adults, mature adults, and very old animals all may have different behavioral niches to fill in the population's organization. In general, the complexity of social organ-ization is reflected in the number of specifically different age roles that an individual assumes during its life. Among invertebrate animals the greatest complexity is found in the social insects. In a honey bee colony, for example, a single worker is likely to engage in a series of as many as seven different activities during the first fifteen to twenty days of life.

For the first five days after she emerges from her cell she provides food for larvae which are older than four days; for approximately the next five days she feeds younger larvae; then she becomes a receiver of nectar that has been brought into the hive by the forager bees, and at about this time she may also become involved in cleaning debris from the hive or in building comb. Also, at some time between the ages of six to thirteen days, she may tend the queen bee. As we have already seen in Chapter 8, this behavior plays an important part in one of the basic communication systems operating within the hive. Although she may take her first flight from the hive at any time between the ages of five and fifteen days, the worker does not forage for food until she is twenty days old. And, as a bee of foraging age, she may alternate food gathering with guarding the hive entrance against other insects or bees from other hives (Ribbands, 1964). Age-related changes in behavior such as those observed in the honey bee are sometimes referred to as *age polyethism*.

Among nonhuman vertebrates the greatest complexity in age roles is found in the Old World monkeys and the apes. Infant and juvenile monkeys and apes do not "work" in their social community as newly emerged honey bees do. Nevertheless their presence as expressed by size, coloration, and behavior, serves definite social functions. For example, the position of a female baboon in the troop's dominance hierarchy changes markedly when she is carrying or tending a newborn infant (DeVore, 1963). Expressed in terms of a communication system, this means that the social unit "mother-infant" sends and receives different messages than either the female or infant alone. Moreover, infant and juvenile primates must *learn* the nature of their group's organization: "appropriate" social behavior is not given to them as it is to the worker honey bee. This aspect of primate development produces characteristics of social organization not present in groupings of social insects. Mitchell has presented material related to this topic in Chapter 4.

Gender and age roles account for the major features of animal social organization—but by no means all of them. Within a group of animals of the same gender and approximate age, definite relationships may be established and maintained. Among adult male baboons, for example, a dominance hierarchy is established in regard to feeding and copulation. This hierarchy is more complicated than the linear "peck-order" of chickens, because two or three males may successfully cooperate in acting against another male. An individual's status relative to other adult males depends upon both his individual dominance and his success in soliciting support from other males during an aggressive encounter (Hall and DeVore, 1965). This is a rather sophisticated form of social organization among nonhuman animals; it clearly entails the need for relatively precise modes of communication.

In summary, social organization is the result of differentiation of social roles among members of a group. As roles become more diverse, social organization becomes more complex. Social organization succeeds to the extent that it meets the challenges posed by the environment for the group. Complex social organization demands social maintenance. In other words, energy must be expended to maintain the necessary distinctions among the various social roles. One consequence of this is that animal groups with complex social organizations may be expected to relate to each other not only in ways that achieve immediate beneficial results in the physical environment, but also in ways that sustain the distinctions of the various social roles.

These considerations allow us to sharpen our intuitive understanding of animal communication by characterizing animal communication systems as *social adaptations of behavior that meet and exploit the challenges and advantages of group life.* Like all adaptation, communication is an inevitable outcome of the evolutionary process. It is a social adaptation because it is directly tied to, and partly defines, the structure and maintenance of the group's social organization. It is a behavioral adaptation because it is through an animal's gestures, postures, vocalizations, and other responses that his social roles are defined. Communication is not one feature of social organization; it is the proof of it.

There are many cases in which animal communication systems meet the advantages and challenges that group life presents. We will now consider some examples of these systems in regard to several aspects of group life: sexual reproduction, care of the young, cooperative foraging, and population density control.

Challenges and Advantages of Group Life

Sexual Reproduction

In the introduction to Chapter 7 we outlined the behavioral requirements operating on animals that practice sexual reproduction. Taken together, courtship and mating behaviors serve to promote the joining of the right gametes at the right time in the right place. Numerous problems confront animals who engage in this task, and it is where the problems are the most serious that we may most clearly construe what the animals are doing in terms of a communication system.

The problem of reproductive isolation is an obvious example. The genetic advantages of sexual reproduction cannot be fully realized if there is a large degree of gamete wastage due to interspecific mating. We might expect, therefore, that communication in the service of species identification during courtship would be most precise in those circumstances wherein the danger of gamete wastage is greatest.

There is evidence to support this expectation, for example, in the mating calls of male narrow-mouthed frogs (genus *Microhyla*). Two species of this genus have a small zone of overlapping home ranges in eastern Texas; the range of one species (*M. carolinensis*) extends east to Florida while the range of the other (*M. olivacea*) spreads west to Arizona. Blair (1955) studied the physical characteristics of the mating calls of both species, both in areas where they live apart and in the zone of overlap. The average pitch and duration of these calls, as they were made by frogs living at the furthermost extremes of the ranges (Arizona and Florida), were quite similar. Presumably an *M. olivacea* from Arizona, if placed in a pond in Florida, could emit a call that would attract a female *M. carolinensis;* everything else being equal, interspecific mating, and hence gamete wastage, would result. In the zone of overlap in Texas, however, where the danger of interspecific mating is real, the calls of the two species are not the same (Figures 10-3 and 10-4): *M. carolinensis* calls become somewhat lower and briefer and *M. olivacea* calls become markedly higher and longer. Thus in Arizona the average duration of an *M. olivacea* call is approximately 1.5 seconds and its average frequency approximately 3500 Hz, while in the zone of overlap the average is close to 2.1 seconds and its average frequency approximately 4400 Hz. By contrast, the average frequency used by *M. carolinensis* in the zone of overlap is 3100 Hz and the average duration is 1.1 seconds.

This example highlights the importance of coadaptation in the evolution of communication systems. Changes in the males' calls in the zone of overlap will not be of use in reproductive isolation unless the females of the two species are differentially responsive to these calls. It is the communication system within the population that must evolve to

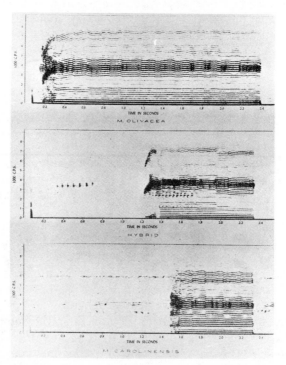

From "Mating call and stage of speciation in the *Microhyla olivacea—M. carolinensis* complex" by W. F. Blair, in *Evolution* **9**, 1955.

Figure 10-3 Representative sound spectrograms of *Microhyla* mating calls in eastern Texas. Sound intensity depicted as darkness of markings. Note emphasized frequency is at about 3000 Hz in *M. carolinensis* and at about 4000 Hz in *M. olivacea*.

meet the evolutionary challenge posed by the possibility of gamete wastage. In this example, the marked difference in the quantitative "operating characteristics" of the system can be maintained over generations only if the hybrids that arise from occasional interspecific matings are at a disadvantage to members of the parent population in regard to achieving reproductive status (Blair, 1955).

A given portion of a courtship display may serve more than one biological function. In the example of the frogs just described, the characteristics of a male's call serve both to increase the probability of intraspecific mating and decrease the probability of interspecific mating. The dual function of a display is also evident in the "semaphore signaling" employed by male fiddler crabs (Crane, 1941; 1957). As described in Chapter 7, each species of crab (genus *Uca*) is characterized by the unique fashion with which the male moves his outsized claw in front of the female during courtship. Several species of crab may live sympatrically on a single beach and go through their courtship rituals at the same time; the species-specificity of the male's visual signals are effective in preventing interspecific breeding. However, even after a female has located herself directly in front of a male of her species, she does not enter with him

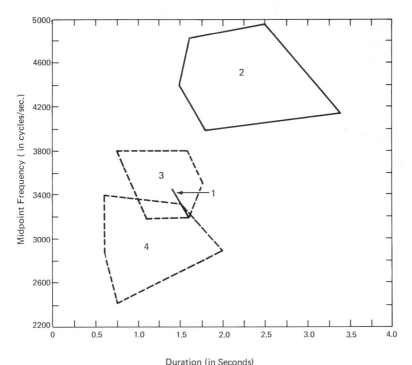

Duration (in Seconds)

Modified from "Mating call and stage of speciation in the *Microhyla olivacea—M. carolinensis* complex" by W. F. Blair, in *Evolution* **9**, 1955.

Figure 10-4 Duration and frequency of *Microhyla carolinensis* and *M. olivacea* mating calls in different geographical regions. (1) *M. olivacea* from Arizona; (2) *M. olivacea* from overlap zone in eastern Texas; (3) *M. carolinensis* from Florida; (4) *M. carolinensis* from overlap zone.

into the sand burrow where copulation occurs until his repeated signals produce physiological changes that result in her behavioral sexual receptivity. From the viewpoint of a communication system, this means that there is more than one "message" contained in the male's signal and that the signal has more than one "destination." And again, the importance or effectiveness of the signal may not be understood except by reference to the rest of the system or systems of which it is a part.

Care of the Young

Mitchell (Chapter 4) and Lott (Chapter 8) have dealt in detail with the behavior of immature animals and their relations with their parents. A good deal of the exemplary material presented in those chapters might be reconsidered from the perspective of communication systems. One clear example, described by Lott, consists of the interactions between mother goats and their newborn kids that involve three modalities or channels; visual, auditory, and olfactory. A critical feature of these signals is their relationship to the kid's identity, which is initially established and subsequently monitored by the mother by means of olfactory signals. Under natural ·conditions the first five minutes of interaction between doe and kid establish the doe's ability to discriminate the odor of her own kid from the odors of other kids in the flock. The kid's odor becomes the key that unlocks subsequent nurturant behavior in the doe's repertoire.

The successful establishment of the kid's olfactory identity with a doe does not depend solely upon the chemical composition of the olfactory stimulus; it depends as well upon the condition of the doe at the time she is first exposed to the odor. As described in Chapter 8, experimentally induced inability of the doe to smell the kid at the time of parturition prevents the establishment of a unique bond between the two animals without eliminating the doe's general tendencies toward maternal behavior. Thus the characteristics of the "receiver" at the time the "message" is first transmitted can affect the response to identical and similar messages much later in time. This is generally the case with the phenomena of early learning or imprinting: responsiveness to species-typical signals in adulthood depends upon having particular experiences during infancy or early childhood.

What is initially surprising to us about these goats is not that the doe recognizes her own kid but that she does so by its smell and that the timing of their first interactions is critical for success. We tend to take for granted that parents will recognize their own young and behave appropriately to them regardless of the young animal's behavior. This assumption is generally invalid for animals; parental care often depends upon the performance of particular signaling movements or vocalizations (sign stimuli). Biologically inappropriate objects, e.g., animals of another species or offspring of other parents, will, by producing the proper signals, elicit the same behavior as the parent's own young. A clear example of this is found in the parental behavior of the English robin. Robin parents recognize each other and are apparently also able to identify an egg in the nest which is not their own. However, after the eggs hatch, the control of the adult's care-giving behavior becomes dependent upon the gaping response (beak pointed up, wide open) of the young; fledglings that gape are fed, those that do not are not. The gaping response is such a powerful signal that any bird of approximately the right size may use it to elicit feeding behavior from robins that have recently hatched their own young. Linnets, blackbirds, song-thrushes, cuckoos, meadow pipits, and wrens are some of the species reported to have been fed by parent robins. However, the susceptibility of robins to the gaping signal is highly dependent upon their position in the breeding cycle. Except for the time between hatching and fledging, gaping itself elicits no feeding behavior from them. However, it is interesting to note that the female can elicit feeding from the male by exhibiting a crouching, wing-fluttering display accompanied by a characteristic vocalization (Figure 10-5). This so-called

From *The Life of the Robin* by D. Lack. London: Penguin Books, 1963.

Figure 10-5 Forms of feeding in the robin. Above, the cock feeds the hen during courtship. Below, the hen feeds a fledgling.

"courtship feeding" occurs with varying frequency from the time of initial pairing until approximately one week after the young are fledged. The function of the display and the male's response to it appears to be to strengthen the bond between cock and hen (Lack, 1953).

Cooperative Foraging

One of the most fascinating examples of communication among animals involves the behavior of worker honey bees as they forage out from the hive for pollen and nectar. The details of this behavior have been analyzed intensively and elegantly by Professor Karl von Frisch of the University of Munich. Von Frisch (1950) became interested in communication among bees when he noticed that the number of bees appearing at a rich food source grew very rapidly soon after the first bee appeared there; this was not to be expected if the bees were foraging solitarily, without interacting during trips between food source and hive. In order to study the interactions of the foraging workers von Frisch constructed a glass-walled bee hive. The next step was to place feeding stations (tables with dishes of sugar water on them) at some distance from the hive. When bees arrived at the feeding station they were marked by small drops of paint; a code using several different paint colors allowed a large number of bees to be identified individually. Then, when the marked bees had collected sugar water from the feeding station and had returned to the hive, their behavior could be observed. Sometimes the sugar water was cleverly arranged with flowers on the feeding station so that the bee would have to stand on the flower, thereby picking up some of its odor, while ingesting the sugar water.

In a series of ingenious experiments based on this general method, von Frisch was able to demonstrate close relationships between the location of the feeding station and the behavior of the returning bee in the hive. To begin with, the returning bee engaged in one of two forms of "dance": the round dance or the tail-wagging dance; these are illustrated in Figure 10-6.

When the feeding station was fifty meters or less away from the hive, returning workers always performed the round dance; at distances between fifty and one-hundred meters some bees used round dances and some used tail-wagging dances; and at distances greater than one-hundred meters all the bees used waggle-tail dances. Moreover, the speed at which the workers performed the waggle-tail dance was inversely related to the distance of the food source: the farther the source the more slowly they performed the dance. The relationship between speed (number of turns per unit time) and distance is shown in Figure 10-7.

This ability of the honey bee to relate the distance of a food source to an arbitrary act on the wall of its hive is truly an amazing behavioral adaptation; but it was not the only incredible feature of the forager's

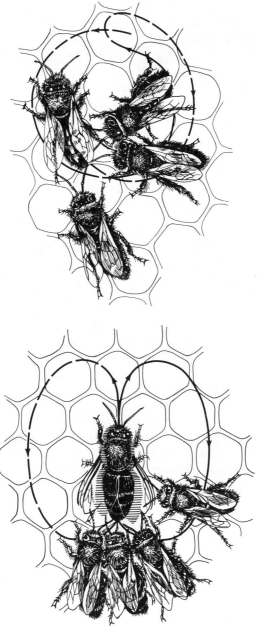

Figure 10-6 a) The round dance of the worker honey bee. The round-dance is performed when the food source is relatively close to the hive. b) The waggle dance of the worker honey bee. The waggle dance is performed upon return from relatively distant feeding sources. The rate and direction of the waggles encode the distance and relative position of the food source.

behavior that von Frisch discovered. He and his colleagues observed that the orientation of the bee on the vertical wall of the hive during waggle-tail dances changed during the course of the day, as the bee went back and forth to the same feeding station. Repeated observations and analysis revealed that these changes in orientation related to the changing position of the sun with respect to the food source and the hive. If the food source was directly between the hive and the sun, then the dancer waggled directly upward on the wall of the hive. But if the food source was directly away from the sun, then the dancer waggled directly down. In general, the bee waggled at an angle to the vertical that equaled the angle between the food source and the sun. This relationship is shown graphically in Figure 10-8.

These relationships between the location (distance and direction) of the food source and the forager's behavior inside the hive are, in the opinion of many, the most sophisticated behavioral adaptation to be found among invertebrate animals. Some authors have gone so far as to call this behavior a "language." In fact it is not a language, for reasons which we will make clear later. Moreover, in our discussion so far, we have not provided convincing evidence that the forager's dancing is part of a communication system, that is to say, that other bees in the hive locate the food source more efficiently as a result of their exposure to the dance. We must not forget that the concept of the "message" depends upon the behavior of the "destination" as well as that of the "source." It is not sufficient that *we, people,* could find the food more efficiently ("read the message"); we have to prove that other bees in the hive do so, in fact.

In his initial experiments von Frisch provided some evidence that other bees responded to the dance by going straight to the food source. The key issue in these experiments is whether naive bees, i.e., bees who have not previously been to a particular food source, get to that food source more efficiently as a result of being exposed to the dance, and that only, than they would have otherwise. It turns out to be quite difficult to design and conduct experiments which support this hypothesis without also supporting alternative hypotheses. For example, Wenner and his colleagues (e.g., Wenner, 1967; Wenner, Wells, and Johnson, 1969) have argued that the data from von Frisch's experiments, and from studies of their own, may be interpreted to mean that bees locate food sources by a combination of random search patterns and olfactory cues gained from the bodies of returning foragers. These authors are critical of the idea that the dance is utilized by other animals in the hive, although they do not deny that "the message" is contained in the dance in the sense that we humans can decode it. Critics of Wenner's hypothesis (Gould, Henery, and MacLeod, 1970; Lindauer, 1971) have argued on both theoretical and empirical grounds that the dancing is in fact an effective signal. Lindauer, for example, states that "one of the most fruitful guiding

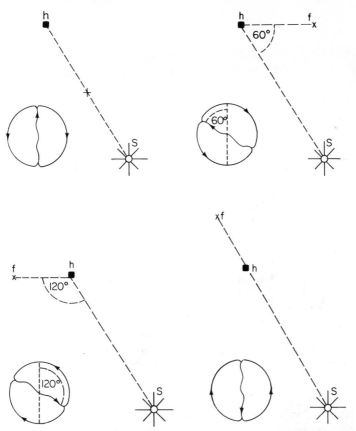

Figure 10-7 Relation between positions of the hive (h), the sun (s), and the food source (f) that determine the angle to gravity that the worker bee takes during the waggle dance. (See the text for more detail.)

principles of biology has been that each morphological structure and behavioral act is associated with a special function. On this basis alone, it would seem highly unlikely that the information contained in the waggle dance of a honey bee is not transmitted to her nest mates. Yet this is precisely the argument that has been advanced by Wenner and his colleagues" (p. 89). Lindauer goes on to report experiments in which olfactory cues from the food source and other nondance related cues are eliminated in so far as possible. Under these conditions, bees are still recruited in relatively great numbers to the feeding station which has been signaled by the returning foragers. The experiments reported by Gould *et al.* (1970) produced similar results. These authors concluded, "Thus under the experimental conditions used, the directional information contained in the dance appears to have been communicated from forager

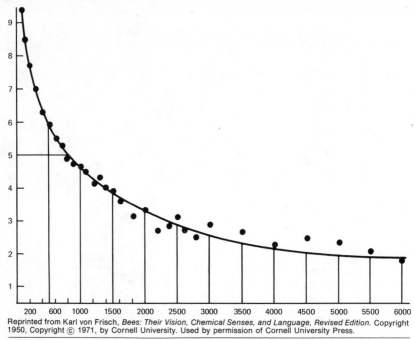

Figure 10-8 Relations between the distance of the food source (in meters) and the number of turns in the waggle dance.

to recruit and subsequently used by the recruit" (p. 553). They confirmed the original interpretation of von Frisch and rejected the alternative hypothesis of Wenner. However, they also cautioned that other, as yet uncontrolled factors might be controlling the recruited bees' path to the food source. In this case as in all examples of complex animal behavior, it is very difficult to design and conduct experiments which confirm one hypothesis absolutely, at the expense of all other hypotheses.

The Control of Population Density

It is often the most basic features of a biological system which escape our attention, because they constitute the apparent background against which the system operates. For example, until relatively recently there was not a great deal of attention paid to the mechanisms by which the sizes of animal populations were regulated. Nevertheless, generally it was agreed that every population had upper and lower limits of size; going beyond these limits could lead to extinction. When populations become too large they utilize food resources faster than the resources can regenerate, and when they become too small they may not contain enough diversity to maintain and increase their numbers in the face of environmental changes. The obvious question, from the behavioral perspective, is how animals behave in different population densities so as to maintain their population size within the limits of tolerance.

In 1962 V. C. Wynne-Edwards published a massive amount of evidence in regard to a theory about the relationship between animal social behavior (communication) and the regulation of population size. This work has generated a good deal of discussion and criticism during the last ten years. In this section we will sketch the outline of Wynne-Edwards' theory and show briefly how he applied it to signaling acts.

We begin by pointing out that there are two processes by which a population may regulate its own density: 1) by encouraging in-migration or out-migration with regard to a particular physical area, and 2) by adjusting the "turnover time" (relation between birth-rate and death-rate) of the population within the area. Death rates are, of course, not under as direct control as birth rates: i.e., it is easier to regulate the number of individuals born by regulating the proportion of the population that breeds than it is to regulate the number dying by mass intraspecific killing. This is particularly obvious when one female is likely to spawn or give birth to hundreds of young. But even reducing the birth rate is not as quick a method of reducing the density within an area as is encouraging emigration.

Given these two general processes for population density regulation, there needs to be some way for individuals to be affected by the density so that their behavior becomes appropriate to restoring optimal conditions. It is central to Wynne-Edwards' viewpoint that the regulation of density is an active process of the population, not simply a result of external environmental forces. In particular, he argues that successful natural populations, especially more complex ones, have evolved control systems which "build up and preserve a favorable balance between population density and available resources. . . ." (p. 9). He likens these systems to the homeostatic (self-balancing) systems of the body, and he calls the self-regulation of animal numbers *population homeostasis.*

The basic problem facing any population is to feed itself. Population homeostasis must operate to keep the number of animals less than the available resources will support. Moreover, it must operate so that individuals within the population are not usually harmed by its operation. One way to insure these outcomes would be to design (evolve) a series of *conventions* within the population so that animals participating in the conventions would, as a result, avoid the direct competition for food which could lead to a destructive outcome.[1] By displacing the direct food competition in favor of some conventional form of competition, and by associating definitive displays and other signaling acts with these conventions, the population can communicate about its density (in relation to available food and other natural resources) without having to engage in

---

[1] See page 351 for a different but related use of "conventional."

destructive competition among group members. The population "plays out its conventions" even when there is an abundance of resources. Because many of the conventions have to do with determining who will breed and with whom, they can serve to regulate the birth rate. Conventions may also serve to promote or retard the rate of exploration or migration, thus affecting population density even more immediately.

Earlier in this chapter we pointed out the intimate relationship between the concepts of social organization and communication. In Wynne-Edwards' view, an animal society is most usefully defined as *"an organization capable of providing conventional competition. . . .* The social organization is originally set up to provide the feedback for [population homeostasis]" (p. 14). Thus he finds the biological function of social organization to reside in its conventional characteristics. These provide forms of competition among individuals which prevent direct competition for food but also allow existing densities to regulate migration or "turnover."

Wynne-Edwards argues that much of the ordinarily observed communication among animals has functional significance with regard to regulating population density. In fact he lumps a good deal of the stereotyped social behavior of higher animals into a category which he calls *epideictic displays:*

Epideictic displays are especially evolved to provide the necessary feedback when the balance of population is about to be restored, or may need to be shifted, either as a seasonal routine or as an emergency measure. They generally involve conventions that have evolved away from the direct primitive contest, which is liable to end in bitter bloodshed and even killing of participants, and have come to assume a highly symbolic quality, not even directly implying threat in many cases, but producing a state of excitation and tension closely reflecting the size and impressiveness of the display. In many cases not only is a special time of day set aside, but also a traditional place, to which all the participants resort for the purpose; and this is undoubtedly the underlying cause of almost all types of communal roosting and hibernation, and many other gregarious manifestations among normally solitary animals. (1962, p. 16–17)

The bulk of Wynne-Edwards' (1962) book is given over to cataloguing displays and showing how they may be interpreted as serving the epideictic function. There are literally hundreds of examples presented there, from which we will describe one, which Wynne-Edwards calls "the perfect example of an epideictic rite" (p. 325): social behavior in the Acorn Woodpecker of California.

Food Storage by the Acorn Woodpecker (Melanerpes formicivorus)

The Acorn Woodpeckers are the most abundant woodpeckers in California. They roost in various species of oak tree, and the oak's acorns are a principal source of food for them. They are gregarious animals;

several adult males and females often roost together in a single hole in the oak and gather acorns together. Normally they do not migrate during the course of the year. The basic social unit larger than the group living in a single hole is called a *settlement*. A single settlement tends to restrict its food-gathering activities to a particular stand of oaks; there is some, but not much, crossing of food-gathering areas by adjacent settlements.

Each settlement possesses one or more "storage trees." These are usually oak but sometimes the softer pine, in which the birds have pecked holes to receive acorns. The birds exhibit remarkable precision in placing and pounding the acorns into the holes. By the time they are finished, the acorn is fitted so flush and tight that only another woodpecker can retrieve it. Neither squirrels nor jays can loosen it easily. One storage tree in the San Jacinto Mountains contains approximately 50,000 holes.

It is important to note that each settlement has one or at most a few storage trees. During the period of autumn harvesting, the birds will by necessity be gathered in rather large groups around the storage trees. "Information" is thus potentially available for the determination of 1) the density of the population at that time of the year, and 2) the abundance of food for the winter. The question then naturally arises, whether the behavior of the birds at that time of year, when gathered together around the storage tree, varies with density and food abundance in such a way as to encourage outmigration when the density is too great for the available food. Similarly, during the breeding season, when males and females spend a good deal of time flying about the storage tree, a behavioral "assessment" of density might lead to a regulation of breeding, thereby controlling population size.

The reports of those who have observed the behavior of the woodpeckers at the time of harvesting suggest that there is much ritualized display and "mock-battle" around the storage tree. Moreover, given the relatively benign California climate, there is seldom if ever an actual danger of the food supply being insufficient. Thus, the behavior around the tree is, in fact, related to the necessities of food storage only *conventionally*. The form of the behavior is relatively arbitrary, given the requirements of harvesting itself; thus it may be serving the epideictic function of segregating animals on the basis of their performances in the species' conventions. Direct competition for food has been replaced by more or less "ceremonial" functions, relatively arbitrary in form, which have evolved to keep the population density well under what the environment will support. Wynne-Edwards comments as follows:

The conclusion is not difficult to reach that the acorn rite is something more than just harvesting: rather it resembles a harvest-festival. Food is ceremonially placed in a ceremonial tree, to an accompaniment of self-display and social competition, in a manner that has about it quite a reminiscent ring of the tribal stage in human evolution. In fact it seems to be the perfect example of an epideictic adaptation. On the one hand, competition is the means of applying stress, having potential consequences in relation to the social hierarchy, which can lead to the elimination of supernumeraries and determine the reproductive output of the community. On the other hand, the harvesting itself is a direct sampling of the food-resource on which the community must subsist until the following summer—the resource that actually determines the safety-limit of numbers. And the whole performance, finally, is focused on a traditional object, the tree, which symbolizes the territorial system. (1962, p. 323-324)

In order for this colorful interpretation to be strengthened we would very much like to see quantitative analysis of the birds' behavior and density. Nevertheless, Wynne-Edwards' theory does allow the organization of large amounts of data about animal social behavior and signaling acts that would otherwise remain diverse.

We have concluded our consideration of how animals exploit conditions of group living. Hopefully this very brief overview gives the reader a feeling for the diversity of animal signaling acts and the functions they serve. Now we will consider the mechanisms upon which animal signaling acts are based.

Mechanisms of Animal Communication

Up to this point we have introduced and developed the idea that communication is a set of adaptations by which an animal population exploits the biological advantages deriving from the coordinated actions of its individual members. These social adaptations range in complexity from the most fundamental interactions involved in sexual reproduction to the enormous intricacies of colony life in the social insects and the group lives of higher primates. Now we will consider the mechanisms by which communication is achieved. As it turns out, what "mechanism" means in this context varies as a function of the perspective of the observer; it can refer to events at the levels of the population, the individual, or a portion of the individual's physiology. We will expand upon this point later. Our first task is to present some new terminology that will help to focus on the mechanisms of communication at the level of the individual organism.

Signaling Acts: Their Reference and Social Significance

Within the conceptual framework of communication described in Figure 10-1, we will define a *signal* as a change in the environment of an animal which can affect its behavior. Second, we will define a *signaling act* as any act by an individual animal that generates a signal. Third, we will define the *social significance* of a signaling act as its effects on the behavior of animals that perceive the resulting signal. Finally, we will define the *reference* of a signaling act as the events and processes that govern its occurrence.

How do these concepts relate to our understanding of communication mechanisms? To begin with, it is clear that an analysis of communication mechanisms at the level of the individual animal must include consideration of both the production and the reception of signals; this is also clearly implied by the notions of "transmitter" and "receiver" in Figure 10-1. But there is also a contrast to be made with the scheme shown in the figure, which is that with these present definitions we are not going to be concerned with the concept of "the message," except in so far as it is represented by empirically determined relationships between the reference and the social significance of a signaling act. In other words, if we can specify both the circumstances under which an animal will perform a signaling act (the reference of the act) and the consequences of that performance on the behavior of other animals (the social significance of the act), and if we can show that there is a regular relationship between the reference and the significance, then that relationship *is* "the message" of the act.

This way of looking at animal communication attempts to account for what we would like to mean by "information transfer" without making unwarranted assumptions about what the animals are intending to do. Even so, there is a great deal that goes on between animals for which this approach fails to account. For example, animals possess many attributes or features which figure in their social interactions as regulators but which do not count as signaling acts. Relatively stable features of an animal such as body size, coloration, tooth size and shape, and special markings of various sorts, take on significance within the context of social interactions. But these are attributes over which the animal has no "control" in the usual sense, hence they are not signaling acts in themselves. However, they often appear as important features in signaling acts, as for example when a male baboon throws his head back and retracts his lips in a way that exposes his giant canine teeth or when a peacock spreads his tail. In these and many other cases, signaling acts serve to expose, emphasize, or exaggerate nonbehavioral attributes of the individual.

Within the ethological vocabulary introduced in Chapter 1, a nonbehavioral attribute that figures prominently in a signaling act is classified as a sign stimulus (releaser) if the social significance of the act is species-specific and relatively fixed within the species. The relationship between the reference and social significance of a signaling act featuring a nonbehavioral attribute, i.e., the "message" of that act, can be accounted for only by taking into account the evolutionary development of the nonbehavioral attribute. As Lorenz (1958) has pointed out, the evolutionary status of species within a taxonomic family can be understood by reference to differences between species in the relationships of signaling acts to nonbehavioral attributes featured in the acts. For example, markings under the wings of ducks are related in interesting ways to the use of wing motions in duck courtship displays. Only by knowing the evolutionary history of the species in question can one have a complete appreciation of the total "message" involved in signaling acts. We will return to the importance of evolutionary development in understanding communication later in this section.

Research Strategies and the Interrelation of Individual and Populational Approaches

We have already suggested that the idea of "mechanism" as it applies to communication systems takes on different meanings with different levels of analysis. To begin with, when we inquire about mechanism we are always asking, "How does it work?" In trying to answer that question there are several alternative approaches available. We will consider three approaches and how they are related.

As with any unknown object or system about which we are curious, we may study the mechanisms of communication simply by observing, unobtrusively, the workings of the entire system in its natural setting. This approach can provide an appreciation of the linkages between social and nonsocial environments, i.e., the ecological factors at work in regulating the references and social significances of signaling acts. In particular, this approach fosters an understanding of how physical features of a signal relate to its reference and social significance. Consider, for example, the vocalizations of some small birds in the presence of a hunting hawk overhead. The sounds emitted by the birds are of intermediate frequency (pitch) with gradual onset and offset. The physical properties of this vocalization are such that localization of its source is difficult. That is to say, even if the hawk should be able to hear the call, it would not be able to tell, easily, where the bird was who was making it. However, this ventriloquial property of the bird's "alarm call" does not prevent other small birds from taking action to evade the hawk. Thus the many thousands of years of selection for this signaling system have led to an optimal relationship among the signaling act's external reference (hawk overhead), its physical form (vocalization of the sort described), and its social significance (evasive action by other small birds) (Marler, 1967a).

If our appreciation of the way communication systems are constructed is restricted to naturalistic observation, we may easily make mistakes in our interpretation of how the system works. In particular we are likely to fall into our human "egocentric trap" and imbue the animals with our own ways of doing things. In order to avoid this problem it is required to do experiments, i.e., to manipulate the system in ways that allow its workings to be shown clearly. For example, when a baby chick is cold or out of sight of its mother it may vocalize repeatedly with a characteristic brief sound of relatively constant loudness and decreasing pitch. If in addition the chick's leg is tied by a string to a post it will struggle and move about; together, the sight and sound of the chick convey definite "distress" to the human observer. The mother hen will normally approach the struggling, vocalizing chick. If a barrier is placed between the hen and the chick the hen will try to get around the barrier. However, if a clear bell jar is placed over the chick to prevent sound from escaping, the hen will not approach or attempt to aid the chick (Tinbergen, 1951). This result is not predictable in advance on the basis of human experience nor would it likely become unequivocally clear during the course of naturalistic observation. Our appreciation of the workings of communication mechanisms has been deepened by the conduct of the experiment: the "distress call" vocalizations of a chick have more social significance for its mother than do the associated movements of its body.

At the physiological level of inquiry we look for the "hardware" that must in some way correlate with the production and reception of signals. It is at this level of analysis that our common usage of "mechanism" seems clearly applicable. Physiological analysis can be very useful in answering questions that have both behavioral and evolutionary relevance. For example, it is now known that one of the reasons attempts to teach chimpanzees to talk have failed is that the vocal apparatus of the chimp is inadequate for the kind of vocal control human language requires; a chimpanzee can do much better at communicating with people, and vice versa, when both species use manual sign language (Gardner and Gardner, 1969).

Thus, the study of mechanisms has several levels of analysis. But even when the understanding furnished by these different approaches is maximized through their combined pursuit, they cannot provide the deepest levels of understanding; for though they can tell us *what* these mechanisms are and how they work, they cannot, in themselves, tell us *why* they exist. For example, it is conceivable that someday we will learn the biochemical details of the physiological processes responsible for a foraging honey bee's signaling actions upon her return to the hive, and perhaps we may achieve a similar account of the tranquilizing effects which result when a dominant male chimpanzee backs up to the elevated rump of an agitated subordinate. But knowing *how* the responsible mechanisms operate will not tell us *why* they are parts of the lives of honey bees or chimpanzees. To account for the existence of these phenomena, and thus to reach the furthest goals of biological inquiry, requires that we be able to construct an adequate understanding of the *development* of the mechanisms, in the two senses of development this term always carries in a biological context: development of individual organisms (ontogeny) and of populations in evolutionary time (phylogeny). Our deepest understanding of communication mechanisms, as of all other biological phenomena, comes from achieving an adequate account of the processes by which these mechanisms have arisen, both in the individual organism and the population of which it is a member; for it is these processes that are ultimately responsible for the structural and functional attributes of mechanisms themselves.

Developmental Questions: The Core of Any Analysis of Communication Mechanisms

We need briefly to explain the emphasis we have placed on the importance of developmental aspects of communication systems. During the last several decades our understanding of the biochemical foundations of genetics has increased to the extent that we may identify the structural arrangements within the DNA molecules of an individual with the strategy for that individual's persistence in the environment. More broadly, the modal or typical features of the genetic organization of a species may be said to constitute to the species' *design for existence.* Evolution consists of changes in this modal design by selection from the variation that the population expresses in the environment.

These ideas are both exposed and beautifully elaborated in the books by Beadle and Beadle (1966) and Monod (1971). For our present concern, the most important point is that the strategies or existence designs contained in genetic material are, by necessity in complex organisms, primarily directed at the processes of individual development. If a species' mode of adaptation is complex, involving a wide array of organ systems integrated within each individual, then necessarily the largest part of the genetic information required for the execution of that life-design will be structured to steer the course of development from the single-celled zygote to the adult reproductive form. Thus, a population's design for existence is first and foremost a design for the development of its individual members. Consequently, evolutionary changes in its design are primarily changes in programs for individual development. This is why the relationship between the history of the species (phylogeny) and of the individual (ontogeny) is so intimate, and also why populational as well as individual perspective must be maintained in order to understand behavior in general and communication in particular.

Variations in Ontogenetic Strategies

In addition to this general point about the evolutionary significance of individual development, there is another reason to emphasize its centrality for the understanding of communication mechanisms. In order to understand communication mechanisms it is imperative to understand how they are acquired through the interaction of the developing individual with its environment. There is a good deal of diversity among species in the extent to which particular environmental events are critical in the development of species-specific signaling acts. Very good examples of this variation may be found among songbirds. In some species, for example the song sparrow (*Melospiza melodia*), the development of the species-typical song does not depend upon the individuals hearing other sparrows; the song sparrow does not need an "external model" upon which to base his own song. Even if a song sparrow has been foster-raised by a canary it will perform a completely normal song sparrow song (Marler, 1967b). Similar independence from the environment appears among pigeons and doves, domestic fowl, and the European cuckoo (*Cuculus canorus*) (Thorpe, 1972). This last case is particularly interesting because the cuckoo is a nest parasite on several other species, yet the male cuckoo's song is not influenced by the calls of his foster parents.

The chaffinch (*Fringilla coelebs*) provides a different sort of example. If a chaffinch is deafened surgically during its first year of life it will never produce the species-typical song. Although a few of its notes will be somewhat "chaffinch-like," the basic structure of its vocalization, its "sound skeleton" (Thorpe, 1972), is distinctly different from the normal. If, however, the bird is allowed to hear and practice its song as an adult before it is deafened, then the deafening has no effect. As an intermediate case, if the bird is deafened after hearing normal chaffinch song but before vocalizing itself, its resulting vocalizations have the basic sound skeleton of the species but lack the fine-grained details of the normal song. Thus the chaffinch presents a case in which there appears to be something like an innate "template" for species-typical song, but in which the template must be activated and refined by particular external stimuli. Finally, we may mention the bullfinch (*Pyrrhula pyrrhula*), whose vocalizations depend virtually entirely upon imitation of the songs of its father. If this finch is foster-raised by canaries it will adopt canary song completely (Thorpe, 1972).

These examples among a few species of birds are sufficient to show that the interaction of genetic programming and particular environmental stimuli in the development of communication mechanisms takes several forms. One important task will be to correlate these differences with other differences among the species; but this work has hardly begun. As the work proceeds it will be very important to distinguish between environmental "fixity" and genetic "fixity" in the production of species-typical signaling acts. In other words, lack of variability in the signaling acts of a population may reflect lack of variability in individual environments during development as well as strict genetic programming of behavior. Only careful experiments will sort these factors out. Also, the reference and social significance of a signaling act may differ in regard to the importance of the environment in their development. Similarly, we may expect that in some cases the signal will develop normally in the absence of normal stimulation but the reference of the signal will be highly dependent upon the environment. A clear example of this is human smiling. Congenitally deaf and blind infants smile, but the final control of smiling in the normal adult depends upon a number of subtle individual and cultural factors. Moreover, the social significance of a smile, i.e. how others respond to it, depends in great detail upon the social and environmental context in which it occurs (Eibl-Eibesfeldt, 1970; Kagan, 1968).

The Channels of Animal Communication

We have already provided examples of signaling acts in the several sensory modalities or channels: visual, auditory, tactile, and olfactory/gustatory. The point we wish to emphasize now is that, in general, communication mechanisms *per se* do not evolve completely separately from the particular sensory specializations and behavioral competencies that figure in the species' nonsocial adaptations. The channels over which social communication takes place in a species will tend to be the same channels that the species uses most effectively in its commerce with the nonsocial environment. Our best conjecture about the evolutionary development of communication mechanisms is that they have derived mainly from nonsocial behavioral adaptations which have been brought into the service of social function but which may still be contributing to their original, nonsocial adaptive purposes (Andrew, 1963). The importance of olfaction in the social and nonsocial adaptations of canines and some other mammals and of vision in the social and nonsocial adaptations of birds provide two basic examples of the correspondence we have in mind.

Other constraints or pressures upon the evolutionary development of communication channels include the nature of the local habitat and predator pressures. For example, we would not expect signals in the visual modality to predominate in animals whose habitats prevent them from seeing each other clearly, e.g., deep or muddy-water fishes, nocturnal animals in general, or animals that move through thick cover. Similarly, auditory signals which would aid detection by a predator will normally be selected against. As we pointed out (p. 330), the alarm calls of some bird species are adapted so as to minimize their localizability.

As a final point about the idea of a communication channel, we need to emphasize that, at least for mammals and in particular primates, signaling acts often operate simultaneously in more than one modality; in other words there is redundancy in the signaling act. Indeed, it is the lack of redundancy in, for example, the distress behavior of the chick, which surprises us; we expect on the basis of our human experience, that *both* the sound *and* the sight of the distressed chick should affect the mother hen. Among nonmammalian vertebrates and invertebrates strict channelization of communication is more likely than it is among mammals. Moreover, within a single channel, and particularly in the visual channel, there is often such a continuity to social interaction that the exact specification of what constitutes *the* signaling act may be difficult. For example, when an aroused subordinate chimpanzee approaches the dominant animal in circumstances appropriate for rump-pressing behavior, many aspects of the subordinate animal's behavior will be correlated with his status and emotional condition. In addition to the general direction of his locomotion, i.e., toward the dominant animal, there may be a halting character to his gait or circuity of his path of approach. His carriage may be more crouched than usual, he may cast only rapid, fleeting glances at the dominant animal during his approach, he may exhibit facial grimaces, and the hair around his neck and shoulders may stand erect. All of these actions can serve as signaling acts within the visual channel. Additionally he will engage in behavior which may have social significance in the auditory channel, and there may be characteristic odors associated with his agitated state. Unless and until the appropriate experiments are conducted to sort out the relative importance of all these potential signals, we cannot be completely certain about which have social significance and which do not. But, in general, we may suspect for adult primates that no single channel or highly specific signal is absolutely essential for the elicitation of appropriate behavior. It is the plasticity of response and the ability to generalize or synthesize signals which characterize primate adaptations.

The "Emotional" Organization of Animal Communication Mechanisms

We have defined the reference of a signaling act to be the events or processes governing the act's occurrence. It is relatively easy to describe the reference existing outside the signaling animal, e.g., a hawk circling overhead as reference for the alarm call of a smaller bird. But in addition we should include the internal condition of the animal prior to its performance of the signaling act as part of the act's reference. Normally, this condition also determines other subsequent behavior by that animal; hence the signaling act also serves as an indicator or predictor of the signaling animal's next action. In fact the signaling act is part of the larger action sequence it serves to index, and its physiological mechanisms are part of the mechanisms underlying the larger action sequence. Brown (1969) has made this point clearly in summarizing a number of studies of auditory signaling by birds: "The controls of vocalization are the controls of sexual, agonistic, and other types of behavior in which vocalization plays a role" (p. 93). And Ploog (1971) has made the same point in regard to the vocalizations and sexual behavior of the spider monkey: ". . . almost all [neural] structures which are known to mediate vocalization are also known to yield other behavioral or autonomic responses including genital responses" (p. 110).

These generalizations should not be taken to imply that signaling acts are related to other behavior as parts of a completely fixed system. Among primates in particular, the relations between the mechanisms controlling signaling and other concurrent actions are often quite fluid; Robinson (1967) has shown this clearly using macaque monkeys. Nevertheless, the work of Delgado (1966; 1969), which has involved remote stimulation and recording of activity in the brains of freely moving monkeys, has shown there is a general thematic relationship between signaling acts and accompanying behavior. Delgado interprets his data to mean that the signaling acts and accompanying behavior are expressions of a superordinate emotional state produced by the immediate environmental situation. In other words, these vocalizations index (have their reference in) the emotional status of the animal. The emotional character of animal signals has been a long-favored interpretation (Darwin, 1872), but Delgado has been foremost in advancing a detailed proposal for "some special cerebral mechanism [which] must choose, link, and direct the sequences of appropriate fragments and structure the complete pattern of emotional response" (Delgado, 1966, p. 30).

One value of this approach is that it allows us to discuss the emotional components of the animal's response without becoming overly anthropocentric. However, when we attempt to use these ideas to understand the communication between animals phylogenetically far removed from ourselves, we may become somewhat uneasy. What can we say without straining, for example, about the emotional qualities of a male spider's courtship dance? What superordinate emotional state controls the dance of the worker honey bee on the wall of her hive? These notions also pose problems in dealing with our own interpersonal communication, for what our emotions *are* depends partially upon our cognitive constructions of them; human emotions are not defined entirely in terms of physiological states (Schacter, 1971). Nevertheless, no basic idea presently available serves as well to capture the nature of the organization of animal communication mechanisms, at least for vertebrate populations, as does the idea of emotional expression.

As a first approximation, then, we may summarize our understanding of *communication acts in vertebrates* by saying that they *are those components of the individual's expressed emotional arousal which produce significant changes in the social environments of other members of the population.*

The reference of the signaling act includes certain key features of the physiological condition of the signaling animal as well as the external events which stimulated that condition. Furthermore, the social significance of a signaling act is likely to be emotionally organized in much the same way as was the signaling act itself. In this way the social significance of a signaling act may itself become a signaling act. For example, an antelope may see or smell a lion and move briskly away. A second antelope may respond to the movement of the first by moving with it. Other antelopes in the vicinity may respond to the sight of the second animal by running themselves. The reference of the first antelope's action included the sight or spoor of a lion. Its social significance, the behavior of the second antelope, became the signaling act producing running in the other animals.

This example points out another difficulty with treating animal communication in terms of information transfer. For while it might seem reasonable to construe the running behavior of the first animal as transmitting the message "Run—there's a lion around here," that message certainly cannot be construed in the behavior of the second animal; its signaling act does not include the lion in its reference.

In concluding this section on mechanisms of communication then, we may say that animal communication "works" by the stimulation of emotionally organized responses. On the behavioral level of analysis, the notion of "information transfer" is more likely to be misleading than helpful. It can be replaced by the analysis, at several levels, of the references and social significances of signaling acts.

Animal Signaling Acts and Human Linguistic Performances

In the previous section we emphasized that when examples of animal communication are analyzed on the behavioral level, taking note of the particular actions of each participant in specific episodes, the socially significant actions are usually only part of the total flow of each participant's activity. Animal signaling acts provide indices or clues to the signaler's current behavioral dispositions, and often they elicit specific changes in the patterns of action of their recipient's behavior. This altered behavior may itself result in signals reciprocally affecting the initial signaller or other animals in the population. We have concluded that the most adequate interpretation of the relationship between the behavior of the signaler and the recipient of the signal is in terms of the effective integration of the two animal's behavior to meet and exploit the challenges and advantages of group life. Neither the signaler nor the recipient need benefit individually from the signaling exchange; the benefit accrues to the population.

There is relatively little in the ongoing stream of behavior of animals that bears resemblance to such signaling acts as hoisting a flag or posting a sign. Where such parallels do occur, for example in the marking of territorial borders, the interactions are usually between animals isolated from each other over space or time. But even these instances resist our ordinary interpretation of what, on the surface, appears to be their counterparts in human affairs, particularly in regard to the kinds of information that might be said to be conveyed by such signals. Even the waggle dance of the bees, as impressive an evolutionary accomplishment as it is, lacks the plasticity and inventiveness of human conversation. However, we ought to be able to get very clear on what we mean by "plasticity" and "inventiveness" in this context. Just how does ordinary human talk differ most importantly from a worker bee's dance on the comb of her hive, or from any other nonhuman signal? In this final section of the chapter we will develop some ideas which we hope will make these differences clear. Before, proceeding, however, we need to emphasize that our concern is with the contrast between ordinary human languages and ordinary animal communications; we will not deal with human attempts to teach other animals to utilize forms of human language (Gardner and Gardner, 1969; Hayes, 1951; Kellogg and Kellogg, 1933; Premack, 1971). It is important for our present purposes to distinguish between what animals do in their normal environments and what they can be brought to do after prolonged association with clever people. In other words, we will be concerned here to contrast human linguistic behavior with animal communication systems as they have naturally evolved. Several other authors have discussed differences between the conduct of human affairs and the social interactions of other animals (see for example Adler, 1967, and Eisenberg and Dillon, 1971). Our approach to the problem will be based primarily on the ideas developed already in this chapter, in particular the section on mechanisms. We will begin with an examination of some common sense ideas people have about each other.

The Concept of Understanding

Most of us believe that the adults around us are responsible for their own behavior most of the time. Another way to put this is to say that we typically assume that each person is, somehow, himself the controlling agency of his own acts. It is this aspect of the behavior of other people that we have in mind, for example, when we ask someone whether he "meant to do" or "intended to do" what he did. If you have just stepped on my toe I am interested to know whether you did it "on purpose" or "by accident." Normally, then, for things people do which are not accidental, we assume that they are aware of their actions. To put the matter slightly differently, we assume that people *understand what they are doing*. People who never or seldom possess this kind of ordinary understanding are considered "irrational," and we treat such people very differently than we do "ordinary" people. For example, we do not hold them responsible for their actions as we would someone whose understanding or rationality is not in question.

Now we need to inquire a little more deeply into this notion of "everyday understanding." What are we assuming about people when we assume that they possess this understanding, and what kinds of information from and about people do we rely upon to justify our assumption? To begin with, we need to emphasize that understanding is not, itself, behavior; rather it is a feature or characteristic of a person who behaves in particular ways. It is a feature of the person's mental or cognitive life. Moreover, we ordinarily assume that action is *guided by* the understanding, that it plays some role in determining what we do. We act as we do partly *because* of our understanding of the situation we are in. If our understanding were different, our behavior would change.

We must also note that this understanding is not thought to be present and participating in determining everything we do. Sometimes a person might say, for example, in apologizing for an action of his, that he had "just not been thinking" about what he was doing. Thus, we ordinarily allow that even though one possesses the requisite understanding of the significance of one's actions, and normally one's actions are guided by that understanding, occasions will nevertheless arise in which this understanding does not participate in determining what one does. Thus, I may say of a particular incident that I should have acted differently but I simply could not help myself.

In general, then, our ordinary appreciation of human conduct includes the assumption that adults act with cognitive understanding of their actions. The presence of understanding is what separates rational action from nonrational, nonresponsible action.

When we are in doubt about whether someone possesses ordinary understanding of his actions at a particular time ("knows what he is doing"), we tend to ask questions like "Why are you doing that?" or "Just what do you think you are doing?" If his answer gives us what we take to be a *meaningful interpretation* of the behavior and its context, then we accept that the person has acted with understanding. It is important to note that our acceptance of understanding in another person does not necessarily depend upon our agreeing with his interpretation. In this sense an interpretation can be meaningful without being correct, deep, or, in general, in line with our own understanding of the situation. It need only be generally coherent and not wildly at variance with reality as we have conceived it. The critical feature of the other person's interpretation is that it is in a form which we can share. We rely upon the *sharability* of the other person's interpretations as an indication that the other person is acting with understanding.

Our inference that someone is behaving with understanding depends upon more than the behavior itself, regardless of how effectively the behavior accomplishes some end or goal. We may have reason to doubt the soundness of a person's own interpretation of what he is doing and why he is doing it if he does not persuade us that he is aware of some feature or consequence of his action which is significant to us. We may decide that, however effective the person's actions, and in spite of his own testimony, the person's own understanding is so faulty in a particular instance that it could not have been involved in guiding his actions in the way he tells us.

Here is the other side of the same coin: we may possess understanding of some behavioral performance without being skilled in that performance. For example, I may know the theory of the proper golf swing perfectly well, and be able to tell you all about it and even correct your swing, and yet not be able myself to perform the swing correctly. In general, there is no *logically binding* relationship between our behavior and our understanding of it; if there were, we would not need to ask each other, ever, if we understand what we are doing. It is our talk about nonverbal behavior, not the behavior itself, that is indicative of our understanding.

Let us summarize what has been said about the concept of ordinary understanding. Understanding is constituted by an interpretation of one's present situation and of the consequences of the action one might take in that situation. It is not logically linked to the behavior which it interprets. It is fallible, i.e., it can be incorrect. But it must be sharable, i.e., it must be in a form which is potentially communicable to other people.

There ought to be nothing in the foregoing description of understanding that strikes the reader as novel. Indeed, we have tried to describe the fundamental, common-sense view of the relationship between how we behave and how we talk about how we behave. These common-sense notions have received enough confirmation in the experience of many, many human generations to have become deeply ingrained parts of all human cultures. In fact, the persuasiveness and pervasiveness of these ideas have prompted a strong tendency to extend them to account for other natural phenomena. Many of us are quite prone to personify natural objects and events such as the wind, rivers, the sea, trees, and so on. Indeed, our ordinary Western idea of "cause" as a push, pull, or other kind of force, is sometimes used in everyday language as if impersonal events "knew what they were doing" in producing effects. For example, the sentences "The wind knocked over my mailbox" and "The vandals knocked over my mailbox" have externally the same form. But these sentences are in fact very different, as can be appreciated by considering legitimate answers to the question "Why?" in each case. A cogent discussion of the relationship between "natural" and "personal" causation may be found in de Charms (1968).

"Knowing How"

The tendency to personify nonhuman objects and events is particularly obvious in our dealings with nonhuman animals and their behavior. We have dealt with this topic earlier in this chapter and also in Chapter 1. At this point we want to concentrate on the concept of knowledge, and in particular the sort of knowledge that we have in mind when we say of someone that he "knows how" to do something. We will call this sort of knowledge *performance knowledge*. We will be more explicit about the meaning of this term in a little while. For now, it is enough to rely upon the common understanding of the phrase "knowing how" to proceed with the argument.

To what extent might we be justified in attributing performance knowledge to nonhuman animals? There is a sense in which the behavior of animals might be considered to indicate performance knowledge. We might say, for example, that spiders know how to spin webs and that birds know how to build nests. We would make these statements on the basis of the fact that spiders do in fact spin webs and birds do in fact build nests. Moreover, we might allow that failures to perform adequately are indicative of a lack of performance knowledge, particularly if the failure occurs in an immature animal. Thus we might say of a lion cub that it does not yet know how to hunt effectively or of a child that it does not yet know how to feed itself.

A problem with using the phrase "know how" in this way arises when the only reason one uses it is that the animal or object in question does in fact perform adequately. If adequate performance itself is the sole criterion, then one can also say that people know how to breathe, DNA molecules know how to replicate themselves, computers know how to add numbers, and rain knows how to fall. In other words, if the inference about the existence of performance knowledge depends completely upon the success of the performance itself, as usually observed, then it is a gratuitous inference. Under these conditions, to say that something or some organism possesses performance knowledge is to say no more than that it performs reasonably well under certain circumstances. This is an empty use of the word "knowledge," but it is harmless enough.

Throughout the history of the study of animal behavior there have been several attempts to make more substantive claims for animals as possessors of knowledge. In general, these attempts start from the assumption that the individual animal (usually conceived as the "typical" mammal) needs somehow to represent the external world, or map it in some way, within its nervous system, and to be able to "work with" this representation, if it is to survive. The details of these theories of inner representations of outer events, and the technical languages in which they are couched, are what separate the various schools of theorists.

For example, one of the earliest papers of the famous neo-behaviorist Clark L. Hull (1884–1952) began as follows:

One of the oldest problems with which thoughtful persons have occupied themselves concerns the nature and origin of knowledge. How can one physical object become acquainted with the ways of another physical object and of the world in general? In approaching this problem from the point of view of habit, it is important to recognize that knowledge is mediated by several fairly distinct habit mechanisms. (Hull, 1930, p. 511)

Following this introduction Hull went on to develop a theory of one such habit mechanism, namely the mechanism by which the early portion of a series of environmental events can come to elicit an inner representation of the whole series of events. In other words, Hull emphasized the importance of the nervous system's ability to remember a sequence of stimulation by repeating its earlier responses to the entire sequence when the first part of the sequence is presented alone. The nervous system's ability to perform this operation, which is called *redintegration,* was in Hull's opinion the foundation for the animal's knowledge of the world. It was on the basis of this process that animals could learn to anticipate future events and behave adaptively with regard to them.

Hull attempted to build this theory of knowledge in animals without reference to the everyday vocabulary of knowledge and understanding that we typically use in describing human activities. Other psychological theorists have been decidedly more "cognitive" in their approach to nonhuman animals, i.e., they have freely used terms ordinarily applied only to the mental lives of people. Edward C. Tolman (1886–1961), a contemporary of Hull's and for some years his major competitor for the leadership of neo-behaviorism, entitled his major work *Purposive Behavior in Animals and Men* (Tolman, 1932). Tolman and others who worked in his tradition were very willing to apply terms like "expectancy," "hypothesis testing," "vicarious trial and error," and other decidedly cognitive terms to operations presumably occurring within the brains or minds of white rats as they traversed mazes. Contemporary critics of Tolman criticized his theorizing on the grounds that it left animals "lost in thought" in the middle of the maze. Several other brands of theory were also devised to account for changing conduct of rats in this and similar circumstances. Everyone agreed that rats did in fact reduce the number of errors and increase their speed through the maze; there was far less agreement about the best way to account for this improved performance, this "learning how." But all of the serious theories presumed some form of internal representation of the external world, whether labeled "stimulus-response chains" or "cognitive maps." Further, it was the potential flexibility or generality of these inner representations that were generally assumed to be the mark of the animal's intelligence.

A modern version of this way of approaching animal conduct has been provided by Mason (1970), who uses the term "schema" (plural "schemata") to describe an animal's internal representation of its present environmental situation. According to Mason, an animal's schemata provide it with a basis for anticipating the consequences of its own actions; hence they serve to organize the animal's activity. If a particular schema is adequate, then the behavior that it organizes will be adaptive.

The reader will recognize that up to this point the notion of schema might fall under the same charge of emptiness that "performance knowledge" does, if all we mean by the term is that animals behave appropriately under particular circumstances. Although the postulation of some entity called a schema might lead to research aimed at characterizing its neurophysiological properties, this research could equally well be generated by a concern to describe the physiological correlates of adaptive behavior; the word "schema" itself really adds nothing at all except perhaps a slogan around which to gather. However, Mason's notion of schema contains more than its simple relationship to behavior, and hence it is saved from emptiness. In particular he suggests that schemata are characterized by greater or lesser degrees of "openness," which he defines as " . . . the ease with which [a schema] can be modified by experience, the extent to which [it] can incorporate new information" (Mason, 1970, p. 30).

"Openness" may operate in several ways, for example by the elaboration or reorganization of a schema, or changes in the activity which the schema organizes, or a process by which several schemata become organized into a single superordinate schema. What is most important for our present concerns is Mason's proposal that one indication of "openness" is an animal's ability to achieve the same end by more than one behavioral means. According to this proposal, which is similar to the concept of "multiple means-end readiness" suggested many years ago by Tolman (1932), we may infer the presence of a schema whenever an animal demonstrates skill in effecting a particular environmental change in more than one way. If an animal can achieve the same outcome by two or more behavioral pathways, and particularly if the animal can meet the challenge posed by a change in the environment so that novel means of "solving the problem" must be found, then there are grounds for claiming that the animal possesses a schema for achieving an adaptive result in the environment. In this case we would feel justified in claiming, without fear of logical circularity, that the animal "knew how" to achieve a particular result.

There is a very large experimental literature that could be combed to determine whether various animal species do in fact operate with such "openness" that we want to claim that they "know how" to achieve certain results; for an introduction to the literature on primates in particular, see Riopelle (1967). It is neither appropriate nor necessary to review that literature here; but it is important to point out the apparent similarities between cognitively-based theories of animal behavior, which we will call theories of performance knowledge, and the common-sense notion of human understanding which we introduced earlier. First, both sets of ideas suppose that the determinants of behavior include neurophysiological conditions corresponding to cognitive states which register significant features of the environment (including feedback from one's own behavior) and provide bases for anticipating the consequences of action. Second, both sets of ideas suppose that these cognitive states are subject to elaboration and transformation as a result of experience, and that, because of these elaborations and transformations, they acquire increasingly greater influence in the organization of behavior.

These similarities between the two concepts are not sufficient to show that they are identical. We will now consider some differences between them.

"Knowing How" and "Understanding"

To begin with, let us review briefly what is involved in the idea of performance knowledge. We infer the existence of performance knowledge from the observation of successful or competent performance.

"Success" implies that the performance is efficiently executed under at least some variation in environmental conditions. Thus, if a young bird's attempts at flight keep it aloft only in gentle winds, or if a child can ride its bicycle only in a straight line but cannot yet turn without falling, then we say that the bird does not yet know how to fly and the child does not yet know how to ride its bike. Success, then, as a criterion of performance knowledge, is at least the ability to "do the same thing" under several sets of circumstances.

It is this consistency of performance under varying conditions which leads us to infer the existence of a guiding or controlling cognitive scheme, i.e., performance knowledge. In particular, our inference of performance knowledge is strongest when the chain of action is completed even under markedly different environmental conditions. Extreme examples of this flexibility of performance by animals have in the past been called "insight," a term not nearly so popular in psychology as it once was. The classical example of this sort of performance is the behavior of the captive chimpanzee named Sultan who was studied by the Gestalt psychologist Wolfgang Köhler during the years of the First World War. Sultan was accustomed to gathering pieces of fruit into his cage by putting a stick out through the bars and dragging the fruit back. Köhler wanted to determine how flexible Sultan could be in acquiring different ways to bring fruit into the cage. He gave Sultan two sticks, both hollow, with one slightly slimmer than the other so that the two might be fitted snugly together to make a long pole. Then he put the fruit so far away from the cage that Sultan could drag it in only if he put the sticks together. When first he encountered this new situation, Sultan stuck one stick, then the other, through the bars toward the fruit; of course neither would reach it. His next maneuver was to push one stick toward the fruit, let it go outside the bars, then push the other stick against it until the tip of the first stick touched the fruit; this led to gesture and motions on Sultan's part which Köhler interpreted as a sense of satisfaction. At a certain point, when it appeared Sultan would get no further in the solution of the problem, Köhler "gave him a hint" by himself sticking a finger into the end of the larger stick; he did this while the Sultan was looking at him. However, an hour after this attempt at instruction, Sultan still was without fruit. But he did not leave the sticks alone; he sat in his cage and manipulated them in various ways. Finally, after some time of this activity, Sultan put the two sticks loosely together. Shortly thereafter he stuck his new pole out after the fruit. The pole fell apart before the fruit could be dragged in; after several attempts, however, Sultan succeeded in bringing in the fruit. From that point on for the rest of the day, Sultan dragged in many objects with the longer pole, including stones and sticks as well as fruit. The next day, after a brief period in which he repeated the ineffective maneuvers of the previous day, he once again put the sticks together and got the fruit (Köhler, 1925).

Extended laboratory investigations of chimpanzees could very well demonstrate that the animals can master several additional means of achieving the goal of bringing the fruit from outside the cage to within it. Fortunately, however, we now have information based on the behavior of chimpanzees in natural habitats that bears on very much the same point. Van Lawick-Goodall (1968) has reported on extensive use by chimps of sticks and blades of stiff grass as implements for the collection of ants and termites to be eaten. In addition to being able to use the stick once they have acquired it, chimp adults also strip vines or leaves to produce an implement of the proper size. Van Lawick-Goodall also reports on the behavior of young chimpanzees in this regard. She observed them on many occasions to make clumsy approximations of the successful behavior of adults. The animals may start with this kind of behavior as early as during their third year, but they do not show complete competence for several years thereafter. In her words, it takes the chimpanzee several years to "learn how" to gather ants and termites using an implement. Adult chimpanzees may be said to "know how" to do this. This kind of "know how" leads us to believe that the chimp possesses some kind of scheme, which guides or directs its insect-gathering activity. Moreover, the process by which young chimpanzees acquire this performance knowledge involves activity apparently similar to what we normally call *play* in human children. As we shall see, the concept of play figures strongly in our later discussion.

Now we need to ask how the performance knowledge of the chimpanzee, and other nonhuman animals, differs from ordinary human understanding. We can get at that by recalling the properties of understanding given above and contrasting them with the properties of performance knowledge.

First, consider the relationship between the performance knowledge and the relevant performances: *performance knowledge is logically related to the success of the performances which it controls.* We have already pointed out that the concept of performance knowledge is empty if it refers only to an animal's competent performance of one act in one way. If, however, the animal reaches the same end point in different ways at different times, then we may legitimately credit the animal with performance knowledge in regard to achieving that end point. Nevertheless, it is, *always and necessarily,* behavior instrumental to achieving that end point which is used as the criterion for performance knowledge. This is what we mean when we say there is a *logical* connection between the success of the performance itself and performance knowledge.

This is one of the ways in which performance knowledge differs from understanding. For as we have pointed out, *understanding is not logically related to the performances which it guides.* The criteria for understanding do not necessarily include the achievement of particular behavioral end-points. Thus we can all agree that someone can understand some activity, for example unpowered flight, without performing it success-fully.[2]

A second, related distinction between performance knowledge and understanding may be made in the following way: *performance knowledge is infallible, while understanding is fallible.* That performance knowledge is never "wrong" follows from its logical relationship to behavior; precisely because the criterion for the possession of performance knowledge *is* behaving successfully, the performance knowledge *cannot* be incorrect. By contrast, our understanding can be inadequate, shallow, misguided, etc., and still be credited as understanding.

Third and finally, *understanding is always potentially sharable; performance knowledge is not.* There is no necessary relationship between knowing how to do something and being able to interpret your performance meaningfully to someone else, or even to yourself. In order to be a championship sprinter it is not required to be able to talk about sprinting so that someone else can understand it; nor need the champion be able to talk to himself about it. Potential sharability is a definitive characteristic of understanding, but is not a characteristic of performance knowledge.

These three differences between performance knowledge ("knowing how") and understanding are the heart of the difference between animal signaling acts and human conversation. Our argument is that understanding is unique to our own species, and that it derives from human capacities to develop the cognitive schemes which provide for the competent use of languages. In other words, we shall argue that the understanding which permits us to act knowingly and intentionally is made possible through the development of cognitive schemes that underlie our participation in linguistic interactions with other persons. Thus, although we

---

[2]It is interesting to note how these distinctions, and failures to make them, arise when psychiatrists and psychologists are challenged to "heal themselves." The experimental psychologist B. F. Skinner placed a discussion of this issue into the dialogue of his utopian novel *Walden Two.* The designer and chief of the community, Frazier, has been challenged for not having, in his own conduct, the characteristics which he prescribes for others in the community. Frazier responds to the challenge as follows:

> How much can you ask of a man? . . . . Isn't it enough that I've made other men likable and happy and productive? Why expect me to resemble them? Must I possess the virtues which I've proved to be best suited to a well-ordered society? Must I exhibit the interests and skills and untrammeled spirit which I've learned how to engender in others? Must I wear them all like a damned manikin? After all, emulation isn't the only principle in education—all the saints to the contrary. Must the doctor share the health of his patient? Must the icthyologist swim like a fish? Must the maker of firecrackers pop? (Skinner, 1948, p. 207)

We sense the problem being raised here: on at least some occasions we *understand* more than we are able to do, whatever the reasons for our inabilities may be. And, as we hope is becoming increasingly clear, understanding and linguistic performance (being able to talk) are intimately related.

have just demonstrated that understanding and performance knowledge are not the same sorts of cognitive schemes, we will now try to show how understanding may arise from a particular kind of performance knowledge, namely from our knowledge of how to participate effectively in our linguistic communities.

With the differences between performance knowledge and understanding clearly in mind, we need to examine how linguistic performances evolved so as to permit the development of understanding. In other words, how did the capacity for the development of linguistic competence arise? What are the properties of human linguistic performances that permit them to form the basis of human understanding? In the following section we suggest some answers to these questions.

The Evolution of Linguistic Performances

In our search for the sources of understanding we will begin by emphasizing some aspects of human zoology which are often underemphasized or ignored in ethologically oriented discussions of human behavior. First, we must never overlook the central importance of cultural factors in our ways of life and their evolution. *The single most generally valid characterization of our zoological status is that our survival has always depended upon cultural modes of adaptation.* We can no more leave this out of our consideration of any aspect of human existence than we could undertake a discussion of the zoology of whales and dolphins without reference to their status as aquatic mammals. At the risk of some overstatement, we will put the matter this way: cultures are the seas in which our species evolves.

A second human characteristic of prime importance is the set of physical adaptations on which is based our ability to manufacture and use tools and instruments. These implemental or instrumental adaptations, as we may call them, require particular consideration because we believe they are intimately related to our capacities for linguistic competence. A more complete discussion of these adaptations than we can supply here will be found in Tobias (1971); our discussion will be limited to considerations particularly important for the understanding of linguistic skills. Thus, human tool-making-and-using tendencies have demanded and been furthered by the evolution of a precision grip, high-resolution eye-hand coordination, and, very importantly, forms of behavioral organization which allow tools to be fashioned from raw materials having very little resemblance to the completed artifacts, and which provide for the transportation of such products from the place of their manufacture to the place of their use. For example, if a cutting instrument is to be used to open the carcass of a food animal, then a particular kind of raw material must be secured which has properties that permit it to be worked to form a cutting edge. Those actions involved in finding the raw material and those involved in the various stages of the work must be in part controlled by some current conception of the properties of the final product. That is, the many different actions required in finding suitable raw material and working it appropriately must be governed, over perhaps quite lengthy stretches of time, by factors which are not present in the immediate environmental circumstances in which those actions occur. Similarly, either the instrument must be brought to the animal or the animal to the instrument. The behavior involved in either case is different from the behaviors involved in killing the animal, making the tool, or using it.

From what we know about our evolutionary history, it is clear that there were often substantial distances between the location of material for tool building (stone of various sorts) and locations suitable for the tool's use. Of course these distances translate into a substantial time gap between the time of a tool's manufacture and the time of its use. Hence, in addition to the specialized behavior of carrying the tool from one place to the other, there needed to arise some mechanism for insuring that tool making would be maintained under circumstances in which it would not be "reinforced" except after a long delay, and not always then. One may contrast this situation facing early man with the tool making described above for modern chimpanzees, for in the case of the chimpanzee the lag in space and time between tool making and tool using is relatively short.

The only behavior patterns exhibiting comparable organization in other species are the activities of immature mammals, which apparently relate to functional forms of adult behavior. When we see this behavior in the young animal we call it "play," partly in recognition of the fact that it is not then functioning as similar movements normally do when they are performed by an adult; the behavior seems to be "just for fun." Possibly the development of play, which permits the expression of certain behaviors in "biologically idle" situations, was related historically to the development of instrumental adaptations. Several considerations lead us to offer this speculation. First, we know that a great prolongation of infantile and juvenile periods of development occurred in our history on the same evolutionary horizon as tool use and manufacture (Clark, 1967; Schultz, 1963). Second, we know that effective human tool use depends upon information transmitted across generations. The passage of traditions of all sorts is begun and in many cases completely accomplished before the new generation is mature. And third, the period of immaturity, with its attendant "play" under the relatively secure protection and nurturance of adults, offers the opportunity for practice of the refined movement that the eventual skilled use of tools requires. There is thus a close relationship among the several features of infancy and adolescence, including play, affectional development, observational learning, and imitation, and the skills required for tool manufacture and use (Aronfreed, 1969; Hall, 1963).

The next step in our argument is to relate these ideas about instrumental adaptations and play to the development of our linguistic competence, i.e., the way we come to "do things with words."

Similarities between Instrumental Adaptations and Linguistic Competence

The mechanisms subserving instrumental adaptations and linguistic competence share several fundamental properties. First, at the physiological level, both involve specializations of the nervous system which permit the organization of highly refined actions of groups of small muscles into intricate temporal patterns. For instrumental adaptations, the specializations involve fine-grained control of the wrists and fingers combined with eye-hand coordination, while linguistic competence requires the coordinated action of lips, tongue, diaphragm, etc. In both cases the neuromuscular coordination is complex to a degree found only within our species.

Second, at the cultural level, criteria of successful performance of instrumental acts (tool manufacture and use in the broadest sense) and linguistic acts are transmitted across generations as traditions. We acquire our manual and linguistic competencies as part of our cultural heritage, though it is our biological heritage which equips us to participate in the culture.

Third, the successful acquisition of manual and linguistic acts depends upon rather extended periods of "practice" during which the exercise of the acts need not fulfill the functional requirements demanded in adulthood. In other words, there is an extended period in which the vocalizations of infants and children are allowed to be "play"; over time the adult members of the community gradually increase the responsibility they attribute to the vocalizations of younger members. A person becomes responsible for what he says as he comes more fully to understand what he says.

Fourth and finally, the biological advantages of both our instrumental adaptations and our linguistic competence derive from their uses in the environment. This is obvious enough in the case of tool manufacture and use. The uses of our linguistic competence are also obvious in certain ways. But it is worth emphasizing that a major use of language is the encoding and transmission of information about arbitrary features of the environment: features which are as they are due to cultural traditions. The details of each natural language are arbitrary in this sense. For example, there is no necessary relationship between a kind of object (e.g., a horse) and the sound which "stands for" that object in the language ("a horse"). To the extent that traditions are arbitrary features of the environment, to that extent, at least, languages must be open to arbitrary manipulation. We will return to this important point later.

These considerations suggest that there might have been a direct evolutionary linkage between the behavioral specializations underlying technological and linguistic capacities. In both cases patterns of nervous action must be organized to afford a much greater "disengagement" of ongoing behavior from control by the immediate environment than appears in the organization of other kinds of behavior. We conjecture that the evolution of instrumental adaptations set the evolutionary stage in a way that made possible the subsequent specializations of linguistic capacities. This supposition is particularly appealing when we consider how the exploitation of the incipient development of these linguistic specializations could have amplified the biological advantages provided by tool use and manufacture as, for example, in making possible or more effective the kinds of social coordination required in the hunting of large animals during Middle and Late Pleistocene times.

We have so far been considering instrumental adaptations and linguistic competence as if they were similar, parallel categories; this approach has allowed us to clarify certain fundamental features of linguistic competence. Now we need to emphasize the interactions between these two features of human evolution. On the populational level of analysis, where it is possible to discern the biological functions of adaptations in a species' existence strategy, it becomes overwhelmingly apparent that the mechanisms associated with our linguistic competence and traditions serve to foster and strengthen cultural adaptations. In other words, language creates our cultures and binds us to them. It is the medium of transmission within and across generations of those arbitrary environmental features which in general define the capacity for human culture, and which in detail define differences between the various cultures that have evolved on different parts of the earth. Tool manufacture and use is one feature which is held in some form by all cultures but which differs greatly among cultures. The traditions associated with these instrumental adaptations are transmitted via language; language is the glue which binds generations of tool-users together and enables the cultural, and hence biological, possibilities of these technological capacities to be more fully exploited.

Characteristics of Linguistic Traditions and Actions

We have just emphasized linguistic activity as a critical feature of the processes which maintain tool manufacture and use over generations. The maintenance of these technological traditions, their diffusion into the repertoires of wider circles of peoples, their effective application in specific instances, and to at least some degree their invention, all depend upon the utilization of linguistic competence. From these obvious points there follows an important conclusion: *in order to accomplish cultural functions, linguistic activity must share the open, inventive, plastic quality that inheres in the invention, manufacture, and utilization of tools.*

Of course linguistic action supports and maintains more in the culture than technology. Indeed, much that we take to have the highest value in our culture, for example our religions, political philosophies, and so on, are completely dependent upon linguistic action for their expression and transmission. Much that is contained in traditions can take form *only* in language, in that no nonverbal behavior is suggested by it or follows necessarily from it. Linguistic activity thus "takes on a life of its own," quite apart from its specifically descriptive or other instrumental functions.

Many of the properties of linguistic actions that we have been describing could be brought together and summarized by saying that linguistic actions are *conventional.* This means that a particular vocal sound counts as a linguistic act if and only if speakers of a language count it as such. From this conclusion it follows that a description of muscular movements of the jaws, tongue, lips, etc., will not itself serve to characterize the linguistic acts of a culture. It is in part for this reason that we have, up to this point, used the relatively awkward phrase "linguistic acts" rather than the commonly used term "language." We want to save the word "language" to refer to the set of possible products of speakers which has conventionally accepted status. Linguistic acts, then, are the vocal acts which produce events conventionally accepted as "parts of the language."

One of the most obvious features of languages is that they change over time. New words and expressions come into the language and others drop out. These conventionally accepted linguistic products have a status which is different from the vocal acts which produce them, just as a stone axe has a status which is different from the movements of the manufacturer who created it. However, to a much greater degree than other artifacts of culture, the values of products of linguistic productions depend entirely on the conventions governing their usage.

Another fundamental characteristic of linguistic acts is their *hierarchical organization.* The products of linguistic acts do not possess obvious structural integrity; if you look at an oscilloscope tracing or sound spectrograph of human speech, you will see that the peaks and troughs of physical energy in the speech signal do not correspond accurately to the "units of speech" that you hear: parts of words, words, phrases, and so on. These units are, by convention, nested within each other in a systematic hierarchy.

Linguists have found that the most convenient unit by which to describe spoken language is the *phoneme:* a unit of speech sound that speakers of a language normally respond to as unique. The number of acoustically different or phonetically different sounds within a single phoneme may be substantial; for a simple example, compare the "p" sound in "pin" and "spin." The "p" in "pin" has an explosive quality (try blowing out a match with it) that the "p" in "spin" does not. However, in English *this difference is never used to encode different meanings in the language.* Hence, both "p" sounds are included under the single phoneme "p." In another language this difference may be important, and two "p" phonemes would have to be recorded (Miller, 1951).

It is important to emphasize the conventional nature of hierarchical organization. Although our speech may be conveniently analyzed into phonemes, as typical speakers of the language we do not construct our utterances by first picking out phonemes and then stringing them together. Neither do we analyze what people say to us by hearing all the phonemes. For example, one recent experiment demonstrated that the perception of phonemes is a more difficult task than the perception of larger language units such as syllables (Savin and Bever, 1970). Savin and Bever concluded from their results and other considerations that phonemes have a status which is nonperceptual and nonmotoric: we do not speak in phonemes nor do we hear them as single speech units. Yet the description of languages, both of their current states and of their cultural evolutions, proceeds most smoothly by means of phonemic analysis. Hence, in Savin and Bever's view, phonemes are *psychologically real but abstract.* In terms of the theme of this chapter, we can see that the analysis of human linguistic performances depends upon the utilization of a concept that is qualitatively different in its logical status than any required so far for the analysis of nonhuman communication systems.

So far we have characterized human linguistic actions, and hence languages, as conventional and hierarchically organized. We will introduce a third characteristic by pointing out that linguistic actions are *unlimited in regard to their possible combinations.* Because linguistic acts are hierarchically organized, smaller units can be reorganized to create novel larger units. Syllables can be placed together in different combinations to make different words, words can be placed together in different combinations to make different sentences, and so on. In any linguistic community, there is no limit to the number of such combinations that can have clearly recognized status within the community's conventions. This lack of limitation is actually most puzzling because of the very evident limitations on human perceptual capacities. If each of us had to learn our language by remembering everything we heard and associating it with the context in which we heard it, so that we could use it appropriately later on, we would soon confuse different utterances and the different contexts for their appropriate use. There have been many attempts to account for the means by which linguistic performance capacities somehow manage to bypass these limitations (e.g., Bastian, 1964; 1965; Lenneberg, 1967). Though none of these efforts have proved to be completely satisfactory, all agree in supposing that the key to the puzzle is related to the fact that only a few of all the conceivable combinations of linguistic units will be recognized as legitimate.

On at least a superficial level of description, it appears that there are always severe restrictions on the combinatorial privileges enjoyed by each linguistic unit, so that only some kinds of combinations are permitted. Those which are permitted provide for infinitely many combinations.

It is most difficult to adequately specify the basics of such restrictions, because they rest on quite abstract features of the ways the permitted combinatorial elements are interrelated. Further, these features cannot be discerned in the items considered in isolation. For example, we might be led to say that a verb may never be directly combined with another verb, thereby ruling out a combination such as "Some birds sleep eat together." But if we consider another, apparently very similar, combination such as "Some birds sleep walk together," we might save our initial description by deciding that there is one verb when we combine these particular words. However, this illustrates that we cannot decide which items are verbs by considering items by themselves. It also indicates that we cannot decide which parts of the combination are indeed combinatorial elements without considering the ways combinations may be interrelated to one another. In addition, examples of this sort should enable us to see the quite abstract nature of the relationships between utterances—relationships which we must be able to appreciate in order to recognize utterances that do not conform to basic linguistic practices.

At least some of the conceptual schemes underlying linguistic competence must incorporate logical complexities of this kind. But the most convenient and effective way to characterize such abstract regularities in the organization of linguistic utterances is, in general, to suppose that they are organized in terms of *rules* which specify 1) the ways in which abstract classes of linguistic units may be formed into permissable combination and 2) the ways in which one or more of such combinations may be transformed to yield still other varieties of permitted combinations. The convenience of accounting for the regularities of linguistic utterances in terms of such rules stems from the enormous range of linguistic organizations which may be encompassed in a most economical fashion. For this kind of account to be effective, the rules must be systematically related to one another so that the conditions under which a given rule is applicable are specified by other rules. With such a system, a fairly small inventory of rules may be sufficient to account for the regular features in the organization of the infinitely many utterances which are permitted in the language.

Following this general way of treating the regularities in the organization of linguistic utterances, we may thus suppose that one of the challenges which must be successfully met in learning how to participate in a linguistic community is the development of cognitive schemes which incorporate an appropriate set of rules for the formation and transformation of utterances. Once attained, such cognitive schemes would enable their possessor to appreciate the regular features in the organization of all the infinitely many utterances he might ever produce or hear in his language. It is this infinitely open character of the performance knowledge underlying linguistic competence, called its *systematic productivity,* which is the third attribute of linguistic actions that must be recognized in order to trace human understanding to its sources.

But before proceeding with that task we should point out that there is much that is unsatisfactory about this way of treating the systematic production of linguistic performance knowledge. In particular, we should be aware of the implicit or tacit nature of the rules which underlie the systematic productivity of linguistic capabilities. While this lack of explicitness conforms to our previous discussion of the nature of the cognitive schemes involved in any sort of performance knowledge, it makes most difficult the conception of additional properties that might be possessed by these cognitive schemes. It is not at all easy to suggest what the psychological and physiological attributes of these schemes might be, e.g., how they are learned and how they operate. These implicit rules in terms of which we have described the systematic productivity of our linguistic capabilities must not be confused with the explicitly sharable and logically simpler rules involved in games and polite conduct, even though there may be some reasons for thinking the two kinds of rules are somehow related.

Performance Knowledge, Understanding, and Language

Full participation in a linguistic community requires the utilization of the systematic rules governing which among the infinitely many possible combinations of linguistic units will be uttered. These rules constitute a schema for the production of language. We *know how* to speak our language; language depends upon performance knowledge in this way. However, the rules are "open" in very important ways: they do not limit the lengths of utterances nor the circumstances in which the utterances occur. Although the rules govern the forms of production, they do not limit the circumstances of production. The rules set only minimal requirements for our utterances. There are, of course, *other* social rules which govern linguistic acts: rules of etiquette, "common sense," scientific accuracy, and so on. But the separation of basic linguistic rules from other social rules "unties" or otherwise frees language from the kinds of behavioral criteria that define performance knowledge. In other words, there is an enormous difference between ascertaining *that* or *how well* someone speaks English and evaluating *what he says* in English.

Meaning and Reference in Language

Our ability to respond to what has been said by saying something appropriate ourselves is one important indication that we "understand the meaning" of what has been said. We claim to understand meaning when we can supply an appropriate linguistic act in response to someone else's linguistic act. *Meaning*, or more specifically *linguistic meaning*, is a term having to do with relations between linguistic units. The meaning of a linguistic unit is the set of the endlessly many utterances that may refer to, or comment on, that unit.

We can distinguish the meaning of a linguistic unit from its *reference,* by which we mean whatever nonlinguistic objects, events, or processes the linguistic act conventionally designates. That is, speaking loosely, we can say that linguistic meaning is a matter of the relations that prevail among linguistic units within the practices of a linguistic community, whereas the reference of a linguistic unit is a matter of the conventional relations it has to the *non*-linguistic objects, events, or processes designated by it.

The reader will have noticed that the term *reference* has been introduced for the second time. In the section on mechanisms we defined the reference of a signaling act as the events or processes that govern its occurrence. We also pointed out that an animal's signaling act could not be interpreted to designate a given object or event; while one antelope's running might have a lion as its reference, the running of other antelopes in the herd need not. The important distinction between the reference of a linguistic act and the reference of a signaling act is that the former is completely conventional and the latter is not. Cultural conventions hold linguistic references in place, but they also allow rapid changes.

These distinctions help us to see a feature of linguistic conventions of the utmost consequence for human conduct. Whereas the practices of a linguistic community provide meaning for all well-formed linguistic events, these events may have only the most remote and indirect reference to things external; they may even be completely devoid of any extralinguistic reference. The basis of these possibilities is found in the ways linguistic competence is attained; to acquire interpretive and productive control over a linguistic event does not require any commerce with its referents. After the child has gained some knowledge of the systematic productivity of the linguistic conventions of its community, increments to its competence become cumulative, so that its present understanding contributes to its subsequent acquisitions. To put the matter colloquially, if somewhat misleadingly, we could say that a child, after the first stages of linguistic development have been reached, increasingly learns new vocabulary in terms of his already established vocabulary. This is at least partly responsible for the exponential nature of his initial vocabulary growth. The child will ordinarily begin to acquire control over new linguistic events by acquiring an understanding of their meaning and by developing some conception of what can be appropriately said with and about these new linguistic entities, but *not* primarily through any experiential discrimination of their referents.

In this way, later acquisition to the child's competence increasingly incorporates linguistic events such as "pain," "accident," "mammal," or "the day before yesterday," which are linked to the nonlinguistic circumstances to which they refer only indirectly through a network of intermediate linguistic events formed by their meanings. And, finally, we acquire conventional control of such events as "proof," "adverbial phrase," and "the square root of minus one," all of which have no extralinguistic reference at all. Our linguistic performances with these units of speech are constrained only by intralinguistic factors. And these factors are, as we have seen, entirely conventional. Linguistic performances are qualitatively different from other human performances and the performances of animals, for they have no necessary reference to the extralinguistic environment. They are not characterized by "knowing how," but by "understanding" within the framework of cultural convention.

Conclusions

Linguistic meanings, and the understandings which are based upon them, are only weakly constrained by the nature of the nonlinguistic environment. The constraints imposed by the linguistic community are also generally weak. Because linguistic acts usually have only diffuse reference to the nonlinguistic environment, the maintenance of their status within the practices of the linguistic community must be based on their meanings. However, in a number of different circumstances, the linguistic community tolerates extensive variability in the meanings assigned to a given linguistic act by different individuals. Thus, not only is much of our conversation biologically idle, it is idle in idiosyncratic ways. Because the assignment of meaning is an extremely open process, different individuals come to focus on different transformations of what might have been initially a common understanding. The cultural receptions that these variations receive depend upon the particular cultural context in which they appear; the rules for changing meanings in mathematics are different from those in poetry. Thus, although our performances are not generally tightly linked to nonlinguistic experiences (i.e., most of our conversation does not consist of "precise descriptions"), and although even such performances are by no means infallible, our understandings remain generally sharable within our culture. Indeed, habitually accepted cultural understandings are exploited by individuals whose depth of understanding is minimal. For example, children often participate in a wide range of conversations with only a superficial understanding of what is said.

The key feature of human language is that its form produces no limitations on the development of new meanings. The systematic, conventional character of our linguistic acts insures that there is an infinity of potentially meaningful linguistic events. The individual learns from his culture, but the openness inherent in the assignment of meanings provides a mechanism by which the individual may feed new meanings back into the culture. Cultures and individuals teach each other. There are no limitations on the development of human understanding produced by the nature of human languages.

We have come to the end of our long and complicated argument. Most optimistically, we have tried to describe the relationship between biological and cultural factors in the development and control of human linguistic acts. Innately equipped to acquire the performance knowledge upon which our linguistic productions are based, we eventually acquire the core of meanings with which our culture interprets the world. It is virtually inevitable that each individual's set of understandings will be unique to some degree. These understandings serve as a map to mark the culturally significant features of the topography of our lives, including those for which we have no direct experience. These maps change as we acquire new or refined referents and meanings based on our linguistic and nonlinguistic experiences. In all of this activity we are being distinctly human, for the openness of our meanings is directly due to the conventional, hierarchical, systematic nature of our linguistic acts. No naturally occurring signaling acts of nonhuman animals possess these characteristics. But these distinctly human characteristics must not be taken as signs that we have "escaped nature" or otherwise avoided our evolutionary history. On the contrary, our cultural mode of adaptation, which is completely linked to our physiological capacity for the acquisition of linguistic competence, *is* our species-specific evolutionary heritage.

References

Abraham, K. (1926). The psychological relations between sexuality and alcoholism. *International Journal of Psychoanalysis, 7,* pp 2–10.

Adler, M. J. (1967). *The Difference of Man and the Difference It Makes.* New York: Holt, Rinehart and Winston.

Adler, N. (1969). Effects of the male's copulatory behavior on successful pregnancy of the female rat. *Journal of Comparative and Physiological Psychology, 69,* pp 613–622.

Adler, N. and Bermant, G. (1966). Sexual behavior of male rats: effects of reduced sensory feedback. *Journal of Comparative and Physiological Psychology, 61,* pp 240–243.

Adler, N. and Zoloth, S. R. (1970). Copulatory behavior can inhibit pregnancy in female rats. *Science, 168,* pp 1480–1482.

Adrian, E. D. (1964). *The Basis of Sensation.* New York: Hafner Publishing Co.

Adrian, E. D. and Matthews, R. (1927a). The action of light on the eye. I: The discharge of impulses in the optic nerve and its relation to the electric change in the retina. *Journal of Physiology (London), 63,* pp 378–414.

Adrian, E. D. and Matthews, R. (1927b). The action of light on the eye. II: The processes involved in retinal excitation. *Journal of Physiology (London), 64,* pp 279–301.

Alcock, J. (1970). Punishment levels and the response of black-capped chickadees (*Parus atricapillus*) to three kinds of artificial seed. *Animal Behaviour, 18,* pp 592–599.

Alcock, J. (1971). Interspecific differences in avian feeding behavior and the evolution of batesian mimicry. *Behaviour, 40,* pp 1–9.

Andrew, R. J. (1963). Evolution of facial expression. *Science, 142,* pp 1034–1041.

Ardrey, R. (1961). *African Genesis.* New York: Atheneum.

Ardrey, R. (1966). *The Territorial Imperative.* New York: Atheneum.

Arling, G. L. and Harlow, H. F. (1967). Effects of social deprivation on maternal behavior of rhesus monkeys. *Journal of Comparative and Physiological Psychology, 64,* pp 371–378.

Armstrong, E. A. (1958). *The Folklore of Birds.* London: Collins.

Aronfreed, J. (1969). The problem of imitation. In Lipsitt, L. P. and Reese, H. W. (Eds.), *Advances in Child Development and Behavior, 4.* New York: Academic Press.

Aronson, L. and Cooper, M. (1968). Desensitization of the glans penis and sexual behavior in cats. In Diamond, M. (Ed.), *Perspectives in Reproduction and Sexual Behavior.* Bloomington: University of Indiana Press, pp 51–82.

Aronson, L. and Cooper, M. (1969). Mating behaviour in sexually inexperienced cats after desensitization of the glans penis. *Animal Behaviour, 17,* pp 208–212.

Aronson, L., Tobach, E., Lehrman, D., and Rosenblatt, J. (1970). *Development and Evolution of Behavior—Essays in Memory of T. C. Schneirla.* San Francisco: W. H. Freeman.

Ashby, E. (1922). Notes on the mound-building birds of Australia, with particulars of features peculiar to the Mallee-fowl, *Leipoa ocellata* Gould, and a suggestion as to their origin. *Ibis, 4,* pp 702–709.

Ashida, S. (1964). Modification by early experience of the tendency toward gregariousness in rats. *Psychonomic Science, 1,* pp 343–344.

Atz, J. W. (1970). The application of the idea of homology to behavior. In Aronson, L. *et al.* (Eds.), *Development and Evolution of Behaviour.* San Francisco: W. H. Freeman. pp 53–74.

Bard, P. (1940). The hypothalamus and sexual behavior. *Research Publication of the Association for Research of Nervous and Mental Disease, 20,* pp 551–579.

Barfield, R. J. (1969). Activation of copulatory behavior by androgen implanted into the preoptic area of the male fowl. *Hormones and Behavior, 1,* pp 37–52.

Barfield, R. J. (1971a). Activation of sexual and aggressive behavior by androgen implanted into the male ring dove brain. *Endocrinology, 89,* pp 1470–1476.

Barfield, R. J. (1971b). Gonadotropic hormone secretion in the female ring dove in response to visual and auditory stimulation by the male. *Journal of Endocrinology, 49,* pp 305–310.

Barnett, S. A. (1963). *The Rat, a Study in Behavior.* Chicago: Aldine Publishing Company.

Barnett, S. A. (1967). *Instinct and Intelligence: The Behavior of Animals and Man.* Englewood Cliffs, New Jersey: Prentice-Hall.

Barrington, E. (1971). Evolution of hormones. In Schoffeniels, E. (Ed.), *Biochemical Evolution and the Origin of Life.* Amsterdam: North Holland Publishing Co., pp 174–190.

Bastian, J. (1964). The biological background of man's languages. *Georgetown Monograph Series on Languages and Linguistics, 17,* pp 141–148.

Bastian, J. (1965). Primate signaling systems and human languages. In DeVore, I. (Ed.), *Primate Behavior: Field Studies of Monkeys and Apes.* New York: Holt, Rinehart and Winston, pp 585–606.

Bastock, M. (1967). *Courtship: An Ethological Study.* Chicago: Aldine Publishing Company.

Bateson, M. C. (1972). *Our Own Metaphor.* New York: Alfred A. Knopf.

Beach, F. A. (1950). The snark was a boojum. *American Psychologist, 5,* pp 115–124.

Beach, F. A. (1956). Characteristics of masculine "Sex drive." In Jones, M. R. (Ed.), *Nebraska Symposium on Motivation, 1956.* Lincoln, Nebraska: University of Nebraska Press.

Beach, F. A. (1958). Normal sexual behavior in male rats isolated at fourteen days of age. *Journal of Comparative and Physiological Psychology, 51,* pp 37–38.

Beach, F. A. (1965). Retrospect and prospect. In Beach, F. A. (Ed.), *Sex and Behavior.* New York: Wiley, pp 535–569.

Beach, F. A. (1966). Ontogeny of "coitus-related" reflexes in the female guinea pig. *Proceedings of the National Academy of Sciences of the United States of America, 56,* pp 526–533.

Beach, F. A. (1967). Cerebral and hormonal control of reflexive mechanisms involved in copulatory behavior. *Physiological Reviews, 47,* pp 289–316.

Beach, F. A. (1968). Coital behavior in dogs: III. Effects of early isolation on mating in males. *Behaviour, 30,* pp 218–238.

Beach, F. A. (1970). Coital behavior in dogs: VI. Long-term effects of castration upon mating in the male. *Journal of Comparative and Physiological Psychology Monograph, 70:3, Part 2,* pp 1–32.

Beach, F. A. and Holz, M. (1946). Mating behavior in male rats castrated at various ages and injected with androgen. *Journal of Experimental Zoology, 101,* pp 91–142.

Beach, F. A. and Pauker, R. (1949). Effects of castration and subsequent androgen administration upon mating behavior in the male hamster (*Cricetus auratus*). *Endocrinology, 45,* pp 211–221.

Beach, F. A. and Rabedeau, R. (1959). Sexual exhaustion and recovery in the male hamster. *Journal of Comparative and Physiological Psychology, 52,* pp 56–61.

Beach, F. A. and Wilson, J. R. (1963). Effects of prolactin, progesterone and estrogen on reactions of nonpregnant rats to foster young. *Psychological Reports, 13,* pp 231–239.

Beadle, G. and Beadle, M. (1966). *The Language of Life.* Garden City, New York: Doubleday and Company.

Beamer, W., Bermant, G., and Clegg, M. (1969). Copulatory behavior of the ram, *Ovis aries:* II. Factors affecting copulatory satiation. *Animal Behaviour, 17,* pp 706–711.

Bell, R. H. V. (1971). A grazing system in the Serengeti. *Scientific American, 225:7,* pp 86–93.

Bell, R. Q. and Costello, N. S. (1964). Three tests for sex differences in tactile sensitivity in the newborn. *Biology of the Neonate, 7,* pp 335–347.

Berkson, G. (1967). Abnormal stereotyped motor acts. In Zubin, J. and Hunt, H. F.

(Eds.), *Comparative Psychopathology.* New York: Grune and Stratton, pp 76–94.

Bermant, G. (1963). Intensity and role of distress calling in chicks as a function of social contact. *Animal Behaviour, 11,* pp 514–517.

Bermant, G. (1965). Rat sexual behavior: photographic analysis of the intromission response. *Psychonomic Science, 2,* pp 65–66.

Bermant, G., Anderson, T., and Parkinson, S. (1969). Copulation in rats: Relations among intromission curation, frequency, and pacing. *Psychonomic Science, 17,* pp 293–294.

Bermant, G., Clegg, M., and Beamer, W. (1969). Copulatory behavior of the ram, *Ovis aries:* I. A normative study. *Animal Behaviour, 17,* pp 700–705.

Bermant, G. and Davidson, J. (1973). *Biological Bases of Sexual Behavior.* New York: Harper and Row.

Bermant, G. and Taylor, T. (1969). Interactive effects of experience and olfactory bulb lesions in male rat copulation. *Physiology and Behavior, 4,* pp 13–18.

Bermant, G. and Westbrook, W. (1966). Peripheral factors in the regulation of sexual contact by female rats. *Journal of Comparative and Physiological Psychology, 61,* pp 244–250.

Birdsell, J. B. (1953). Some environmental and cultural factors influencing the structuring of Australian aboriginal populations. *American Naturalist Supplement, 87,* p 834.

Birns, B., Blank, M., and Bridger, W. (1966). The effectiveness of various soothing techniques on human neonates. *Psychosomatic Medicine, 28,* pp 316–322.

Bitterman, M. E. (1965). The evolution of intelligence. In McGaugh, J. L., Weinberger, N. M., and Whalen, R. E. (Eds.), *Psychobiology: The Biological Bases of Behavior.* San Francisco: W. H. Freeman, pp 150–157.

Bitterman, M. E. (1969). Thorndike and the problem of animal intelligence. *American Psychologist, 24,* pp 444–453.

Blair, W. F. (1955). Mating call and stage of speciation in the *Microhyla olivacea–M. carolinenis* complex. *Evolution, 9,* pp 469–480.

Blanchard, R. J., Dielman, T. E., and Blanchard, D. C. (1968). Prolonged after effects of a single foot shock. *Psychonomic Science, 10,* pp 327–328.

Bloch, G. and Davidson, J. (1968). Effects of adrenalectomy and experience on postcastration sex behavior in the male rat. *Physiology and Behavior, 3,* pp 461–465.

Boling, J. L., Blandau, R. J., Wilson, J. G., and Young, W. C. (1939). Postparturitional heat responses of newborn and adult guinea pigs. Data on parturition. *Proceedings of the Society for Experimental Biology and Medicine, 42,* pp 128–132.

Bolles, R. C. (1969). Avoidance and escape learning: simultaneous acquisition of different responses. *Journal of Comparative and Physiological Psychology, 68,* pp 355–358.

Bolles, R. C. (1970). Species-specific defense reactions and avoidance learning. *Psychological Review, 77,* pp 32–48.

Bolles, R. C. and McGillis, D. B. (1968). The non-operant nature of the bar-press escape response. *Psychonomic Science, 11,* pp 261–262.

Bolles, R. C., Rapp, H. M., and White, G. C. (1968). Failure of sexual activity to reinforce female rats. *Journal of Comparative and Physiological Psychology, 65,* pp 311–313.

Bolles, R. C. and Seelbach, S. E. (1964). Punishing and reinforcing effects of noise onset and termination for different responses. *Journal of Comparative and Physiological Psychology, 58,* pp 127–131.

Bolles, R. C. and Stokes, L. W. (1965). Rat's anticipation of diurnal and a-diurnal feeding. *Journal of Comparative and Physiological Psychology, 60,* pp 290–294.

Boring, E. (1950). *A History of Experimental Psychology.* 2nd Edition. New York: Appleton-Century-Crofts.

Boring, E. (1969). Titchenev, meaning, and behaviorism. In Krantz, D. L. (Ed.), *Schools of Psychology: A Symposium.* New York: Appleton-Century-Crofts, pp 21–34.

Brazier, M. A. (1959). The historical development of neurophysiology. In Field, J. (Ed.), *Handbook of Physiology. Neurophysiology. Vol. 1.* Washington, D.C.: American Physiological Society, pp 1–58.

Breder, C. M. and Halpern, F. (1946). Innate and acquired behavior affecting the aggregation of fishes. *Physiological Zoology, 9,* pp 154–190.

Breland, K. and Breland, M. (1961). The misbehavior of organisms. *American Psychologist, 16,* pp 681–684.

Breland, K. and Breland, M. (1966). *Animal Behavior.* New York: Macmillan.

Broadhurst, P. L. (1960). Experiments in psychogenetics: applications of biometrical genetics to the inheritance of behavior. In Eysenck, H. J. (Ed.), *Experiments in Personality: Vol. I., Psychogenetics and Psychopharmacology.* London: Routledge and Kegan Paul.

Broadhurst, P. L. and Bignami, G. (1965). Correlative effects of psychogenetic selection: a study of the Roman high and low avoidance strains of rats. *Behavioral Research and Therapy, 2,* pp 273–280.

Broadhurst, P. L. and Eysenck, H. J. (1964). *Emotionality in the Rat: A Problem of Response Specificity.* London: University of London.

Bronson, F. and Eleftheriou, B. (1965). Adrenal response to fighting in mice: separation of physical and psychological causes. *Science, 147,* pp 627–628.

Brown, J. L. (1964). The evolution of diversity in avian territorial systems. *Wilson Bulletin, 76,* pp 160–169.

Brown, J. L. (1969). The control of avian vocalization by the central nervous system. In Hinde, R. A. (Ed.), *Bird Vocalization.* Cambridge: Cambridge University Press.

Brown, J. L. and Hunsperger, R. W. (1963). Neurothology and the motivation of agonistic behaviour. *Animal Behaviour, 11,* pp 439–448.

Brown, P. L. and Jenkins, H. M. (1968). Auto-shaping of the pigeon's key-peck. *Journal of the Experimental Analysis of Behavior, 11,* pp 1–8.

Buechner, H. K. and Schloeth, R. (1965). Ceremonial mating behavior in Uganda Kob (*Adenota kob thomasi* Neumann). *Zeitschrift für Tierpsychologie, 22,* pp 209–225.

Buerger, A. A. and Dawson, A. M. (1968). Spinal kittens: long-term increases in electromyograms due to a conditioning routine. *Physiology and Behavior, 3,* pp 99–103.

Buerger, A. A. and Dawson, A. M. (1969). Spinal kittens: effect of clamping of the thoracic aorta on long-term increases in electromyograms due to a conditioning routine. *Experimental Neurology, 23,* pp 457–464.

Buerger, A. A. and Fennessy, A. (1970). Learning of leg position in chronic spinal rats. *Nature, 225,* pp 751–752.

Bullough, W. S. (1961). *Vertebrate Reproductive Cycles.* New York: Wiley.

Butler, C. G. (1955). *The World of the Honeybee.* New York: Macmillan.

Butterfield, H. (1965). *The Origins of Modern Science* (revised edition). New York: The Free Press.

Caggiula, A. (1970). Analysis of the copulation-reward properties of posterior hypothalamic stimulation in male rats. *Journal of Comparative and Physiological Psychology, 70,* pp 399–412.

Calhoun, J. B. (1963). *The Ecology and Sociology of the Norway Rat.* Bethesda, Maryland: U.S. Department of Health, Education, and Welfare, Public Health Service.

Callan, H. (1970). *Ethology and Society.* Oxford, England: Clarendon Press.

Capranica, R. R. (1965). *The Evoked Vocal Response of the Bullfrog: A Study of Communication by Sound.* Cambridge, Massachusetts: Massachusetts Institute of Technology Press.

Capretta, P. J. (1961). An experimental modification of food preference in chickens. *Journal of Comparative and Physiological Psychology, 54,* pp 238–242.

Capretta, P. J. and Moore, M. J. (1970). Appropriateness of reinforcement to cue in the conditioning of food aversions in chickens (*Gallus gallus*). *Journal of Comparative and Physological Psychology, 72,* pp 85–89.

Carlsson, S. and Larsson, K. (1964). Mating in male rats after local anesthetization of the glans penis. *Zeitschrift für Tierpsychologie, 21,* pp 854–856.

Carran, A. B., Yeudall, L. T., and Royce, J. R. (1964). Voltage level and skin resistance in avoidance conditioning of inbred strains of mice. *Journal of Comparative and Physiological Psychology, 58,* pp 427–430.

Cate, J. T. (1940). Quelques observations sur la locomotion des chiens dont la moelle épinière est sectionée transversalement. *Archives Néerlandaises de Physiologie, 24,* pp 476–485.

Chamberlain, T. J., Halick, P., and Gerard, R. W. (1963). Fixation of experience in the rat spinal cord. *Journal of Neurophysiology, 26,* pp 662–673.

Chomsky, N. (1965). *Aspects of the Theory of Syntax.* Cambridge: Massachusetts Institute of Technology Press.

Chomsky, N. (1967). The general properties of language. In Millikan, C. H. and Darley, F. L. (Eds.), *Brain Mechanisms Underlying Speech and Language.* New York: Grune and Stratton.

Clark, D. L. *The Social Isolation of Animals.* Unpublished manuscript.

Clark, D. L. (1968). Immediate and Delayed Effects of Early, Intermediate, and Late Social Isolation in the Rhesus Monkey. Unpublished doctoral dissertation, University of Wisconsin.

Clark, W. E. L. (1959). *The Antecedents of Man.* Edinburgh: Edinburgh University Press.

Clark, W. E. L. (1967). *Man-Apes or Ape-Men? The Study of Discoveries in Africa.* New York: Holt, Rinehart and Winston.

Clegg, M., Beamer, W., and Bermant, G. (1969). Copulatory behavior of the ram, *Ovis aries:* III. Effects of pre- and postpubertal castration and androgen replacement therapy. *Animal Behaviour, 17,* pp 712–717.

Clegg, M., Santolucito, J., Smith, J., and Ganong, W. (1958). The effect of hypothalamic lesions on sexual behavior and estrous cycles in the ewe. *Endocrinology, 57,* pp 790–797.

Cold Spring Harbor Laboratory of Quantitative Biology. (1965). *Sensory receptors. Cold Spring Harbor Symposia on Quantitative Biology, Vol. XXX.* Cold Spring Harbor, Long Island, New York: Author.

Cole, H. and Cupps, P. (1969). *Reproduction in Domestic Animals.* 2nd Edition. New York: Academic Press.

Cole, H. and Ronning, M. (1973). *Animal Agriculture.* San Francisco: W. H. Freeman.

Collias, N. E. (1956). The analysis of socialization in sheep and goats. *Ecology, 37,* pp 228–239.

Collins, R. L. (1964). Inheritance of avoidance conditioning in mice: a diallel study. *Science, 143,* pp 1188–1190.

Cooper, K. K. (1970). An electrophysiological study of the effects of castration on the afferent system of the glans penis of the cat. *Dissertation Abstracts International, 30,* p 3509b.

Corso, J. F. (1967). *The Experimental Psychology of Sensory Behavior.* New York: Holt, Rinehart and Winston.

Cott, H. B. (1957). *Adaptive Coloration in Animals.* New York: Barnes & Noble.

Crane, J. (1941). Eastern Pacific expeditions of the New York zoological society: crabs of the genus *Uca* from the west coast of Central America. *Zoologica, 26,* pp 145–207.

Crane, J. (1957). Basic patterns of display in fiddler crabs *Ocypodidae* (genus *Uca*). *Zoologica, 42,* pp 69–82.

Crook, J. H. (1965). The adaptive significance of avian social organization. *Symposium of the Zoological Society, London, 14,* pp 181–218.

Crook, J. H. (1970). The socio-ecology of primates. In Crook, J. H. (Ed.), *Social Behaviour in Birds and Mammals: Essays on the Social Ethology of Animals and Man.* New York: Academic Press, pp 103–166.

Cross, H. A. and Harlow, H. F. (1965). Prolonged and progressive effects of partial isolation in the behavior of macaque monkeys. *Journal of Experimental Research in Personality, 1,* pp 39–49.

Dale, P. S. (1972). *Language and Development.* Hinsdale, Illinois: Dryden Press.

D'Amato, M. R. and Schiff, D. (1964). Long-term discriminated avoidance performance in the rat. *Journal of Comparative and Physiological Psychology, 57,* pp 123–126.

Darwin, C. (1859). *The Origin of Species by Means of Natural Selection.* London: J. Murray.

Darwin, C. (1860). *The Voyage of the Beagle.* Engel, L. (Ed.), (1962). Garden City, New York: Doubleday & Company.

Darwin, C. (1871). *The Descent of Man, and Selection in Relation to Sex.* New York: Appleton.

Darwin, C. (1872). *The Expression of the Emotions in Man and the Animals.* London: J. Murray.

Davidson, J. (1966). Characteristics of sex behaviour in male rats following castration. *Animal Behaviour, 14,* pp 266–272.

Davidson, J. (1969). Hormonal control of sexual behavior in adult rats. *Advances in Bioscience, 1,* pp 119–169.

Davidson, J. and Bloch, G. (1969). Neuroendocrine aspects of male reproduction. *Biology of Reproduction, 1, Suppl. 1,* pp 67–92.

Davis, D. E. (1964). Physiological analysis of aggressive behavior. In Etkin, W. (Ed.), *Social Behavior and Organization Among Vertebrates.* Chicago: University of Chicago Press.

DeCharms, R. (1968). *Personal Causation.* New York: Academic Press.

DeFries, J. C. (1967). Quantitative genetics and behavior: overview and perspective. In Hirsch, J. (Ed.), *Behavior-genetic Analysis.* New York: McGraw-Hill.

Delgado, J. (1966). Emotions. In Vernon, J. A. (Ed.), *Introduction to Psychology: A Self-selection Text Book.* Dubuque, Iowa: Wm. C. Brown Co. Publishers.

Delgado, J. (1967). Aggression and defense under cerebral radio control. In Clemente, C. D. and Lindsley, D. B. (Eds.), *Aggression and Defense: Neural Mechanisms and Social Patterns (Brain Function), Vol. V.* Los Angeles: University of California Press, pp 171–193.

Delgado, J. (1969). *Physical Control of Mind.* New York: Harper and Row.

Dember, W. N. (1965). The new look in motivation. *American Scientist, 53,* pp 409–427.

Dember, W. N. and Earl, R. W. (1957). Analysis of exploratory, manipulative, and curiosity behaviors. *Psychological Review, 64,* pp 91–96.

Demerec, M. and Kaufmann, B. P. (1962). *Drosophila Guide.* Washington, D.C.: Carnegie Institution.

Denenberg, V. H. (1962). The effects of early experience. In Hafez, E. S. E. (Ed.), *The Behavior of Domestic Animals.* Baltimore: Williams and Wilkins, pp 109–138.

Denenberg, V. H. (1967). Stimulation in infancy, emotional reactivity, and exploratory behavior. In Glass, D. C. (Ed.), *Neurophysiology and Emotion.* New York: Russell Sage Foundation and Rockefeller University Press, pp 161–189.

Descartes, R. (1972). *Treatise of Man.* French Text with Translation and Commentary by T. S. Hall. Cambridge: Harvard University Press.

Dethier, V. G. (1962). *To Know a Fly.* San Francisco: Holden-Day.

Dethier, V. G. (1971). A surfeit of stimuli: a paucity of receptors. *American Scientist, 59,* pp 706–715.

Dethier, V. G. and Stellar, E. (1964). *Animal Behavior.* 2nd Edition. Englewood Cliffs, New Jersey: Prentice-Hall, Inc.

DeVore, I. (1963). Mother-infant relations in free-ranging baboons. In Rheingold, H. (Ed.), *Maternal Behavior in Mammals.* New York: Wiley, pp 305–335.

DeVore, I. (1965). Male dominance and mating behavior in baboons. In Beach, F. A. (Ed.), *Sex and Behavior.* New York: Wiley, pp 266–289.

Dewsbury, D. A. (1967a). A quantitative description of the behavior of rats during copulation. *Behaviour, 29,* pp 154–178.

Dewsbury, D. A. (1967b). Effect of alcohol ingestion on copulatory behavior of male rats. *Psychopharmacologia (Berl.), 11,* pp 276–281.

Dewsbury, D. A. (1969). Copulatory behavior of rats (*Rattus norvegicus*) as a function of prior copulatory experience. *Animal Behaviour, 17,* pp 217–223.

Diamond, M. (1970). Intromission pattern and species vaginal code in relation to induction of pseudopregnancy. *Science, 169,* pp 995–997.

Diamond, M. (1972). Vaginal stimulation and progesterone in relation to pregnancy and parturition. *Biology of Reproduction, 6,* pp 281–287.

Diamond, S., Balvin, R. S., and Diamond, F. R. (1963). *Inhibition and Choice.* New York: Harper and Row.

Dicks, D., Myers, R. E., and Kling, A. (1969). Uncus and amygdala lesions: effects on social behavior in the free-ranging rhesus monkey. *Science, 165,* pp 69–71.

DiGiorgio, A. M. (1929). Persistenza nell'animale spinale, di asimmetrie posturali e matorie de origine cerebellare. Nota I–III. *Archivio di Fisiologia, 27,* pp 518–580.

DiGiorgio, A. M. (1943). Rickerche sulla persistenza dei fenomeni cerebellari nell'animale spinale. *Archivio di Fisiologia, 43,* pp 47–63.

Dilger, W. C. (1961). Changes in nest-material carrying behavior of F_1 hybrids between *Agapornis roseicollis* and *A. personata fischeri* during three years. *American Zoologist, 1,* p 350. (Abstract.)

Dilger, W. C. (1962). The behavior of lovebirds. *Scientific American, 206:1,* pp 88–98.

Dobrzecka, C., Szwejkowska, G., and Konorski, J. (1966). Qualitative versus directional cues in two forms of differentiation. *Science, 153,* pp 87–89.

Dobzhansky, T. (1950). Mendelian populations and their evolution. *American Naturalist, 84,* pp 401–413.

Dobzhansky, T. (1962). *Mankind Evolving.* New Haven: Yale University Press.

Dobzhansky, T. (1968). Genetics and the social sciences. In Glass, D. (Ed.), *Genetics.* New York: Russell Sage Foundation, Social Science Research Council, and Rockefeller University Press, pp 129–142.

Dobzhansky, T. (1972). Genetics and the diversity of behavior. *American Psychologist, 27,* pp 523–530.

Dodge, R. and Benedict, F. G. (1915). *Psychological Effects of Alcohol.* Washington, D.C.: Carnegie Institution Publication No. 232.

Dröscher, V. B. (1969). *The Magic of the Senses.* New York: E. P. Dutton and Company.

Dykman, R. A. and Shurrager, P. S. (1956). Successive and maintained conditioning in spinal carnivores. *Journal of Comparative and Physiological Psychology, 49,* pp 27–35.

Efron, R. (1966). The conditioned reflex: a meaningless concept. *Perspectives in Biology and Medicine, 9,* pp 488–514.

Egerton, P. (1962). The cow-calf relationship and rutting behavior in the American bison. Masters thesis, University of Alberta.

Ehrlich, P. and Ehrlich, A. (1970). *Population, Resources, Environment; Issues in Human Ecology.* San Francisco: W. H. Freeman.

Eibl-Eibesfeldt, I. (1970). *Ethology—the Biology of Behavior.* New York: Holt, Rinehart and Winston.

Eisenberg, J. F. and Dillon, W. S. (Eds.) (1971). *Man and Beast: Comparative Social Behavior.* Washington, D.C.: Smithsonian Institution Press.

Eisenberg, L. (1972). The *human* nature of human nature. *Science, 176,* pp 123–128.

Ellenberger, H. (1970). *The Discovery of the Unconscious.* New York: Basic Books.

Ellis, P. (1953). Social aggregation and gregarious behavior in hoppers of *Locusta migratoria migratorioides* (R. and F.). *Behaviour, 5,* pp 225–260.

Ellis, P. (1959). Learning and social aggregation in locust hoppers. *Animal Behaviour, 7,* pp 91–106.

Ellis, P. (1963a). The influence of some environmental factors on learning and aggregation in locust hoppers. *Animal Behaviour, 11,* pp 142–151.

Ellis, P. (1963b). Changes in the social aggregation of locust hoppers with changes in rearing conditions. *Animal Behaviour, 11,* pp 152–160.

Erickson, C. J. (1970). Induction of ovarian activity in female ring doves by androgen treatment of castrated males. *Journal of Comparative and Physiological Psychology, 71,* pp 210–215.

Erlenmeyer-Kimling, L., Hirsch, J., and Weiss, J. M. (1962). Studies in experimental behavior genetics: III. Selection and hybridization analyses of individual differences in the sign of geotaxis. *Journal of Comparative and Physiological Psychology, 55,* pp 722–731.

Esplin, D. W. and Woodbury, D. M. (1961). Spinal reflexes and seizure patterns in the two-toed sloth. *Science, 133,* pp 1426–1427.

Etkin, W. (1964). Cooperation and competition in social behavior. In Etkin, W. (Ed.), *Social Behavior and Organization Among Vertebrates.* Chicago: University of Chicago Press, pp 35–52.

Evans, H. E. (1966). *The Comparative Ethology and Evolution of the Sand Wasps.* Cambridge, Massachusetts: Harvard University Press.

Evans, H. E. (1968). *Life on a Little Known Planet.* New York: E. P. Dutton Company.

Ewer, R. F. (1968). *Ethology of Mammals.* New York: Plenum Press.

Falconer, D. S. (1960). *Introduction to Quantitative Genetics.* New York: Ronald.

Faure, J. C. (1932). The phases of locusts in South Africa. *Bulletin of Entomology Research, 23,* pp 293–405.

Fearing, F. (1930, 1970). *Reflex Action: A Study in the History of Physiological Psychology.* Baltimore: Williams and Wilkens (1930); Cambridge: M.I.T. Press (1970).

Fisher, A. E. (1955). The effects of differential early treatment on the social and exploratory behavior of young puppies. Unpublished doctoral dissertation, Pennsylvania State University.

Fleay, D. H. (1937). Nesting habits of the brush turkey. *Emu, 36,* pp 153–163.

Flynn, J. P. (1967). The neural basis of aggression in cats. In Glass, D. C. (Ed.), *Neurophysiology and Emotion.* New York: Russell Sage Foundation and Rockefeller University Press, pp 40–60.

Folman, Y. and Drori, D. (1965). Normal and aberrant copulatory behaviour in male rats (*R. norvegicus*) reared in isolation. *Animal Behaviour, 13,* pp 427–429.

Forbes, A. and Mahan, C. (1963). Attempts to train the spinal cord. *Journal of Comparative and Physiological Psychology, 56,* pp 36–40.

Fox, M. W. and Stelzner, D. (1966). Behavioral effects of differential early experience in the dog. *Animal Behaviour, 14,* pp 273–281.

Fox, M. W. and Stelzner, D. (1967). The effects of early experience on the development of inter- and intraspecific social relationships in the dog. *Animal Behaviour, 15,* pp 377–386.

Franzisket, L. (1951). Gewohnheitsbildung und bedingte Reflexe bei Rückenmarksfröschen. *Zeitschrift für vergleichende Physiologie, 33,* pp 142–178.

Fraser, A. (1968). *The Reproductive Behavior of Ungulates.* New York: Academic Press.

Friedman, M. C. (1966). Physiological conditions for the stimulation of prolactin secretion by external stimuli in the ring dove. *Dissertation Abstracts, 27,* pp 312–B.

Frith, H. J. (1962). *The Mallee-Fowl.* London: Angus and Robertson.

Fuller, J. L. (1960). Behavior genetics. *Annual Review of Psychology, 11,* pp 41–70.

Fuller, J. L. (1961). Effects of experiential deprivation upon behaviour in animals. *Third World Congress of Psychiatry Proceedings, Vol. III,* pp 223–227.

Fuller, J. L. (1964). The K-puppies. *Discovery, 25:2,* pp 18–22.

Fuller, J. L. and Clark, L. D. (1966). Genetic and treatment factors modifying the postisolation syndrome in dogs. *Journal of Comparative and Physiological Psychology, 61,* pp 251–257.

Fuller, J. L. and Thompson, W. R. (1960). *Behavior Genetics.* New York: Wiley.

Galtsoff, P. (1961). Physiology of reproduction in molluscs. *American Zoologist, 1,* pp 273–289.

Gantt, W. H., Newton, J. E. O., Royer, F. L., and Stephens, J. H. (1966). Effect of person. *Conditional Reflex, 1,* pp 18–35.

Garcia, J., Ervin, F. R., and Koelling, R. A. (1967). Bait-shyness: a test for toxicity with N = 2. *Psychonomic Science, 7,* pp 245–246.

Garcia, J., Kimeldorf, D. J., and Hunt, E. L. (1961). The use of ionizing radiation as a motivating stimulus. *Psychological Review, 68,* pp 383–395.

Garcia, J., Kimeldorf, D. J., and Koelling, R. A. (1955). Conditioned aversion to saccharin resulting from exposure to gamma radiation. *Science, 122,* pp 157–158.

Garcia, J. and Koelling, R. A. (1966). Relation of cue to consequence in avoidance learning. *Psychonomic Science, 4,* pp 123–124.

Gardner, R. A. and Gardner, B. T. (1969). Teaching sign language to a chimpanzee. *Science, 165,* pp 664–672.

Gause, G. J. (1942). The relation of adaptability to adaptation. *Quarterly Review of Biology, 17,* pp 99–114.

Gerall, A. (1958). Effect of interruption of copulation on male guinea pig sexual behavior. *Psychological Reports, 4,* pp 215–221.

Gerall, A. (1963). An exploratory study of the effect of social isolation variables on the sexual behaviour of male guinea pigs. *Animal Behaviour, 11,* pp 274–282.

Gerall, H., Ward, I., and Gerall, A. (1967). Disruption of the male rat's sexual behaviour induced by social isolation. *Animal Behaviour, 15,* pp 54–58.

Gibb, J. A. (1960). Populations of Tits and Goldcrests and their food supply in pine plantations. *Ibis, 102,* pp 63–208.

Gilliard, E. T. (1963). The evolution of bowerbirds. *Scientific American, 209:2,* pp 38–46.

Glickman, S. E. and Schiff, B. B. (1967). A biological theory of reinforcement. *Psychological Review, 74,* pp 81–109.

Goldfarb, W. (1945). Effects of psychological deprivation in infancy and subsequent stimulation. *American Journal of Psychiatry, 102,* pp 18–33.

Goodrich, H. B. and Taylor, H. C. (1934). Breeding reactions in *Betta splendens. Copeia, 4,* pp 165–166.

Gottlieb, G. (1972). Zing-Yang Kuo: radical scientific philosopher and innovative experimentalist (1898–1970). *Journal of Comparative and Physiological Psychology, 80,* pp 1–10.

Gould, J., Henerey, M., and MacLeod, M. (1970). Communication of direction by the honeybee. *Science, 169,* pp 544–554.

Gould, R. A. (1969). *Yiwara: Foragers of the Australian Desert.* New York: Charles Scribner's Sons.

Graham, C. H. (1965). *Vision and Visual Perception.* New York: Wiley.

Green, D. M. and Swets, J. A. (1966). *Signal Detection Theory and Psychophysics.* New York: Wiley.

Griffin, J. P. and Pearson, J. A. (1967). Habituation of the flexor reflex in the rat. *Journal of Physiology, London, 190,* pp 3P–5P.

Griffin, J. P. and Pearson, J. A. (1968). The effect of bladder distension on habituation of the flexor withdrawal reflex in the decerebrate spinal cat. *Brain Research, 8,* pp 185–192.

Groves, P. M., DeMarco, R., and Thompson, R. F. (1969). Habituation and sensitization of spinal interneuron activity in acute spinal cat. *Brain Research, 14,* pp 521–525.

Groves, P. M., Lee, D., and Thompson, R. F. (1969). Effects of stimulus frequency and intensity on habituation and sensitization in acute spinal cat. *Physiology and Behavior, 4,* pp 383–388.

Grunt, J. and Young, W. (1953). Consistency of sexual behavior patterns in individual male guinea pigs following castration and androgen therapy. *Journal of Comparative and Physiological Psychology, 46,* 138–144.

Haber, R. N. (1968). *Contemporary Theory and Research in Visual Perception.* New York: Holt, Rinehart and Winston.

Haber, R. N. (1969). *Information Processing Approaches to Visual Perception.* New York: Holt, Rinehart and Winston.

Hafez, E. (Ed.). (1962). *The Behaviour of Domestic Animals.* Baltimore: Williams and Wilkins.

Hafez, E. (1968). *Reproduction of Farm Animals.* 2nd Edition. Philadelphia: Lea and Febiger.

Hafez, E. (Ed.). (1969). *The Behaviour of Domestic Animals.* 2nd Edition. Baltimore: Williams and Wilkins.

Hailman, J. P. (1967). The ontogeny of an instinct. *Behaviour Supplement 15,* pp 1–159.

Hailman, J. P. (1969). How an instinct is learned. *Scientific American, 221:6,* pp 98–106.

Hall, C. S. (1947). Genetic differences in fatal audiogenic seizures between two inbred strains of house mice. *Journal of Heredity, 38,* pp 2–6.

Hall, G. S. (1904). *Adolescence; Its Psychology and Its Relation to Physiology, Anthropology, Sociology, Sex, Crime, Religion, and Education.* New York: Appleton.

Hall, K. (1963). Observational learning in monkeys and apes. *British Journal of Psychology, 54,* pp 201–226.

Hall, K. and DeVore, I. (1965). Baboon social behavior. In DeVore, I. (Ed.), *Primate Behavior.* New York: Holt, Rinehart and Winston, pp 55–110.

Hamilton, W. D. (1964). The genetical evolution of social behaviour. I and II. *Journal of Theoretical Biology, 7,* pp 1–16, 17–52.

Hansen, E. W. (1966a). Squab induced crop growth in ring dove foster parents. *Journal of Comparative and Physiological Psychology, 62,* pp 120–122.

Hansen, E. W. (1966b). The development of maternal and infant behavior in the rhesus monkey. *Behaviour, 27,* pp 107–149.

Hardin, G. (1956). Meaning of the word protoplasm. *Scientific Monthly, 82,* p 112.

Harlow, H. F. (1958a). The evolution of learning. In Row, A. and Simpson, G. G. (Eds.), *Behavior and Evolution.* New Haven: Yale University Press, pp 269–290.

Harlow, H. F. (1958b). The nature of love. *American Psychologist, 13,* pp 673–685.

Harlow, H. F. (1962). The heterosexual affectional system in monkeys. *American Psychologist, 17,* pp 1–9.

Harlow, H. F. (1964). Effects of radiation on the central nervous system and on behavior—general survey. In Haley, T. J. and Snider, R. S. (Eds.), *Response of the Nervous System to Ionizing Radiation: Second International Symposium.* Boston: Little, Brown, pp 627–644.

Harlow, H. F. (1965). Sexual behavior in the rhesus monkey. In Beach, F. A. (Ed.), *Sex and Behavior.* New York: Wiley, pp 234–265.

Harlow, H. F. (1966). The primate socialization motives. *Transactions and Studies of the College of Physicians of Philadelphia, 33,* pp 224–237.

Harlow, H. F. and Harlow, M. K. (1965). The affectional systems. In Schrier, A. M., Harlow, H. F., and Stollnitz, F. (Eds.), *Behavior of Nonhuman Primates, 2.* New York: Academic Press.

Harlow, H. F., Harlow, M. K., Dodsworth, R. O., and Arling, G. L. (1966). Maternal behavior of rhesus monkeys deprived of mothering and peer associations in infancy. *Proceedings of the American Philosophical Society, 110,* pp 58–66.

Harlow, H. F., Harlow, M. K., and Hansen, E. W. (1963). The maternal affectional system of rhesus monkeys. In Rheingold, H. L. (Ed.), *Maternal Behavior in Mammals.* New York: Wiley.

Harlow, H. F., Harlow, M. K., and Suomi, S. J. (1971). From thought to therapy: lessons from a primate laboratory. *American Scientist, 59,* pp 538–549.

Harlow, H. F. and Kuenne, M. (1949). Learning to think. *Scientific American, 181:2,* pp 36–39.

Harlow, H. F. and Suomi, S. J. (1970). Nature of love—simplified. *American Psychologist, 25,* pp 161–168.

Harlow, H. F. and Zimmerman, R. R. (1959). Affectional responses in the infant monkey. *Science, 130,* pp 421–432.

Harrington, T. and Merzenich, M. M. (1970). Neural coding in the sense of touch: human sensations of skin indentation compared with the responses of slowly adapting mechanoreceptive afferents innervating the hairy skin of monkeys. *Experimental Brain Research, 10,* pp 251–264.

Hart, B. L. (1967a). Sexual reflexes and mating behavior in the male dog. *Journal of Comparative and Physiological Psychology, 64,* pp 388–399.

Hart, B. L. (1967b). Testosterone regulation of sexual reflexes in spinal male rats. *Science, 155,* pp 1283–1284.

Hart, B. L. (1968a). Sexual reflexes and mating behavior in the male rat. *Journal of Comparative and Physiological Psychology, 65,* pp 453–460.

Hart, B. L. (1968b). Role of prior experience in the effects of castration on sexual behavior of male dogs. *Journal of Comparative and Physiological Psychology, 66,* pp 719–725.

Hart, B. L. (1968c). Alteration of quantitative aspects of sexual reflexes in spinal male dogs by testosterone. *Journal of Comparative and Physiological Psychology,* pp 726–730.

Hart, B. L. (1968d). Neonatal castration: influence on neural organization of sexual reflexes in male rats. *Science, 160,* pp 1135–1136.

Hart, B. L. (1968e). Effects of alcohol on sexual reflexes and mating behavior in the male dog. *Quarterly Journal of Studies on Alcohol, 29,* pp 839–844.

Hart, B. L. (1969a). Gonadal hormones and sexual reflexes in the female rat. *Hormones and Behavior, 1,* pp 65–71.

Hart, B. L. (1969b). Effects of alcohol on sexual reflexes and mating behavior in the male rat. *Psychopharmacologia (Berl.), 14,* pp 377–382.

Hart, B. L. (1970). Mating behavior in the female dog and the effects of estrogen on sexual reflexes. *Hormones and Behavior, 1,* pp 93–104.

Hart, B. L. (1971a). Facilitation by strychnine of reflex walking in spinal dogs. *Physiology and Behavior, 6,* pp 932–934.

Hart, B. L. (1971b). Facilitation by estrogen of sexual reflexes in female cats. *Physiology and Behavior, 7,* pp 675–678.

Hart, B. L. and Haugen, C. M. (1968). Activation of sexual reflexes in male rats by spinal implantation of testosterone. *Physiology and Behavior, 3,* pp 735–738.

Hatch, A., Wiberg, G. S., Balzas, T., and Grice, H. C. (1963). Long-term isolation stress in rats. *Science, 142,* p 507.

Hayes, C. (1951). *The Ape in Our House.* New York: Harper and Row.

Heath, R. G. (1964). Pleasure response of human subjects to direct stimulation of the brain: physiologic and psychodynamic considerations. In Heath, R. G. (Ed.), *The Role of Pleasure in Behavior.* New York: Harper and Row, pp 219–243.

Hebb, D. O. and Thompson, W. (1954). The social significance of animal studies. In Lindzey, G. (Ed.), *Handbook of Social Psychology.* Reading, Massachusetts: Addison–Wesley.

Hediger, H. (1965). Environmental factors influencing the reproduction of zoo animals. In Beach, F. A. (Ed.), *Sex and Behavior.* New York: Wiley, pp 319–354.

Heidbreder, E. (1969). Functionalism. In Krantz, D. L. (Ed.), *Schools of Psychology: A Symposium.* New York: Appleton-Century-Crofts, pp 35–50.

Heinroth, O. (1910). Beiträge zur Biologie, namentlich Ethologie und Physiologie der Anatiden, *5 International Ornithologisches Kongress,* (*Verh.*), *5,* pp 589–702.

Heinroth, O. and Heinroth, K. (1958). *The Birds.* Ann Arbor, Michigan: The University of Michigan Press.

Held, R. and Hein, A. (1958). Adaptation of disarranged hand-eye coordination contingent upon re-afferent stimulation. *Perceptual and Motor Skills, 8,* pp 87–90.

Held, R. and Hein, A. (1963). Movement-produced stimulation in the development of visually guided behavior. *Journal of Comparative and Physiological Psychology, 56,* pp 872–876.

Henderson, N. D. (1967). Prior treatment effects on open field behavior of mice: a genetic analysis. *Animal Behaviour, 15,* pp 364–376.

Herrick, C. J. (1924). *Neurological Foundations of Animal Behavior.* New York: Holt, Rinehart and Winston.

Hersher, L., Richmond, J., and Moore, A. (1963a). Maternal behavior in sheep and goats. In Rheingold, H. (Ed.), *Maternal Behavior in Mammals.* New York: Wiley.

Hersher, L., Richmond, J. B., and Moore, A. U. (1963b). Modifiability of the critical period for the development of maternal behavior in sheep and goats. *Behaviour, 20,* pp 311–320.

Hess, E. (1962). Ethology: an approach toward the complete analysis of behavior. In Brown, R. *et al., New Directions in Psychology.* New York: Holt, Rinehart and Winston, pp 157–266.

Hess, E. H. (1964). Imprinting in birds. *Science, 146,* pp 1128–1139.

Hilgard, E. R. (1948, 1956). *Theories of Learning.* 2nd Edition. New York: Appleton-Century-Crofts.

Hinde, R. A. (1953). The conflict between drives in the courtship and copulation of the Chaffinch. *Behaviour, 5,* pp 1–31.

Hinde, R. A. (1959). Some recent trends in ethology. In Koch, S. (Ed.), *Psychology: A Study of a Science, Vol. 2.* New York: McGraw-Hill, pp 561–610.

Hinde, R. A. (1962). The relevance of animal studies to human neurotic disorders. In Richter, D. *et al.* (Eds.), *Aspects of Psychiatric Research.* London: Oxford University Press.

Hinde, R. A. (1966). *Animal Behavior: A Synthesis of Ethology and Comparative Psychology.* New York: McGraw-Hill.

Hinde, R. A. (1970). *Animal Behavior: A Synthesis of Ethology and Comparative Psychology.* 2nd Edition. New York: McGraw-Hill.

Hinde, R. A., Rowell, T. E., and Spencer-Booth, Y. (1964). Behavior of socially living rhesus monkeys in their first six months. *Proceedings of the Zoological Society, London, 143,* pp 609–659.

Hineline, P. N. and Rachlin, H. (1969). Escape and avoidance of shock by pigeons pecking a key. *Journal of the Experimental Analysis of Behavior, 12,* pp 533–538.

Hirsch, J. (1959). Studies in experimental behavior genetics: II. Individual differences in geotaxis as a function of chromosome variations in synthesized *Drosophila* populations. *Journal of Comparative and Physiological Psychology, 52,* pp 304–308.

Hirsch, J. (1963). Behavior genetics and individuality understood. *Science, 142,* pp 1436–1442.

Hirsch, J. (1967a). Behavior-genetic or "experimental" analysis: the challenge of science versus the lure of technology. *American Psychologist, 22,* pp 118–130.

Hirsch, J. (1967b). Behavior-genetic analysis. In Hirsch, J. (Ed.), *Behavior-Genetic Analysis.* New York: McGraw-Hill.

Hirsch, J. (1970). Behavior-genetic analysis and its biosocial consequences. *Seminars in Psychiatry, 2,* pp 89–105.

Hirsch, J., and Erlenmeyer-Kimling, L. (1961). Signs of taxis as a property of the genotype. *Science, 134,* pp 835–836.

Hirsch, J. and Erlenmeyer-Kimling, L. (1962). Studies in experimental behavior genetics: IV. Chromosome analyses for geotaxis. *Journal of Comparative and Physiological Psychology, 55,* pp 732–739.

Hirsch, J. and Tryon, R. C. (1956). Mass screening and reliable individual measurement in the experimental behavior genetics of lower organisms. *Psychological Bulletin, 53,* pp 402–410.

Hodos, W. and Campbell, C. B. G. (1969). *Scala Naturae:* why there is no theory in comparative psychology. *Psychological Review, 76,* pp 337–350.

Hostetter, R. C. and Hirsch, J. (1967). Genetic analysis of geotaxis in *Drosophila melanogaster:* complementation between forward and reverse selection lines. *Journal of Comparative and Physiological Psychology, 63,* pp 66–70.

Hulet, C., Ercanbrack, S., Price, D., Blackwell, R., and Wilson, L. (1962a). Mating behavior of the ram in the one-sire pen. *Journal of Animal Science, 21,* pp 857–864.

Hulet, C., Ercanbrack, S., Blackwell, R., Price, D., and Wilson, L. (1962b). Mating behavior of the ram in the multi-sire pen. *Journal of Animal Science, 21,* pp 865–869.

Hulet, C., Blackwell, R., Ercanbrack, S., Price, D., and Wilson, L. (1962c). Mating behavior of the ewe. *Journal of Animal Science, 21,* pp 870–874.

Hull, C. L. (1930). Knowledge and purpose as habit mechanisms. *Psychological Review, 37,* pp 511–525.

Hull, C. L. (1952). *A Behavior System.* New Haven: Yale University Press.

Hutchison, J. B. (1964). Investigations on the neural control of clasping and feeding in *Xenopus laevis* (Daudin). *Behaviour, 24,* pp 47–66.

Hutchison, J. B. (1967). A study of the neural control of sexual clasping behaviour in *Rana angolensis* bocage and *Bufo regularis* reuss with a consideration of self-regulatory hindbrain systems in the anura. *Behaviour, 28,* pp 1–57.

Hutchison, J. B. (1971). Effects of hypothalamic implants of gonadal steroids on courtship behaviour in Barbary doves (*Streptopelia risoria*). *Journal of Endocrinology, 50,* pp 97–113.

Hutchison, J. B. and Poynton, J. C. (1963). A neurological study of the clasp reflex in *Xenopus laevis* (Daudin). *Behaviour, 22,* pp 41–63.

Huxley, J. (Ed.). (1966). A discussion on ritualization of the behaviour in animals and man. *Proceedings of the Zoological Society, London, 251,* Part B.

Johnson, A. S. (1972). Man, grizzly, and national parks. *National Parks and Conservation Magazine, 46,* pp 10–15.

Jones, J. (1967). Fishes. In *The Tarousse Encyclopedia of Animal Life.* New York: McGraw-Hill, pp 207–267.

Kagan, J. (1968). On cultural deprivation. In Glass, D. C. (Ed.), *Environmental Influences.* New York: Russell Sage Foundation and Rockefeller University Press.

Kalat, J. W. (1968). The salience effect in taste-aversion learning. Unpublished paper, University of Pennsylvania.

Kalat, J. W. and Rozin, P. (1970). "Salience:" a factor which can override temporal contiguity in taste-aversion learning. *Journal of Comparative and Physiological Psychology, 71,* pp 192–197.

Kaplan, W. D. and Trout, W. E. (1969). The behavior of four neurological mutants of *Drosophila. Genetics, 61,* pp 399–409.

Kellogg, W. N., Deese, J., and Pronko, N. H. (1946). On the behavior of the lumbo-spinal dog. *Journal of Experimental Psychology, 36,* pp 503–511.

Kellogg, W. N., Deese, J., Pronko, N. H., and Feinberg, M. (1947). An attempt to condition the chronic spinal dog. *Journal of Experimental Psychology, 37,* pp 99–117.

Kellogg, W. N. and Kellogg, L. A. (1933). *The Ape and the Child.* New York: Whittlesey House.

Kellogg, W. N., Pronko, N. H., and Deese, J. (1946). Spinal conditioning in dogs. *Science, 103,* pp 49–50.

Kenshalo, D. R. (1968). *The Skin Senses.* Springfield, Illinois: Charles C. Thomas.

Kimble, G. (1961). *Hilgard and Marquis' Conditioning and Learning.* New York: Appleton-Century-Crofts.

Kimeldorf, D. J., Garcia, J., and Rubadeau, D. O. (1960). Radiation-induced conditioned avoidance behavior in rats, mice, and cats. *Radiation Research, 12,* pp 710–718.

Klein, D. B. (1970). *A History of Scientific Psychology.* New York: Basic Books.

Klinghammer, E. and Hess, E. H. (1964). Parental feeding in ring doves (*Streptopelia roseogrisea*): innate or learned? *Zeitschrift für Tierpsychologie, 21,* pp 338–347.

Klopfer, P. H. (1962). *Behavioral Aspects of Ecology.* Englewood Cliffs, New Jersey: Prentice-Hall.

Klopfer, P. H. (1968). Is heartrate an indicator of imprinted preferences and affect? *Developmental Psychobiology, 1,* pp 205–209.

Klopfer, P. H. (1969). *Habitats and Territories.* New York: Basic Books.

Klopfer, P. H., Adams, D. and Klopfer, M. (1964). Maternal "imprinting" in goats. *Proceedings of the National Academy of Sciences, 52,* pp 911–914.

Klopfer, P. H. and Gamble, J. (1966). Maternal imprinting in goats: the role of chemical senses. *Zeitschrift für Tierpsychologie, 23,* pp 588–592.

Klopfer, P. H. and Klopfer, M. S. (1968). Maternal imprinting in goats: fostering of alien young. *Zeitschrift für Tierpsychologie, 25,* pp 862–866.

Knipling, E. F. (1960). The eradication of the screw-worm fly. *Scientific American, 203:7,* pp 54–61.

Koch, S. (1964). Psychology and emerging conceptions of knowledge as unitary. In Wann, T. W. (Ed.), *Behaviorism and Phenomenology.* Chicago: University of Chicago Press, pp 1–41.

Köhler, W. (1925). *The Mentality of Apes.* New York: Harcourt, Brace & Company.

Komisaruk, B. R. (1967). Effects of local brain implants of progesterone on reproductive behavior in ring doves. *Journal of Comparative and Physiological Psychology, 64,* pp 219–224.

Konishi, M. and Nottebohm, (1969). Experimental studies in the ontogeny of avian vocalizations. In Hinde, R. A. (Ed.), *Bird Vocalizations.* Cambridge: Cambridge University Press.

Konorski, J. (1967). *Integrative Activity of the Brain.* Chicago: University of Chicago Press.

Kozak, W., MacFarlane, W. V., and Westerman, R. (1962). Long-lasting reversible changes in the reflex responses of chronic spinal cats to touch, heat, and cold. *Nature, 193,* pp 171–173.

Krafka, J. (1920). The effect of temperature upon facet number in the bar-eyed mutant of *Drosophila:* Part I. *Journal of General Physiology, 2,* pp 409–432.

Krasner, L. and Ullmann, L. (1965). *Research in Behavior Modification.* New York: Holt, Rinehart and Winston.

Krech, D., Rosenzweig, M. R., and Bennett, E. L. (1962). Relations between brain chemistry and problem-solving among rats raised in enriched and impoverished environments. *Journal of Comparative and Physiological Psychology, 55,* pp 801–807.

Kruuk, H. (1972). *The Spotted Hyena.* Chicago: University of Chicago Press.

Kuffler, S. W. (1953). Discharge patterns and functional organization of mammalian retina. *Journal of Neurophysiology, 16,* pp 37–68.

Kugelberg, E. (1952). Facial reflexes. *Brain, 75,* pp 385–396.

Kugelberg, E. (1962). Polysynaptic reflexes of clinical importance. *Electroencephalography and Clinical Neurophysiology, Suppl. 22,* pp 385–396.

Kuo, Z. Y. (1921). Giving up instincts in psychology. *Journal of Philosophy, 18,* pp 645–664.

Kuo, Z. Y. (1960a). Studies on the basic factors in animal fighting: V. Interspecies coexistence in fish. *Journal of Genetic Psychology, 97,* pp 181–194.

Kuo, Z. Y. (1960b). Studies on the basic factors in animal fighting: VI. Interspecies coexistence in birds. *Journal of Genetic Psychology, 97,* pp 195–209.

Kuo, Z. Y. (1960c). Studies on the basic factors in animal fighting: VII. Interspecies coexistence in mammals. *Journal of Genetic Psychology, 97,* pp 211–225.

Kuo, Z. Y. (1967). *The Dynamics of Behavior Development.* New York: Random House.

Lack, D. (1953). *The Life of the Robin.* London: Penguin Books.

Landauer, T. K. and Whiting, J. W. (1964). Infantile stimulation and adult stature of human males. *American Anthropologist, 66,* pp 1007–1028.

Lane, F. (1960). *The Kingdom of the Octopus.* New York: Sheridan Press.

Leblond, C. P. (1940). Nervous and hormonal factors in the maternal behavior of the mouse. *Journal of Genetic Psychology, 57,* pp 327–344.

LeBoeuf, B. and Peterson, R. (1969). Social status and mating activity in elephant seals. *Science, 163,* pp 91–93.

LeGrand, Y. (1967). *Form and Space Vision.* Bloomington: Indiana University Press.

Lehrman, D. S. (1953). A critique of Konrad Lorenz's theory of instinctive behavior. *Quarterly Review of Biology, 28,* pp 337–363.

Lehrman, D. S. (1955). The physiological basis of parental feeding behavior in the ring dove (*Streptopelia risoria*). *Behaviour, 7,* pp 241–286.

Lehrman, D. S. (1965). Interaction between internal and external environments in the regulation of the reproductive cycle of the ring dove. In Beach, F. A. (Ed.), *Sex and Behavior.* New York: Wiley.

Lehrman, D. S. (1970). Semantic and conceptual issues in the nature-nurture problem. In Aronson, L. *et al.* (Eds.), *Development and Evolution of Behavior.* San Francisco: W. H. Freeman, pp 17–52.

Lehrman, D. S. and Brody, P. (1957). Oviduct response to estrogen and progesterone in the ring dove (*Streptopelia risoria*). *Proceedings of the Society for Experimental Biology and Medicine, 95,* pp 373–375.

Lenneberg, E. H. (1967). *Biological Foundations of Language.* New York: Wiley.

Lerner, I. M. (1958). *The Genetic Basis of Selection.* New York: Wiley.

Lerner, I. M. (1968). *Heredity, Evolution and Society.* San Francisco: Freeman.

Lettvin, J., Maturana, H., McCulloch, W., and Pitts, W. (1959). What the frog's eye tells the frog's brain. *Proceedings of the Institute of Radio Engineers, 47,* pp 1940–1951.

Levine, S. (1962). The effects of infantile experience on adult behavior. In Bachrach, A. J. (Ed.), *Experimental Foundations of Clinical Psychology.* New York: Basic Books, pp 139–169.

Levine, S. (1968). Hormones and conditioning. In Arnold, W. (Ed.), *Nebraska Symposium on Motivation: 1968.* Lincoln: University of Nebraska Press, pp 85–102.

Lindauer, M. (1971). The functional significance of the honeybee waggle dance. *American Naturalist, 105,* pp 89–96.

Lindburg, D. G. (1969). Rhesus monkeys: mating season mobility of adult males. *Science, 166,* pp 1176–1178.

Lindsley, D. L. and Grell, E. H. (1968). *Genetic Variations of Drosophila Melanogaster.* Washington: Carnegie Institution.

Lockard, R. B. (1971). Reflections on the fall of comparative psychology: is there a message for us all? *American Psychologist, 26,* pp 168–179.

Locke, E. (1971). Is "behavior therapy" behavioristic? (An analysis of Wolpe's psychotherapeutic methods.) *Psychological Bulletin, 76,* pp 318–327.

Lorenz, K. Z. (1937). The companion in the bird's world. *Auk, 54,* pp 245–273.

Lorenz, K. Z. (1941). Vergleichende Bewegungsstudien an Anatiden. *Suppl., Journal für Ornithologie, 89,* pp 194–293.

Lorenz, K. Z. (1952). *King Solomon's Ring.* New York: Thomas Y. Crowell.

Lorenz, K. Z. (1958). The evolution of behavior. Reprinted in McGaugh, J., Weinberger, N., and Whalen, R. (Eds.), *Psychobiology: The Biological Bases of Behavior.* San Francisco: W. H. Freeman, 1966, pp 33–44.

Lorenz, K. Z. (1965a). *Evolution and Modification of Behavior.* Chicago: University of Chicago Press.

Lorenz, K. Z. (1965b). Preface. In Darwin, C., *The Expression of the Emotions in Man and Animals.* Chicago: University of Chicago Press, pp ix–xiii.

Lorenz, K. Z. (1966). *On Aggression.* New York: Harcourt Brace Jovanovich.

Lorenz, K. Z. (1969). Innate bases of learning. In Pribram, K. (Ed.), *On the Biology of Learning.* New York: Harcourt Brace Jovanovich.

Lorenz, K. Z. (1970). *Studies in Animal and Human Behavior, Vol. 1.* Cambridge: Harvard University Press.

Lorenz, K. Z. and Tinbergen, N. (1938). Taxis und Instinkthandlung in der Eirollbewegung der Graugans. *Zeitschrift für Tierpsychologie, 2,* pp 1–29.

Lorenz, K. Z. and Tinbergen, N. (1957). Taxis and instructive action in the egg-retrieving behavior of the greylag goose. In Schiller, C. H. (Ed.), *Instinctive Behavior.* New York: International Universities Press.

Lotka, A. J. (1956). *Elements of Mathematical Biology.* New York: Dover Publications.

Lott, D. F. (1962). The role of progesterone in the maternal behavior of rodents. *Journal of Comparative and Physiological Psychology, 55,* pp 610–613.

Lott, D. F. and Comerford, S. (1968). Hormonal initiation of parental behavior in inexperienced ring doves. *Zeitschrift für Tierpsychologie, 25,* pp 71–75. *25,* pp 71–75.

Lott, D. F. and Fuchs, S. S. (1962). Failure to induce retrieving by sensitization or the injection of prolactin. *Journal of Comparative and Physiological Psychology, 55,* pp 1111–1113.

Lott, D. F. and Rosenblatt, J. S. (1969). Development of maternal responsiveness during pregnancy in the laboratory rat. In Foss, B. M. (Ed.), *Determinants of Infant Behavior IV.* London: Methuen & Company.

Lovejoy, A. O. (1936). *The Great Chain of Being.* Cambridge: Harvard University Press.

Lovejoy, A. O. (1936, reprinted 1960). *The Great Chain of Being.* New York: Harper and Row.

Lovejoy, A. O. (1961). *Reflections on Human Nature.* Baltimore: The Johns Hopkins Press.

Lush, J. L. (1945). *Animal Breeding Plans.* Ames, Iowa: Collegiate Press.

Maas, J. W. (1962). Neurochemical differences between two strains of mice. *Science, 137,* pp 621–622.

Maatsch, J. L. (1959). Learning and fixation after a single shock trial. *Journal of Comparative and Physiological Psychology, 52,* pp 408–410.

MacArthur, R. H. (1965). Patterns of species diversity. *Biological Reviews of the Cambridge Philosophical Society, 40,* pp 510–533.

Maffei, L. and Rizzolatti, G. (1965). Effect of synchronized sleep on the response of lateral geniculate units to flashes of light. *Archives Italiennes de Biologie, 103,* pp 609–622.

Marler, P. (1956). Behaviour of the chaffinch, *Fringilla coelebs. Behaviour Supplement 5,* pp 1–184.

Marler, P. (1961). The filtering of external stimuli during instinctive behavior. In Thorpe, W. H. and Zangwill, O. L. (Eds.), *Current Problems in Animal Behaviour.* Cambridge: Cambridge University Press, pp 150–166.

Marler, P. (1967a). Animal communication signals. *Science, 157,* pp 769–774.

Marler, P. (1967b). Comparative study of song development in emberizine finches. *Proceedings of the 14th International Ornithological Congress.* Oxford: Blackwell, pp 231–244.

Marler, P. and Hamilton, W. (1966). *Mechanisms of Animal Behavior.* New York: Wiley.

Mason, W. A. (1961). The effects of social restriction on the behavior of rhesus monkeys: II. Tests of gregariousness. *Journal of Comparative and Physiological Psychology, 54,* pp 287–290.

Mason, W. A. (1967). Motivational aspects of social responsiveness in young chimpanzees. In Stevenson, H. W., Hess, E. H., and Rheingold, H. L. (Eds.), *Early Behavior: Comparative and Developmental Approaches.* New York: Wiley, pp 103–126.

Mason, W. A. (1968a). Scope and potential of primate research. *Science and Psychoanalysis, 12,* pp 101–112.

Mason, W. A. (1968b). Early social deprivation in nonhuman primates: implications for human behavior. In Glass, D. C. (Ed.), *Environmental Influences.* New York: Russell Sage Foundation and Rockefeller University Press.

Mason, W. A. (1970). Early deprivation in biological perspective. In Denneberg, V. (Ed.), *Education of the Infant and Young Child.* New York: Academic Press, pp 25–50.

Mayr, E. (1963). *Animal Species and Evolution.* Cambridge: Harvard University Press.

McClearn, G. E. (1961). Genotype and mouse activity. *Journal of Comparative and Physiological Psychology, 54,* pp 674–676.

McClearn, G. E. (1967). Genes, generality, and behavior research. In Hirsch, J. (Ed.), *Behavior-Genetic Analysis.* New York: McGraw-Hill.

McClearn, G. E. and Meredith, W. (1964). Dimensional analysis of activity and elimination in a genetically heterogeneous group of mice (*Mus musculus*). *Animal Behaviour, 12,* pp 1–10.

McClearn, G. E. and Meredith, W. (1966). Behavioral genetics. *Annual Review of Psychology, 17,* pp 515–550.

McClearn, G. E. and Rodgers, D. A. (1959). Differences in alcohol preference among inbred strains of mice. *Quarterly Journal of Studies on Alcohol, 20,* pp 691–695.

McCulloch, T. L. and Haslerud, G. M. (1939). Affective responses of an infant chimpanzee reared in isolation from its kind. *Journal of Comparative and Physiological Psychology, 28,* pp 437–445.

McGill, T. (1962). Sexual behavior in three inbred strains of mice. *Behaviour, 19,* pp 341–350.

McGill, T. (1965). *Readings in Animal Behavior.* New York: Holt, Rinehart and Winston.

McGill, T. and Tucker, G. (1964). Genotype and sex drive in intact and castrated male mice. *Science, 145,* pp 514–515.

McHugh, T. (1958). Social behavior of the American buffalo (*Bison bison bison*). *Zoologica, 43,* pp 1–40.

McKinney, F. (1961). An analysis of the displays of the European Eider *Somateria mollissima mollissima* (Linnaeus) and the Pacific Eider *Somateria mollissima v. nigra* Bonaparte. *Behaviour Supplement 7.*

Mech, L. D. (1967). *The Wolf.* Garden City, New York: Natural History Press.

Meier, G. (1961). Infantile handling and development of Siamese kittens. *Journal of Comparative and Physiological Psychology, 54,* pp 284–286.

Melzack, R. (1954). The genesis of emotional behavior, an experimental study of the dog. *Ibid, 47,* pp 166–168.

Melzack, R. and Scott, T. H. (1957). The effects of early experience on the response to pain. *Journal of Comparative and Physiological Psychology, 50,* pp 155–161.

Merzenich, M. M. and Harrington, T. (1969). The sense of flutter-vibration evoked by stimulation of the hairy skin of primates: comparison of human sensory capacity with the responses of mechanoreceptive afferents innervating the hairy skin of monkeys. *Experimental Brain Research, 9,* pp 236–260.

Michelmore, S. (1964). *Sexual Reproduction.* Garden City, New York: The Natural History Press.

Miller, G. A. (1951). *Language and Communication.* New York: McGraw-Hill.

Milne, L. and Milne, M. (1962). *The Senses of Animals and Men.* New York: Atheneum.

Mitchell, G. (1968a). Attachment differences in male and female infant monkeys. *Child Development, 39,* pp 611–620.

Mitchell, G. (1968b). Persistent behavior pathology in rhesus monkeys following early social isolation. *Folia Primatologica, 8,* pp 132–147.

Mitchell, G. (1969). Paternalistic behavior in primates. *Psychological Bulletin,* pp 399–417.

Mitchell, G. (1970). Abnormal behavior in primates. In Rosenblum, L. (Ed.), *Primate Behavior, Vol. 1.* New York: Academic Press.

Mitchell, G. and Clark, D. L. (1968). Long-term effects of social isolation in nonsocially adapted rhesus monkeys. *Journal of Genetic Psychology, 113,* pp 117–128.

Mitchell, G., Raymond, E. J., Ruppenthal, G. C., and Harlow, H. F. (1966). Long-term effects of total social isolation upon behavior of rhesus monkeys. *Psychological Reports, 18,* pp 567–580.

Moan, C. E. and Heath, R. G. (1972). Septal stimulation for the initiation of heterosexual behavior in a homosexual male. *Journal of Behavior Therapy and Experimental Psychiatry, 3,* pp 23–30.

Møller, G. W., Harlow, H. F., and Mitchell, G. (1968). Factors affecting agonistic communication in rhesus monkeys (*Macaca mulatta*). *Behaviour, 31,* pp 339–357.

Monod, J. (1971). *Chance and Necessity.* New York: Knopf.

Montagu, A. (1969). *Sex, Man and Society.* New York: Tower Publications.

Montaigne, M. de (1948). *The Complete Works of Montaigne,* translated by Frame, D. M. Stanford, California: Stanford University Press.

Morgan, C. L. (1894). *An Introduction to Comparative Psychology.* London: W. Scott, Ltd.

Morris, D. (1956, 1970). The function and causation of courtship ceremonies. In Morris, D., *Patterns of Reproductive Behavior.* New York: McGraw-Hill, pp 128–152.

Morris, D. (1967). *The Naked Ape.* New York: McGraw-Hill.

Mountcastle, V. B. (1968). *Medical Physiology.* St. Louis, Missouri: The C. V. Mosby Company.

Mountcastle, V. B., Talbot, W. H., Darian-Smith, I., and Kornhuber, H. H. (1967). Neural basis of the sense of flutter-vibration. *Science, 155,* pp 597–600.

Mowrer, O. H. (1960). *Learning Theory and Behavior.* New York: Wiley.

Moyer, K. E. (1968). Kinds of aggression and their physiological basis. *Communications in Behavioral Biology, Part A, 2,* pp 65–87.

Moynihan, M. (1962). Hostile and sexual behavior patterns of South American and Pacific *Laridae. Behaviour Supplement 8.*

Munn, N. L. (1965). *The Evolution and Growth of Human Behavior.* Second Edition. Boston: Houghton Mifflin Company.

Murphey, R. M. (1965). Sequential alternation behavior in the fruit fly, *Drosophila melanogaster. Journal of Comparative and Physiological Psychology, 60,* pp 196–199.

Murphey, R. M. and Hall, C. F. (1969). Some correlates of negative geotaxis in *Drosophila melanogaster. Animal Behaviour, 17,* pp 181–185.

Murphy, M. and Schneider, G. (1970). Olfactory bulb removal eliminates mating behavior in the male golden hamster. *Science, 167,* pp 302–303.

Nash, J. (1970). *Developmental Psychology: A Psychobiological Approach.* Englewood Cliffs, New Jersey: Prentice-Hall.

National Institute of Child Health and Human Development (1968). *Perspectives on Human Deprivation.* Washington, D.C.: Author.

Nesmeinova, T. N. (1957). The inhibition of the motor reflex in spinal dogs under conditions of chronic experimentation. *Sechenov Physiological Journal of the USSR, 43,* pp 281–288.

Noble, G. (1931). *The Biology of the Amphibia.* New York: McGraw-Hill.

Oldroyd, H. (1965). *The Natural History of Flies.* New York: W. W. Norton.

Orians, G. H. (1969). On the evolution of mating systems in birds and mammals. *American Naturalist, 103,* pp 589–603.

Osgood, C. (1953). *Method and Theory in Experimental Psychology.* New York: Oxford University Press.

Parsons, P. A. (1967). *The Genetic Analysis of Behaviour.* London: Methuen.

Pavlov, I. (1904). Physiology of digestion. In *Nobel Lectures: Physiology or Medicine, 1901–1921.* Amsterdam: Elsevier, 1967, pp 141–155.

Payne, A. P. and Swanson, H. H. (1970). Agonistic behaviour between pairs of hamsters of the same and opposite sex in a neutral observation area. *Behaviour, 36,* pp 259–269.

Peakall, D. B. (1970). Pesticides and the reproduction of birds. *Scientific American, 222:4,* pp 72–83.

Perdeck, A. C. (1958). The isolating value of specific song patterns in two sibling species of grasshoppers (*Chorthippus brunneus* Thunb. and *C. biguttulus* L.). *Behaviour, 12,* pp 1–75.

Peters, R. S. (Ed.). (1965). *Brett's History of Psychology.* Cambridge: Massachusetts Institute of Technology Press.

Petrunkevitch, A. (1926). Value of instinct as a taxonomic guide in spiders. *Biological Bulletin, 50,* pp 427–432.

Pfeiffer, J. (1969). *The Emergence of Man.* New York: Harper and Row.

Phoenix, C., Goy, R., and Young, W. (1967). Sexual behavior: general aspects. In Martin, C. and Ganorig, W. (Eds.), *Neuroendocrinology.* New York: Academic Press.

Ploog, D. (1971). Neurological aspects in social behavior. In Eisenberg, J. F. and Dillon, W. S. (Eds.), *Man and Beast: Comparative Social Behavior.* Washington, D.C.: Smithsonian Institution Press.

Pollin, W. and Strabenau, J. R. (1968). Biological, psychological and historical differences in a series of monozygotic twins discordant for schizophrenia. In Rosenthal, D. and Kety, S. S. (Eds.), *The Transmission of Schizophrenia.* New York: Pergamon.

Premack, D. (1971). Language in chimpanzee? *Science, 172,* pp 808–822.

Prescott, J. W. (1967). The psychobiology of maternal-social deprivation and the etiology of violent-aggressive behavior: a special case of sensory deprivation. Paper presented at Stanford Research Institute on July 21.

Prosser, C. L. and Hunter, W. S. (1936). The extinction of startle responses and spinal reflexes in the white rat. *American Journal of Physiology, 117,* pp 609–618.

Ramon y Cajal, S. (1906). The structure and connexions of neurons. In *Nobel Lectures: Physiology or Medicine, 1901–1921.* Amsterdam: Elsevier, 1967, pp 220–253.

Ratliff, F. (1965). *Mach Bands.* San Francisco: Holden-Day.

Resko, J., Feder, H., and Goy, R. (1968). Androgen concentrations in plasma and testes of developing rats. *Journal of Endocrinology, 40,* pp 485–491.

Revusky, S. H. (1968). Aversion to sucrose produced by contingent X-radiation: temporal and dosage parameters. *Journal of Comparative and Physiological Psychology, 65,* pp 17–22.

Revusky, S. H. and Bedarf, E. W. (1967). Association of illness with prior ingestion of novel foods. *Science, 155,* pp 219–220.

Rheingold, H. (Ed.). (1963). *Maternal Behavior in Mammals.* New York: Wiley.

Rheingold, H. and Bayley, N. (1959). The later effects of an experimental modification of mothering. *Child Development, 30,* pp 363–372.

Ribbands, C. (1953, reprinted 1964). *The Behavior and Social Life of Honeybees.* New York: Dover Publications.

Ribble, M. (1944). Infantile experience in relation to personality development. In Hunt, J. McV. (Ed.), *Personality and Behavior Disorders.* New York: Ronald, pp 621–651.

Riddle, O., Lahr, E., and Bates, R. W. (1942). The role of hormones in the initiation of maternal behavior in rats. *American Journal of Physiology, 137,* pp 299–317.

Riesen, A. H. (1950). Arrested vision. *Scientific American, 183,* pp 16–19.

Riopelle, A. J. (Ed.). (1967). *Animal Problem Solving.* Harmondsworth: Penguin Books.

Roberts, R. C. (1967). Some concepts and methods in quantitative genetics. In Hirsch, J. (Ed.), *Behavior-Genetic Analysis.* New York: McGraw-Hill.

Roberts, W. W. (1969). Are hypothalamic motivational mechanisms functionally and anatomically specific? *Brain, Behavior and Evolution, 2,* pp 317–342.

Roberts, W. W., Steinberg, M. L., and Means, L. W. (1967). Hypothalamic mechanisms for sexual, aggressive, and other motivational behaviors in the opossum, *Didelphis virginiana. Journal of Comparative and Physiological Psychology, 64,* pp 1–15.

Robinson, B. W. (1967). Neurological aspects of evoked vocalization. In Altman, S. A. (Ed.), *Social Communication Among Primates.* Chicago: University of Chicago Press.

Rodieck, R. W. and Stone, J. (1965). Responses of cat retinal ganglion cells to moving visual patterns. *Journal of Neurophysiology, 28,* pp 819–832.

Roeder, K. (1963). *Nerve Cells and Insect Behavior.* Cambridge: Harvard University Press.

Roeder, K. (1966). Auditory system of noctuid moths. *Science, 154,* pp 1515–1521.

Romer, A. S. (1958). Phylogeny and behavior with special reference to vertebrate evolution. In Roe, A. and Simpson, G. G. (Eds.), *Behavior and Evolution.* New Haven: Yale University Press, pp 48–76.

Rosen, J. and Hart, F. M. (1963). Effects of early social isolation upon adult timidity and dominance. *Peromyscus Psychological Reports, 13,* pp 47–50.

Rosenblatt, J. S. (1965). Effects of experience on sexual behavior in cats. In Beach, F. A. (Ed.), *Sex and Behavior.* New York: Wiley, pp 416–439.

Rosenblatt, J. S. (1967). Nonhormonal basis of maternal behavior in the rat. *Science, 156,* pp 1512–1514.

Rosenblatt, J. S. and Aronson, L. (1958a). The decline of sexual behavior in male cats after castration with special reference to the role of prior sexual experience. *Behaviour, 12,* pp 285–338.

Rosenblatt, J. S. and Aronson, L. (1958b). The influence of experience on the behavioural effects of androgen in prepuberally castrated male cats. *Animal Behaviour, 6,* pp 171–182.

Rosenblatt, J. S. and Lehrman, D. S. (1963). Maternal behavior in the laboratory rat. In Rheingold, H. (Ed.), *Maternal Behavior in Mammals.* New York: Wiley.

Rosenblatt, J. S., Turkenwitz, G., and Schneirla, T. C. (1962). Development of sucking and related behavior in neonate kittens. In Bliss, E. L. (Ed.), *Roots of Behavior.* New York: Harper.

Rosenblith, W. A. (1961). *Sensory Communication.* New York: Massachusetts Institute of Technology and Wiley.

Rosenzweig, M. R., Krech, D., and Bennett, E. L. (1960). A search for relations between brain chemistry and behavior. *Psychological Bulletin, 57,* pp 476–492.

Rothenbuhler, W. (1964a). Behaviour genetics of nest cleaning in honey bees. I. Responses of four inbred lines to disease-killed brood. *Animal Behaviour, 12,* pp 578–583.

Rothenbuhler, W. (1964b). Behaviour genetics of nest cleaning in honey bees. IV. Response of F₁ and backcross generations to disease-killed broods. *American Zoologist, 4,* pp 111–123.

Rothschild, L. (1956). Unorthodox methods of sperm transfer. *Scientific American, 195:5,* pp 121–132.

Rowell, T. E. (1963). The social development of some rhesus monkeys. In Foss, B. M. (Ed.), *Determinants of Infant Behaviour II.* New York: Wiley, pp 35–49.

Sachs, B. D. and Barfield, R. J. (1970). Temporal patterning of sexual behavior in the male rat. *Journal of Comparative and Physiological Psychology, 73,* pp 359–364.

Sackett, G. P. (1965a). Effects of rearing conditions upon the behavior of rhesus monkeys (*Macaca mulatta). Child Development, 36,* pp 855–868.

Sackett, G. P. (1965b). Manipulatory behavior in monkeys reared under different levels of early stimulus variation. *Perceptual and Motor Skills, 20,* pp 985–988.

Sackett, G. P. (1966). Monkeys reared in isolation with pictures as visual input: evidence for an innate releasing mechanism. *Science, 154,* pp 1468–1472.

Sackett, G. P., Porter, M., and Holmes, H. (1965). Choice behavior in rhesus monkeys: effect of stimulation during the first month of life. *Science, 147,* pp 304–306.

Sade, D. S. (1965). Some aspects of parent-offspring and sibling relations in a group of rhesus monkeys, with a discussion of grooming. *American Journal of Physical Anthropology, 23,* pp 1–18.

Sahakian, W. S. (Ed.). (1968). *History of Psychology—A Source Book in Systematic Psychology.* Itasca, Illinois: F. E. Peacock.

Salk, L. (1962). Mother's heartbeat as an imprinting stimulus. *Transactions of the New York Academy of Sciences, 24,* pp 753–763.

Salzen, E. A. (1962). Imprinting and fear. *Symposium of the Zoological Society, London, 8,* pp 199–218.

Salzen, E. A. (1966). The interaction of experience, stimulus characteristics and exogenous androgen in the behaviour of domestic chicks. *Behaviour, 26,* pp 286–322.

Salzen, E. A. (1967). Imprinting in birds and primates. *Behaviour, 28,* pp 232–254.

Salzen, E. A. (1968). Discussion of Y. *Taketomo's* paper. *Science and Psychoanalysis, 12,* pp 184–189.

Sargent, T. D. and Keiper, R. R. (1967). Stereotypes in caged canaries. *Animal Behaviour, 15,* pp 62–66.

Savin, H. and Bever, T. (1970). The nonperceptual reality of the phoneme. *Journal of Verbal Learning and Verbal Behavior, 9,* pp 295–302.

Schacter, S. (1968). Obesity and eating. *Science, 161,* pp 751–755.

Schacter, S. (1971). *Emotion, Obesity, and Crime.* New York: Academic Press.

Schaeffer, T. (1968). Some methodological implications of the research on "early handling" in the rat. In Newton, G. and Levine, S. (Eds.), *Early Experience and Behavior.* Springfield, Illinois: Thomas, pp 102–141.

Schaffer, H. and Emerson, P. (1964). The development of social attachments in infancy. *Monographs of the Society for Research in Child Development, 29:3.*

Schaller, G. B. (1972). Predators of the Serengeti, Parts I, II, III. *Natural History, 81:2, 3, 4,* pp 38–49, 60–69, 38–43.

Scharrer, E. and Scharrer, B. (1963). *Neuroendocrinology.* New York: Columbia University Press.

Schein, M. and Hale, E. (1965). Stimuli eliciting sexual behavior. In Beach, F. A. (Ed.), *Sex and Behavior.* New York: Wiley, pp 440–482.

Schleidt, W. M., Schleidt, M., and Magg, M. (1960). Störung der Mutter-Kind-Beziehung bei Truthühnern durch Gehörverlust. *Behaviour, 16,* pp 254–260.

Schneiderman, H. A. and Gilbert, L. I. (1964). Control of growth and development in insects. *Science, 143,* pp 325–333.

Schneirla, T. C. (1952). A consideration of some conceptual trends in comparative psychology. *Psychological Bulletin, 49,* pp 559–597.

Schneirla, T. C. (1953). Basic problems in the nature of insect behavior. In Roeder, K. D. (Ed.), *Insect Physiology.* New York: Wiley, pp 656–684.

Schneirla, T. C., Rosenblatt, J. S., and Tobach, E. (1963). Maternal behavior in the cat. In Rheingold, H. (Ed.), *Maternal Behavior in Mammals.* New York: Wiley, pp 122–168.

Schultz, A. H. (1963). Age changes, sex differences, and variability as factors in the classification of primates. In Washburn, S. L. (Ed.), *Classification and Human Evolution*. Chicago: Aldine Publishing Company.

Scobey, R. P. (1968). Dissertation submitted to the Johns Hopkins University, Baltimore, Maryland.

Scobey, R. P. and Horowitz, J. M. (1970). Unpublished data.

Scobey, R. P. and Horowitz, J. M. (1972). The detection of small image displacements by cat retinal ganglion cells. *Vision Research, 12,* pp 2133–2143.

Scotch, N. A. (1967). Inside every fat man. In Glass, D. C. (Ed.), *Neurophysiology and Emotion*. New York: Russell Sage Foundation and Rockefeller University Press, pp. 155–160.

Scott, J. P. (1962). Critical periods in behavioral development. *Science, 138,* pp 949–958.

Scott, J. P. (1967a). Biology and the emotions. In Glass, D. C. (Ed.), *Neurophysiology and Emotion*. New York: Russell Sage Foundation and Rockefeller University Press, pp 190–200.

Scott, J. P. (1967b). Comparative psychology and ethology. *Annual Review of Psychology, 18,* pp 65–86.

Scott, J. P. (1968). *Early Experience and the Organization of Behavior*. Belmont, California: Brooks-Cole.

Scott, J. P., Fredericson, E., and Fuller, J. (1951). Experimental exploration of the critical period hypothesis. *Personality, 1,* pp 162–183.

Scriven, M. (1959). Explanation and prediction in evolution theory. *Science, 130,* pp 477–481.

Searle, J. (1972). Chomsky's revolution in linguistics. *New York Review of Books, 18, No. 12,* pp 16–24, June 29, 1972.

Searle, L. V. (1949). The organization of hereditary maze-brightness and maze-dullness. *Genetic and Psychology Monographs, 39,* pp 279–325.

Seay, B. M., Alexander, B. K., and Harlow, H. F. (1964). Maternal behavior of socially deprived rhesus monkeys. *Journal of Abnormal Social Psychology, 69,* pp 345–354.

Seligman, M. E. P. (1970). On the generality of the laws of learning. *Psychological Review, 77,* pp 406–418.

Senko, M. G. (1966). The effects of early, intermediate and late experience upon adult macaque sexual behavior. Unpublished master's thesis, University of Wisconsin.

Seto, H. (1963). *Studies on the Sensory Innervation (Human Sensibility)*. 2nd edition. Springfield, Illinois: Charles C Thomas.

Sevenster-Bol, A. (1962). On the causation of drive reduction after a consummatory act. *Archives Néerlandaise de Zoologie, 15,* pp 175–236.

Shackleton, D. M. (1968). Comparative aspects of social organization of American bison. Master's thesis, University of Western Ontario.

Shaw, E. (1964). The development of schooling behavior in fishes. In Ratner, S. C. and Denny, M. R. (Eds.), *Comparative Psychology*. Homewood, Illinois: Dorsey Press.

Sherrington, C. S. (1906). The spinal cord. In Sharpey-Schafer, E. A. (Ed.), *Textbook of Physiology*. Edinburgh: Pentland.

Sherrington, C. S. (1910). Flexion-reflex of the limb, crossed extension-reflex, and reflex stepping and standing. *Journal of Physiology, 40,* pp 28–121.

Sherrington, C. S. (1947). *The Integrative Action of the Nervous System*. New Haven: Yale University Press.

Sherrington, C. S. (1961). *The Integrative Action of the Nervous System*. 2nd edition. (Paperbound.) New Haven: Yale University Press.

Shurrager, P. S. and Culler, E. A. (1938). Phenomena allied to conditioning in the spinal dog. *American Journal of Physiology, 123,* pp 186–187.

Shurrager, P. S. and Culler, E. A. (1940). Conditioning in the spinal dog. *Journal of Experimental Psychology, 26,* pp 133–159.

Shurrager, P. S. and Culler, E. A. (1941). Conditioned extinction of a reflex in the spinal dog. *Journal of Experimental Psychology, 28,* pp 287–303.

Shurrager, P. S. and Dykman, R. A. (1950). Walking spinal carnivores. *Journal of Comparative and Physiological Psychology, 44,* pp 252–262.

Shurrager, P. S. and Shurrager, H. C. (1941). Converting a spinal CR into a reflex. *Journal of Experimental Psychology, 29,* pp 217–224.

Sidman, M. (1953). Two temporal parameters of the maintenance of avoidance behavior by the white rat. *Journal of Comparative and Physiological Psychology, 46,* pp 253–261.

Simons, R. C., Bobbitt, R. A., and Jensen, G. D. (1968). Mother monkeys' (*Macaca nemestrina*) responses to infant vocalizations. *Perceptual and Motor Skills, 27,* pp 3–10.

Simpson, G. G. (1951). *Horses.* New York: Oxford University Press.

Simpson, G. G. (1961). *Principles of Animal Taxonomy.* New York: Columbia University Press.

Sinclair, D. (1967). *Cutaneous Sensation.* New York: Oxford University Press.

Singh, S. D. (1969). Urban monkeys. *Scientific American, 221:1,* pp 108–115.

Skinner, B. F. (1938). *The Behavior of Organisms.* New York: Appleton-Century-Crofts.

Skinner, B. F. (1948). *Walden Two.* New York: Macmillan.

Skinner, B. F. (1953). *Science and Human Behavior.* New York: Macmillan.

Skinner, B. F. (1968). *The Technology of Teaching.* New York: Appleton-Century-Crofts.

Skinner, B. F. (1969). *Contingencies of Reinforcement: A Theoretical Analysis.* New York: Appleton-Century-Crofts.

Skinner, B. F. (1971). *Beyond Freedom and Dignity.* New York: Knopf.

Sluckin, W. (1965). *Imprinting and Early Learning.* Chicago: Aldine.

Smart, J. (1943). *A Handbook for the Identification of Insects of Medical Importance.* London: British Museum (Natural History).

Smith, J. C. and Roll, D. L. (1967). Trace conditioning with X-rays as the aversive stimulus. *Psychonomic Science, 9,* pp 11–12.

Soloman, P., Kubzansky, P., Leiderman, H., Mendelson, J., Trumbull, R., and Wexler, D. (1965). *Sensory Deprivation.* Cambridge: Harvard University Press.

Southwick, C. H., Beg, M. A., and Siddiqi, M. R. (1965). Rhesus monkeys in north India. In DeVore, I. (Ed.), *Primate Behavior.* New York: Holt, Rinehart, and Winston, pp 111–159.

Spalding, D. A. (1873). Instinct, with original observations on young animals. *Macmillan's Magazine, 27,* pp 282–293.

Spencer, W. A., Thompson, R. F., and Nielson, D. R. (1966a). Alterations in responsiveness of ascending and reflex pathways activated by iterated cutaneous afferent volleys. *Journal of Neurophysiology, 29,* pp 240–252.

Spencer, W. A., Thompson, R. F., and Nielson, D. R. (1966b). Decrement of ventral root electrotonus and intracellularly recorded PSP's produced by iterated cutaneous afferent volleys. *Journal of Neurophysiology, 29,* pp 253–274.

Spieth, H. T. (1958). Behavior and isolating mechanisms. In Roe, A. and Simpson, G. G. (Eds.), *Behavior and Evolution.* New Haven: Yale University Press, pp 363–389.

Spitz, R. and Cobliner, W. (1965). *The First Year of Life.* New York: International Universities Press.

Srb, A. M., Owen, R. D., and Edgar, R. S. (1965). *General Genetics.* San Francisco: Freeman.

Steiner, G. (1967). *Language and Silence.* New York: Atheneum.

Stephan, F. K., Valenstein, E. S., and Zucker, I. (1971). Copulation and eating during electrical stimulation of the rat hypothalamus. *Physiology and Behavior, 7,* pp 587–593.

Stern, C. (1960). *Principles of Human Genetics.* San Francisco: Freeman.

Stone, L. J. and Church, J. (1968). *Childhood and Adolescence.* 2nd Edition. New York: Random House, pp 164–221.

Storer, J. H. (1968). *Man in the Web of Life.* New York: Signet.

Storrs, E. E. and Williams, R. J. (1968). Study of monozygous quadruplet armadillos in relation to mammalian inheritance. *Science, 160,* p 443.

Stratton, G. M. (1897). Vision without inversion of the retinal image. *Psychological Review, 4,* pp 341–360, 463–481.

Swets, J. A. (1964). *Signal Detection and Recognition by Human Observers.* New York: Wiley.

Terkel, J. and Rosenblatt, J. S. (1968). Maternal behavior induced by maternal blood plasma injected into virgin rats. *Journal of Comparative and Physiological Psychology, 65,* pp 479–482.

Terkel, J. and Rosenblatt, J. S. (1972). Humoral factors underlying maternal behavior at parturition: cross transfusion between freely moving rats. *Journal of Comparative and Physiological Psychology, 80,* pp 365–371.

Terman, C. R. (1963). The influence of differential early social experience upon spatial distribution within populations of deer-mice. *Animal Behaviour, 11,* pp 246–262.

Thomas, E. M. (1959). *The Harmless People.* New York: Random House.

Thompson, R. F. and Spencer, W. A. (1966). Habituation: a model phenomenon for the study of neuronal substrates of behavior. *Psychological Review, 73,* pp 16–43.

Thompson, W. R. and Melzack, R. (1956). Early environment. *Scientific American, 194,* pp 38–42.

Thorndike, E. L. (1911). *Animal Intelligence.* New York: Macmillan.

Thorpe, W. (1972). Vocal communication in birds. In Hinde, R. A. (Ed.), *Non-Verbal Communication.* Cambridge: Cambridge University Press, pp 153–175.

Tiger, L. (1969). *Men in Groups.* New York: Random House.

Tiger, L. and Fox, R. (1971). *The Imperial Animal.* New York: Holt, Rinehart and Winston.

Tinbergen, N. (1951). *The Study of Instinct.* Oxford: The Clarendon Press.

Tinbergen, N. (1952). The curious behavior of the stickleback. *Scientific American, 187,* pp 22–38.

Tinbergen, N. (1953a). *The Herring Gull's World.* London: Collins.

Tinbergen, N. (1953b). *Social Behaviour in Animals.* New York: Wiley.

Tinbergen, N. (1958, reprinted 1968). *Curious Naturalists.* Garden City, New York: Doubleday.

Tinbergen, N. (1960a). The evolution of behavior in gulls. *Scientific American, 203:6,* pp 118–130.

Tinbergen, N. (1960b). *The Herring Gull's World.* Revised Edition. New York: Basic Books.

Tinbergen, N. (1962). The evolution of animal communications: a critical examination of methods. *Symposium of the Zoological Society, London, 8,* pp 1–6.

Tinbergen, N. (1964). The evolution of signaling devices. In Etkin, W. (Ed.), *Social Behavior and Organization among Vertebrates.* Chicago: University of Chicago Press.

Tobias, P. V. (1971). *The Brain in Hominid Evolution.* New York: Columbia University Press.

Tolman, E. C. (1924). The inheritance of maze-learning in rats. *Journal of Comparative and Physiological Psychology, 4,* pp 1–18.

Tolman, E. C. (1932). *Purposive Behavior in Animals and Men.* New York: Appleton-Century.

Tolman, E. C. (1959). Principles of purposive behavior. In Koch, S. (Ed.), *Psychology: A Study of a Science, Vol. 2.* New York: McGraw-Hill, pp 92–157.

Tryon, R. C. (1930). Studies of individual differences in maze ability: I. The measurement of the reliability of individual differences. *Journal of Comparative and Physiological Psychology, 11,* pp 145–170.

Tryon, R. C. (1931). Studies in individual differences in maze ability: II. The determination of individual differences by age, weight, sex, and pigmentation. *Journal of Comparative and Physiological Psychology, 12,* pp 1–22.

Tryon, R. C. (1939). Studies in individual differences in maze learning: VI. Disproof of sensory components: experimental effects of stimulus variation. *Journal of Comparative and Physiological Psychology, 28,* pp 361–415.

Tryon, R. C. (1940a). Studies in individual differences in maze ability: VII. The specific components of maze ability, and a general theory of psychological components. *Journal of Comparative and Physiological Psychology, 30,* pp 283–335.

Tryon, R. C. (1940b). Studies in individual differences in maze ability: VIII. Prediction validity of the psychological components of maze ability. *Journal of Comparative and Physiological Psychology, 30,* pp 535–582.

Tryon, R. C. (1942). Individual differences. In Moss, F. A. (Ed.), *Comparative Psychology.* Revised Edition. New York: Prentice-Hall. pp 409–448.

Turnbull, C. (1961). *The Forest People.* Garden City, New York: Doubleday and Company.

Ulrich, R., Stachnik, T., and Mabry, J. (Eds.). (1966). *Control of Human Behavior.* Chicago: Scott, Foresman and Company.

Valenstein, E. S., Cox, V. C., and Kakolewski, J. W. (1970). Reexamination of the role of the hypothalamus in motivation. *Psychological Review, 77,* pp 16–31.

Valenstein, E. S. and Goy, R. (1957). Further studies of the organization and display of sexual behavior in male guinea pigs. *Journal of Comparative and Physiological Psychology, 150,* pp 115–119.

Valenstein, E. S., Riss, W., and Young, W. (1955). Experiential and genetic factors in the organization of sexual behavior in male guinea pigs. *Journal of Comparative and Physiological Psychology, 48,* pp 397–403.

van der Post, L. (1958). *The Lost World of the Kalahari.* Harmondsworth, England: Penguin Books.

van der Post, L. (1961). *The Heart of the Hunter.* New York: William Morrow.

van Iersel, J. (1953). An analysis of the parental behaviour of the male three-spined stickleback (*Gasterosteus aculeatus* L.). *Behaviour Supplement 3,* pp 1–159.

van Lawick-Goodall, H. and van Lawick-Goodall, J. (1970). *Innocent Killers.* Boston, Massachusetts: Houghton-Mifflin.

van Lawick-Goodall, J. (1968). The behaviour of free-living chimpanzees in the Gombe Stream Reserve. *Animal Behaviour Monographs, 1,* pp 161–311.

von Békésy, G. (1967). *Sensory Inhibition.* Princeton, New Jersey: Princeton University Press.

von Frisch, K. (1950). *Bees: Their Vision, Chemical Senses, and Language.* Ithaca: Cornell University Press.

von Firsch, K. (1971). *Bees: Their Vision, Chemical Senses, and Language. Revised Edition.* Ithaca: Cornell University Press.

von Uexküll, J. (1921). *Umwelt und Innenwelt der Tiere.* Berlin: Springer.

Waddington, C. H. (1960). *The Ethical Animal.* London: George Allen and Unwin.

Waddington, C. H. (1966). *Principles of Development and Differentiation.* New York: Macmillan.

Walker, B. W. (1952). A guide to the grunion. *California Fish and Game, 38,* pp 409–420.

Wallace, B. and Srb, A. (1961). *Adaptation.* Englewood Cliffs, New Jersey: Prentice-Hall.

Walls, G. L. (1942). *The Vertebrate Eye and Its Adaptive Radiation.* Bloomfield Hills, Michigan: Cranbrook Institute of Science.

Walters, R. H. and Parke, R. D. (1965). The role of the distance receptors in the development of social responsiveness. In Lipsitt, L. P. and Spiker, C. C. (Eds.), *Advances in Child Development and Behavior, Vol. 2.* New York: Academic Press.

Ward, J. A. and Barlow, G. W. (1967). The maturation and regulation of glancing off the parents by young orange chromides (*Etrophus maculatus: Pisces-chichlidae*). *Behaviour, 29,* pp 1–56.

Washburn, S. L. and Hamburg, D. A. (1968). Aggressive behavior in old world monkeys and apes: studies in adaptation and variability. In Jay, P. C. (Ed.), *Primates.* New York: Holt, Rinehart, and Winston.

Watson, J. (1913). Psychology as the behaviorist views it. *Psychological Review, 20,* pp 158–177.

Watson, J. B. (1914). *Behavior—An Introduction to Comparative Psychology.* New York: Holt.

Watson, J. B. (1925). *Behaviorism.* New York: W. W. Norton.

Wecker, S. C. (1963). The role of early experience in habitat selection by the prairie deer mouse, *Peromyscus maniculatus bairdi. Ecological Monographs, 33,* pp 307–325.

Weidmann, U. (1956). Verhaltensstudien an der Stockente (*Anas platyrhynchos* L.). I. Das Aktionssystem. *Zeitschrift für Tierpsychologie, 13,* pp 208–271.

Weir, M. W. and DeFries, J. C. (1964). Prenatal maternal influence on behavior in mice: evidence of a genetic basis. *Journal of Comparative and Physiological Psychology, 58,* pp 412–417.

Wells, M. (1968). *Lower Animals.* New York: McGraw-Hill.

Welty, J. (1963). *The Life of Birds.* New York: Alfred Knopf.

Wendt, H. (1965). *The Sex Life of the Animals.* New York: Simon and Schuster.

Wenner, A. (1967). Honey bees: do they use the distance information contained in their dance maneuver? *Science, 155,* pp 847–849.

Wenner, A., Wells, P., and Johnson, D. (1969). Honey bee recruitment to food sources: olfaction or language? *Science, 164,* pp 84–86.

Westheimer, G. (1965). Visual acuity. *Annual Review of Psychology, 16,* pp 359–380.

Whalen, R. (1963). Sexual behavior of cats. *Behaviour, 20,* pp 321–342.

Whalen, R. (1968). Differentiation of the neural mechanisms which control gonadotropin secretion and sexual behavior. In Diamond, M. (Ed.), *Perspectives in Reproduction and Sexual Behavior.* Bloomington: University of Indiana Press, pp 303–340.

Whalen, R., Beach, F. A., and Kuehn, R. (1961). Effects of exogenous androgen on sexually responsive and unresponsive male rats. *Endocrinology, 69,* pp 373–380.

White, T. H. (1951). *The Goshawk.* New York: Viking Press.

Whiting, J. W. M. (1967). Analysis of infant stimulation. In Glass, D. C. (Ed.), *Neurophysiology and Emotion.* New York: Russell Sage Foundation and Rockefeller University Press, pp 201–204.

Whorf, B. L. (1956). *Language, Thought, and Reality.* New York: Wiley.

Wickelgren, B. G. (1967). Habituation of spinal motoneurons. *Journal of Neurophysiology, 30,* pp 1404–1423.

Wiepkema, P. R. (1961). An ethological analysis of the reproductive behaviour of the bitterling (*Rhodeus amarus bloch*). *Archives Néerlandaises de Zoologie, 14,* pp 103–199.

Wiesenfeld, S. L. (1967). A sickle-cell trait in human biological and cultural evolution. *Science, 157,* pp 1134–1140.

Wiesner, B. P. and Sheard, N. M. (1933). *Maternal Behavior in the Rat.* London: Oliver and Boyd.

Wigglesworth, V. B. (1964). *The Life of Insects.* Cleveland, Ohio: World Publishing.

Wilcoxon, H. C., Dragoin, W. B., and Kral, P. A. (1969). Differential conditioning to visual and gustatory cues in quail and rat: illness-induced aversions. Paper presented at Psychonomic Society, St. Louis.

Wilcoxon, H. C., Dragoin, W. B., and Kral, P. A. (1971). Illness-induced aversions in rat and quail: relative salience of visual and gustatory cues. *Science, 171,* pp 826–828.

Wilentz, J. S. (1968). *The Senses of Man.* New York: Thomas Y. Crowell Company.

Wilkins, W. (1971). Desensitization: social and cognitive factors underlying the effectiveness of Wolpe's procedure. *Psychological Bulletin, 76,* pp 311–317.

Williams, D. R. and Williams, H. (1969). Auto-maintenance in the pigeon: sustained pecking despite contingent non-reinforcement. *Journal of the Experimental Analysis of Behavior, 12,* pp 511–520.

Williams, G. C. (1966). *Adaptation and Natural Selection.* Princeton, New Jersey: Princeton University Press.

Williams, R. J. (1969). Heredity, human understanding, and civilization. *American Scientist, 57,* pp 237–243.

Willows, A. O. D. (1971). Giant brain cells in molluscs. *Scientific American, 224:2,* pp 68–75.

Winkelmann, R. K. (1960). *Nerve Endings in Normal and Pathologic Skin.* Springfield, Illinois: Charles C Thomas.

Winston, H. D. (1964). Heterosis and learning in the mouse. *Journal of Comparative and Physiological Psychology, 57,* pp 279–283.

Wise, R. A. (1968). Hypothalamic motivational systems: fixed or plastic neural circuits? *Science, 162,* pp 377–379.

Woodworth, R. (1948). *Contemporary Schools of Psychology.* Revised Edition. New York: Ronald.

Wortis, R. P. (1969). The transition from dependent to independent feeding in the young ring dove. *Animal Behaviour Monographs, 2,* pp 3–54.

Wynne-Edwards, V. C. (1962). *Animal Dispersion in Relation to Social Behaviour.* Edinburgh and London: Oliver and Boyd.

Yarbus, A. L. (1967). *The Eye Movements and Vision.* New York: Plenum Press.

Yates, A. (1970). *Behavior Therapy.* New York: Wiley.

Young, R. M. (1970). *Mind, Brain, and Adaptation in the Nineteenth Century.* Oxford: Clarendon Press.

Young, W. C. (1961). The hormones and mating behavior. In Young, W. C. (Ed.), *Sex and Internal Secretions.* Baltimore: Williams and Wilkins.

Zener, K. and Gaffron, M. (1962). Perceptual experience: an analysis of its relations to the external world through internal processings. In Koch, S. (Ed.), *Psychology: A Study of a Science, Vol. IV.* New York: McGraw-Hill, pp 515–618.

Zinsser, H. (1935). *Rats, Lice and History.* Boston: Little, Brown and Company.

Zotterman, Y. (Ed.). (1967). *Sensory Mechanisms. Progress in Brain Research, 23.* New York: Elsevier Publishing Company.

Subject Index

Name Index